Sustainable and Smart
Spatial Planning in Africa

Sustainable and Smart Spatial Planning in Africa

Case Studies and Solutions

Edited by
Charles Chavunduka, Walter Timo de Vries, and
Pamela Durán-Díaz

CRC Press is an imprint of the
Taylor & Francis Group, an **informa** business

First edition published 2022
by CRC Press
6000 Broken Sound Parkway NW, Suite 300, Boca Raton, FL 33487–2742

and by CRC Press
4 Park Square, Milton Park, Abingdon, Oxon, OX14 4RN

CRC Press is an imprint of Taylor & Francis Group, LLC

© 2022 Taylor & Francis Group, LLC

Reasonable efforts have been made to publish reliable data and information, but the author and publisher cannot assume responsibility for the validity of all materials or the consequences of their use. The authors and publishers have attempted to trace the copyright holders of all material reproduced in this publication and apologize to copyright holders if permission to publish in this form has not been obtained. If any copyright material has not been acknowledged please write and let us know so we may rectify in any future reprint.

Except as permitted under U.S. Copyright Law, no part of this book may be reprinted, reproduced, transmitted, or utilized in any form by any electronic, mechanical, or other means, now known or hereafter invented, including photocopying, microfilming, and recording, or in any information storage or retrieval system, without written permission from the publishers.

For permission to photocopy or use material electronically from this work, access www.copyright.com or contact the Copyright Clearance Center, Inc. (CCC), 222 Rosewood Drive, Danvers, MA 01923, 978–750–8400. For works that are not available on CCC please contact mpkbookspermissions@tandf.co.uk

Trademark notice: Product or corporate names may be trademarks or registered trademarks and are used only for identification and explanation without intent to infringe.

Library of Congress Cataloging-in-Publication Data
Names: Chavunduka, Charles, editor. | Vries, Walter Timo de, editor. |
    Duran Diaz, Pamela, editor.
Title: Sustainable and smart spatial planning in Africa: case studies and
    solutions / edited by Charles Chavunduka, Walter Timo De Vries, and
    Pamela Duran Diaz.
Description: First edition. | Boca Raton: CRC Press, 2022. | Includes
    bibliographical references and index.
Identifiers: LCCN 2021054283 | ISBN 9781032118420 (hbk) |
    ISBN 9781032118437 (pbk) | ISBN 9781003221791 (ebk)
Subjects: LCSH: Smart cities — Africa. | Sustainable urban
    development — Africa. | Land use — Africa.
Classification: LCC TD159.4.S88 2022 | DDC 307.76096 — dc23/eng/20211116
LC record available at https://lccn.loc.gov/2021054283

ISBN: 978-1-032-11842-0 (hbk)
ISBN: 978-1-032-11843-7 (pbk)
ISBN: 978-1-003-22179-1 (ebk)

DOI: 10.1201/9781003221791

Typeset in Times LT Std
by Apex CoVantage, LLC

Contents

Foreword ix
Acknowledgements xi
About the Editors xiii
Contributors List xv

SECTION I Sustainable and Smart Spatial Planning in Africa: Introduction

Chapter 1 Sustainable and Smart Spatial Planning in Africa: Introduction 3

Charles Chavunduka, Walter Timo de Vries and Pamela Durán-Díaz

SECTION II Theory of Sustainable and Smart Spatial Planning

Chapter 2 Stocktaking Frameworks for the Planning and Development of Smart Cities 11

Tafadzwa Mutambisi and Charles Chavunduka

Chapter 3 'Smart City' Concept and Its Implications for Urban Planning Systems in African Cities 25

Ephraim Kabunda Munshifwa and Niraj Jain

Chapter 4 Transformation Pathways to Smart Villages: Lessons from Mulenzhe Village in Limpopo Province of South Africa 41

Emaculate Ingwani, Madilonga Trevor Mukwevho, Trynos Gumbo and Masala Thomas Makumule

Chapter 5 Smart Growth and New Urbanism, a Sustainable Approach towards Urban Redevelopment: Case of Chivhu 55

Monalissa Kaluwa, Chipo Mutonhodza and Leonard Chitongo

Chapter 6 Trans-Border Spatial Planning: Assessing the Musina–Beitbridge Twinning Agreement between South Africa and Zimbabwe 69

Shylet Nyamwanza, Peter Bikam and James Chakwizira

SECTION III   Context and External Drivers of Sustainable and Smart Spatial Planning

Chapter 7 Transnational Land Governance for Sustainable Development: A Comparative Study of Africa and Latin America 89

Pamela Durán-Díaz

Chapter 8 Are We There Yet? Prospects and Barriers to Implementing Smart City Initiatives in Harare, Zimbabwe 107

Abraham R. Matamanda

Chapter 9 Smartness in Developing Liveable Informal Settlements: The Case of Hopley in Harare.. 121

Morgen Zivhave and Collins Dzvairo

SECTION IV   Goals and Practices of Sustainable and Smart Spatial Planning

Chapter 10 The Urban Laboratory: A Case of Data Mining and Management for the Successful Hosting of Smart Settlements in Zimbabwe 141

Innocent Chirisa, Valeria Muvavarirwa and Fungai N. Mukora

Chapter 11 Opportunities and Constraints of Solar Harvesting as a Sustainable and Resilience Strategy in Zimbabwe: The Case of Nyangani Renewable Energy, Mutoko District............................ 153

Enock G. Mukwekwe, Leonard Chitongo and Godwin K. Zingi

Chapter 12 Infrastructure Projects Design Versus Use in Local Authorities: A Case Study of Banket Small and Medium Enterprises Mall in Mashonaland West Province of Zimbabwe.................................. 169

Moses Chundu

Contents   vii

Chapter 13 Towards Responsive Human Smart Cities: Interrogating Street Users' Perspectives on Spatial Justice on Street Spaces in Small Rural Towns in South Africa 183

Wendy Wadzanayi Tsoriyo, Emaculate Ingwani, James Chakwizira and Peter Bikam

Chapter 14 Synchronising the Spatial Planning Legislative and Administrative Frameworks of Mining and Other Human Settlements in Zimbabwe 197

Audrey Ndarova Kwangwama, Willoughby Zimunya and Wiseman Kadungure

SECTION V   Methods and Tools for Sustainable and Smart Spatial Planning

Chapter 15 The Contribution of Spatial Planning Tools towards Disaster Risk Reduction in Informal Settlements in South Africa 213

Juliet Akola, James Chakwizira, Emaculate Ingwani and Peter Bikam

Chapter 16 Geographic Information Systems for Smart Spatial Planning and Management: Managing Urban Sprawl in Harare Metropolitan, Zimbabwe 229

Tendai Sylvester Mhlanga and Fiza Naseer

Chapter 17 Appraisal of E-Waste Management Approaches in Zimbabwean Cities 245

Takudzwa M. Matyatya, Willoughby Zimunya and Tariro Nyevera

Chapter 18 Three-Dimensional Layout Planning in the Context of Zimbabwe's Planning Profession: Scope, Fears and Potentialities 257

Brilliant Mavhima

Chapter 19 From Two-Dimensional to Four-Dimensional Layout Design: A Necessary Leapfrog in Zimbabwean Urban Planning 269

Brilliant Mavhima

Chapter 20 An Evaluation of the Effectiveness of Material Waste Management Techniques in the Construction Industry of Zimbabwe: A Case of Harare and Bulawayo 283

Crytone Kusaziya and Yvonne Munanga

SECTION VI Future of Sustainable and Smart Spatial Planning in Africa

Chapter 21 Furthering Sustainable and Smart Spatial Planning in Africa 299

Charles Chavunduka, Walter Timo de Vries and Pamela Durán-Díaz

Index 305

Foreword

The centrality of spatial planning in all societies is indisputable. Spatial planning is critical to the sustainable development of a nation as it has the ability to touch on environmental and socioeconomic facets that shape how nations develop. All the 17 sustainable development goals (SDGs) have a direct or indirect relation to spatial planning and the New Urban Agenda identifies it as an important lever for promoting sustainable development and improving the quality of life. Sustainable and smart spatial planning is, therefore, a critical focus for the development agenda and related interventions.

Sustainable and smart spatial planning, particularly in sub-Saharan Africa, is important for various reasons. First, the UN-Habitat estimates that by 2030, the region's urban population will account for over half of the total population and will require infrastructure, social services, affordable housing and employment opportunities, which all depend upon effective spatial planning. Second, spatial planning systems inherited from colonial epistemologies of rational spatial planning are not suitable for current African realities of rapid change and increased uncertainty. Third, the UN-Habitat estimates that a staggering 62% of the urban population in sub-Saharan Africa lives in slums characterised by, among other things, overcrowding, lack of access to clean water, health facilities and poor waste disposal. It is, therefore, opportune to reflect on alternative planning paradigms that could better suit current and future realities.

Inadequate spatial planning has dire consequences for human settlements in sub-Saharan Africa. It contributes to informal spaces and spatial inequalities as elites benefit from access to land and capture most of its value while the majority are disenfranchised. It promotes environmental degradation, with weak controls on land use and natural resources. Weak spatial planning also leads to inadequate investments and poor-performing property taxation systems that rob governments of much-needed revenues.

These effects have been worsened by COVID-19, which by its nature has been restrictive to planning processes. The challenge of the virus brings to society a deliberate consciousness that global processes and events are converging (borders are porous) while local embeddedness is being entrenched through practices like lockdowns and confinement. At the same time, the virus brought opportunities that may impact future sustainable and smart development as practitioners and academia adopt new ways of working.

This volume brings together scholarship from different disciplines to discuss sustainable and smart spatial planning from an interdisciplinary perspective. It is pleasing to notice that most of the contributors are African authors. The book is aimed at readers studying spatial planning, international development, development studies, geography, geoinformatics and professionals and policy makers concerned with spatial and land use planning.

Readers will benefit from this book through its elaboration of the link between global initiatives on sustainable and smart human settlements including HABITAT II,

HABITAT III and the New Urban Agenda, SDGs, World Bank's Global Smart City Partnership Program (GSCP), Sendai Framework as well as regional ones such as Africa's Agenda 2030 and 2063 and the African Urban Research Initiative. Themes include sustainable and smart city frameworks, smart growth concepts, renewable energy, smart urban laboratories, liveable informal settlements and adaptive smart solutions.

The book concludes that sustainable and smart spatial planning in Africa needs to build on innovative epistemic conceptualisations and processes, which provide a fit for African realities. These include sustained informality and dynamic use and occupation of space. Handling such realities requires alternative forms of governance, enhanced reliance on big data and advanced data analytics and improved socio-technical infrastructures.

It is hoped that the book will raise awareness and stimulate new interdisciplinary research with particular attention to promoting smart governance, effective information infrastructure and technological innovation. This book is timely for spatial planning professionals, academia, civil society actors and stakeholders in public and private sectors who are keen to promote better spatial planning in a rapidly urbanising Africa.

Professor K. H. Wekwete
Midlands State University
Zimbabwe

Acknowledgements

This book project was motivated by the Government of Zimbabwe's policy pronouncement on the prioritisation of the development of sustainable and smart cities in the country. It is in this context, and the increasing adoption of the smart concept in sub-Saharan Africa, that the University of Zimbabwe's Third Biennial Symposium adopted the theme: 'Developing Sustainable Human Settlements through the Smart Concept: Finding Solutions to Contemporary Urban and Rural Challenges'. After the organising team received symposium papers from participants mostly based in southern African countries, the event did not take place following the outbreak of the COVID-19 pandemic. However, the authors decided to publish the papers in the form of a book.

The authors would like to acknowledge the generous support of the GIZ-funded Advancing Collaborative Research in Responsible and Smart Land Management in and for Africa (ADLAND) Project which made the publication of this book possible. We are grateful for the training of authors on publishing and editing and financial support towards the publication process of the book. We are especially grateful to Walter Timo de Vries and Pamela Durán-Díaz who provided the training and as co-editors helped shape the book and contributed to much of the learning represented here.

We would further like to thank our supportive publishers Irma Shagla Britton and Shannon Welch at CRC Press—Taylor & Francis Group, whose ideas and guidance throughout the publication process helped sharpen the focus of the book.

Most especially, we wish to thank all the contributors to the Third Biennial Symposium—not all of whom could be represented in this collection for their work and persistent commitment both to the Biennial Symposium and, more generally, for contributing customised sustainable and smart spatial planning solutions to sub-Saharan Africa's development challenges.

Charles Chavunduka
University of Zimbabwe
Harare, October 2021

About the Editors

Charles Chavunduka is a senior lecturer and deputy dean in the Faculty of Engineering and the Built Environment at the University of Zimbabwe. Before he joined the academia, he served as a planner in the public sector and as a land administration specialist in international agencies. He is a participating member of the Southern Africa Node under the Network of Excellence for Land Governance in Africa (NELGA). His research interests are agrarian reform, land policy and management and urban development.

Walter Timo de Vries is Chair of Land Management at the Technical University of Munich since 2015. He is Study Dean Geodesy and Geo-information and Director of the Master's and PhD programmes in Land Management and Geospatial Science. He has worked in numerous international projects in Asia, Africa and South America, dealing with land information and land reform, geospatial data infrastructures and professional training and education in land issues, cadastre and information management. His current research interests at TUM include smart and responsible land management, urban and rural development and land consolidation.

Pamela Durán-Díaz is a senior scientific researcher and lecturer at the Chair of Land Management of the Technical University of Munich. She coordinates the Doctoral and Master's programmes in Land Management and Geospatial Science (former Land Management and Land Tenure) since 2016. She is the project manager of ADLAND 'Advancing collaborative research in responsible and smart land management in and for Africa', a capacity building project financed by GIZ as a key academic strategy to strengthen land governance in Africa, working directly with the Network of Excellence for Land Governance in Africa (NELGA). Her research interests include sustainable urban and rural development, the WEF nexus for food sovereignty, cultural landscapes and land governance in the Global South.

Contributors List

Juliet Akola
Department of Urban and Regional Planning, University of Venda, South Africa

Peter Bikam
Department of Urban and Regional Planning, North-West University, South Africa

James Chakwizira
Department of Urban and Regional Planning, North-West University, South Africa

Charles Chavunduka
Department of Architecture and Real Estate, University of Zimbabwe, Zimbabwe

Innocent Chirisa
Dean of Social and Behavioural Sciences, University of Zimbabwe, Zimbabwe

Leonard Chitongo
School of Management, IT & Governance, University of KwaZulu-Natal, South Africa

Moses Chundu
Department of Economics and Development, University of Zimbabwe, Zimbabwe

Walter Timo de Vries
Chair of Land Management, Technical University of Munich, Germany

Pamela Durán-Díaz
Chair of Land Management, Technical University of Munich, Germany

Collins Dzvairo
Department of Architecture and Real Estate, University of Zimbabwe, Zimbabwe

Trynos Gumbo
Department of Urban and Regional Planning, University of Johannesburg, South Africa

Emaculate Ingwani
Department of Urban and Regional Planning, University of Venda, South Africa

Niraj Jain
Department of Real Estate Studies, Copperbelt University, Zambia

Wiseman Kadungure
Mazowe Rural District Council, Zimbabwe

Monalissa Kaluwa
Department of Rural and Urban Development, Great Zimbabwe University, Zimbabwe

Crytone Kusaziya
Department of Quantity Surveying, National University of Science and Technology, Zimbabwe

Audrey Ndarova Kwangwama
Department of Architecture and Real Estate, University of Zimbabwe, Zimbabwe

Masala Thomas Makumule
Department of Planning and Development, Thulamela Local Municipality, Vhembe District, South Africa

Abraham R. Matamanda
Department of Geography, University of the Free State, South Africa

Takudzwa M. Matyatya
Department of Architecture and Real Estate, University of Zimbabwe, Zimbabwe

Brilliant Mavhima
Department of Architecture and Real Estate, University of Zimbabwe, Zimbabwe

Tendai Sylvester Mhlanga
Department of Urban and Regional Planning, Gazi University, Turkey

Fungai N. Mukora
Department of Computer Engineering, University of Zimbabwe, Zimbabwe

Enock G. Mukwekwe
Department of Rural and Urban Development, Great Zimbabwe University, Zimbabwe

Madilonga Trevor Mukwevho
Department of Development Planning, Makhado Local Municipality, Vhembe District, South Africa

Yvonne Munanga
Department of Architecture and Real Estate, University of Zimbabwe, Zimbabwe

Ephraim Kabunda Munshifwa
Department of Real Estate Studies, Copperbelt University, Zambia

Tafadzwa Mutambisi
Department of Architecture and Real Estate, University of Zimbabwe, Zimbabwe

Chipo Mutonhodza
Department of Rural and Urban Development, Great Zimbabwe University, Zimbabwe

Valeria Muvavarirwa
Department of Architecture and Real Estate, University of Zimbabwe, Zimbabwe

Fiza Naseer
Department of Urban and Regional Planning, Gazi University, Turkey

Shylet Nyamwanza
Department of Urban and Regional Planning, University of Venda, South Africa

Tariro Nyevera
Department of Architecture and Real Estate, University of Zimbabwe, Zimbabwe

Wendy Wadzanayi Tsoriyo
Department of Urban and Regional Planning, North-West University, South Africa

Willoughby Zimunya
Department of Architecture and Real Estate, University of Zimbabwe, Zimbabwe

Godwin K. Zingi
Department of Rural and Urban Development, Great Zimbabwe University, Zimbabwe

Morgen Zivhave
Department of Architecture and Real Estate, University of Zimbabwe, Zimbabwe

Section I

Sustainable and Smart Spatial Planning in Africa

Introduction

1 Sustainable and Smart Spatial Planning in Africa
Introduction

Charles Chavunduka, Walter Timo de Vries and Pamela Durán-Díaz

CONTENTS

1.1 Introduction 3
1.2 Sustainable Spatial Planning 5
1.3 Smart Spatial Planning 5
1.4 Justification and Outline of Book Sections 6
References 6

1.1 INTRODUCTION

The idea for this book originated when several events and developments started to align. The first of such events was a public statement in 2017 by the Ministry of Local Government in Zimbabwe to start developing smart cities as a strategy to cope with limited resources and to regenerate run-down residential areas. Gradually, the original idea has become more concrete and in early 2020, the government of Zimbabwe announced to invest 23 million US$ in the development of a new smart city Mount Hampden straddling Mashonaland West and Mashonaland Central Provinces, some 17 km from the capital of Harare. At that time, however, very limited knowledge and experience existed in Zimbabwe and southern Africa on how to construct such smart cities in a sustainable manner. This initiated the idea to dedicate the biannual planning symposium to this very issue in 2019 for which various authors prepared manuscripts. The symposium could, however, not take place due to the COVID pandemic, which left many of the papers unpublished and ideas unshared. A second event was a workshop in Zanzibar in July 2019 organised by the ADLAND project, focusing on enriching and implementing collaborative research and capacity development in responsible and smart land management in and for Africa. During this meeting, a third event was agreed upon to enhance publishing skills and enabling land management and spatial planning researchers to publish in internationally recognised journals and books. This brought together a similar group of authors in an online event on 18 December 2020, dealing with how to edit and publish books and book chapters. This last gathering resulted in a joint decision to publish a book on sustainable and smart spatial planning for southern Africa.

DOI: 10.1201/9781003221791-2

The process through which this book was written started by the collection and selection of original manuscripts for the third bi-annual symposium on planning in southern Africa. Each of these manuscripts underwent a thorough peer review and revision process, which resulted in a selection and classification of chapters. This selection and classification process relied on the assessment of the content, novelty and relevance with the context of the scope and objective of the book, selection of empirical, sample and case materials and selection of references.

In the context of rapid and uncertain developments, which affect space, and the use of land, the book fills a gap in understanding how smart city policies and strategies can strengthen urban functions in developing countries in general and in Africa in particular. Currently, access and application of information communication technologies and usage of open and linked data pose a major constraint to the informed and smart development of human settlements, especially in urban regions. The smart concept can serve as a catalyst in enhancing the capability and competitiveness of the settlements in national economic development. This book draws on and focuses on examples from southern Africa particularly to show that despite the internationally converging definitions and conceptualisation of the smart city concepts, contexts and historical developments influence how, where and when the smart city concept can provide responsive and adaptive solutions to current challenges. Of particular importance are the effects arising from rapid urbanisation, such as those relating to public health, affordable shelter, water supply, solid waste management and energy supply. In addition, novel challenges such as mobility, access to updated information, tenure security and usage of mobile and smart technologies play a role.

The scope and justification of the book is to provide critical assessments of how and where spatial and settlement planning is currently changing, either as a result of or despite technological advancements. Such assessments and new insights can enhance the knowledge and background of current spatial planning practitioners and researchers in their daily work. The specific institutional and societal contexts of many parts of Africa, such as the dual and/or customary systems of land rights, incomplete systems of land registrations, post-colonial institutional legacies of planning laws and procedures, complexities in national and supranational responsibilities and accountabilities and transnational socio-economic dependencies provide particular challenges that are often idiosyncratic. At the same time, multiple cross-national and international networks of professionals and research institutions have emerged, which have generated new opportunities and chances for African nations, administrators, practitioners and experts. The domain of spatial and settlement planning is thereby no exception. Networks such as the Network of Excellence on Land Governance in Africa (NELGA), Global Land Tool Network (GLTN), the African Urban Research Initiative, the Agendas 2063 & 2030, the Association of African Planning Schools and various other regional African networks have all intensified their research and professional exchange and innovation activities. In this sense, there is now more scope than ever to make an inventory of multiple planning activities and impacts and thus to derive current impacts and further requirements.

1.2 SUSTAINABLE SPATIAL PLANNING

There are multiple ways of looking at spatial planning in general and specific to local institutional, development or topographic contexts. Process wise, GIZ (2012) looks at spatial planning as a consecutive process (encompassing phases of definitions of objective and approach, analysis of problems, plan formulation, approval process, implementation process and activities and monitoring). Lagopoulos (2018) and Wagner and de Vries (2019) view it as a more complex and reiterative process with several internal feedback loops. Metternicht (2018) makes a differentiation between different types and classes of spatial planning, including integrated land use planning, participatory land use planning and regional land use planning. Finally, Chigbu et al. (2016) and Filipe and Norfolk (2017) emphasise particular output requirements for land use planning, such as the need for providing tenure security or the need to address the poor and alleviate poverty. However, in the context of Africa, and of southern Africa in particular, there is still the legacy of colonial planning systems and procedures that is not suitable for current African realities (Andres et al., 2021) (Chirisa & Dumba, 2012). To a large degree, it has also contributed to informal spaces and spatial inequities. It is therefore also opportune and appropriate to reflect on which alternative planning processes would better suit current realities.

The global sustainable development agenda, resulting in the 17 sustainable development goals (SDGs), has a significant impact on the spatial planning goals and processes from multiple integrated multidisciplinary perspectives. First, many of the SDGs have a direct or indirect relation to spatial planning. Just to mention a few of these, reducing poverty (SDG1) is clearly a spatial problem. Spatial inequalities exist in relation to access to jobs and economic opportunities, to public service and to infrastructure, health and schools. Making both cities and rural regions more sustainable thus requires a better spatial alignment in socio-economic improvements. Similarly, SDG11, focusing on sustainable cities and communities, requires better spatial plans enabling affordable housing, transportation and energy resources. Finally, life on land (SDG15) aiming to sustainably manage forests, combat desertification, halt and reverse land degradation and halt biodiversity loss requires a careful arrangement of spatially diverse stakes and interests.

1.3 SMART SPATIAL PLANNING

Gradually there has been a shift towards the 'smart' settlement paradigm in planning in Africa (Echendu & Okafor, 2021). Smart in this case not only refers to the adoption of technologies, digitisation and automation in multiple settings and procedures but also to the enhancement of how citizens can contribute and participate in planning processes with mobile and internet-based applications. In sub-Saharan Africa, the smart concept has so far mostly been applied to water supply (Nickum et al., 2020), renewable energy (Sterl et al., 2020), solid waste management (Gelan, 2021) and mobility/transportation sectors (Acheampong, 2021). Yet, the advancements in processing, predicting, simulating and information provision tend to be rather monolithical and have yet to be assessed in terms of its impacts on the transformation and the reorganisation of land rights, participatory governance and sustainable livelihoods. Making spatial planning 'smart' is therefore much broader than limiting it only to technological advancements. It is therefore important to focus research not only on the design of technological innovations but also on the impacts and requirements on spatial planning processes and the degrees to which the technical changes also affect societal and planning necessities.

1.4 JUSTIFICATION AND OUTLINE OF BOOK SECTIONS

The book fills a gap in knowledge about how sustainable-smart city policies and strategies can strengthen urban functions in developing countries where information communication technology and data access pose a major constraint to sustainable urbanisation. It gives an African perspective of sustainable and smart spatial development, thereby broadening the understanding of the African city. In doing that, it explores the potential to which the smart concept is providing responsive and adaptive solutions to challenges arising from rapid urbanisation. This is mostly accomplished through place-specific analyses and solutions.

It begins by examining how existing frameworks for the planning and development of sustainable and smart settlements have evolved over time to inform emerging trends in Africa. It then explores conceptual and theoretical materials on smart concept and smart growth in urban and rural areas. Next, it discusses the African context particularly with respect to technology and data limitations and adaptive smart solutions that have shown potential for replication at scale and in localised contexts. This is followed by case studies on urban laboratories, solar harvesting as a sustainable and resilient source of energy, design of sustainable small- and medium-scale business spaces and smartness with respect to spatial justice in rural towns. Lastly, the mixed methods and spatial planning tools including geographic information systems and geographic positioning systems that have been used in the application of smart concepts are evaluated for their effectiveness and sustainable use.

REFERENCES

Acheampong, R. A., 2021. Societal impacts of smart, digital platform mobility services—an empirical study and policy implications of passenger safety and security in ride-hailing. *Case Studies on Transport Policy*, March, 9(1), pp. 302–314.

Andres, L. et al., 2021. *Planning for sustainable urban livelihoods in Africa*. s.l.: Routhledge.

Chigbu, U. E. et al., 2016. Combining land-use planning and tenure security: A tenure responsive land-use planning approach for developing countries. *Journal of Environmental Planning and Management*, 29 November, 60(9), pp. 1622–1639.

Chirisa, I. & Dumba, S., 2012. Spatial planning, legislation and the historical and contemporary challenges in Zimbabwe: A conjectural approach. *Journal of African Studies*, 4, pp. 1–13.

Echendu, A. J. & Okafor, P. C. C., 2021. Smart city technology: A potential solution to Africa's growing population and rapid urbanization? *Development Studies Research*, 08 March, 8(1), pp. 82–93.

Filipe, E. & Norfolk, S., 2017. Changing landscapes in Mozambique: Why pro-poor land policy matters. *IIED Briefing papers—International Institute for Environment and Development*, January.

Gelan, E., 2021. Municipal Solid waste management practices for achieving green architecture concepts in Addis Ababa, Ethiopia. *Technologies*, 11 July, 9(3).

GIZ, 2012. *Land use planning. Concept, tools and applications.* s.l.: GIZ.

Lagopoulos, A. P., 2018. Clarifying theoretical and applied land-use planning concepts. *Urban Science*, 18 February, 2(1).

Metternicht, G., 2018. *Land use and spatial planning. Enabling sustainable management of land resources.* s.l.: Springer.

Nickum, J. E., Kuisma, S. & Bjornlund, H. S. R. M., 2020. Smart Water Management: The way to (artificially) intelligent water management, or just another pretty name? *Water International*, 3 November, 45(6), pp. 515–519.

Sterl, S. et al., 2020. Smart renewable electricity portfolios in West Africa. *Nature Sustainability*, 25 May, 3, pp. 710–719.

Wagner, M. & de Vries, W. T., 2019. Comparative review of methods supporting decision-making in urban development and land management. *Land*, 7 August, 8(8), p. 123.

Section II

Theory of Sustainable and Smart Spatial Planning

2 Stocktaking Frameworks for the Planning and Development of Smart Cities

Tafadzwa Mutambisi and Charles Chavunduka

CONTENTS

2.1	Introduction	11
2.2	Background and History of the Smart City Concept	12
	2.2.1 Traditional City	13
	2.2.2 Information Age	13
2.3	Theoretical Underpinnings	15
	2.3.1 Urban Systems Theory	15
	2.3.2 A Theory of Smart Cities	15
	2.3.3 SMELTS Smart City Framework	16
	2.3.4 Smart City Transformation Framework	18
2.4	Methodology	19
2.5	Findings and Discussion	19
	2.5.1 Implementation of Smart City Frameworks in Southern African Countries	19
	2.5.2 Zimbabwe	20
	2.5.3 South Africa	20
2.6	Conclusion	21
References		22

2.1 INTRODUCTION

The development of smart cities in developing and developed countries has been instigated by rapid urbanisation and land economics (Kumar et al. 2020). Scarcity of land has pushed policy makers and other actors to come up with various solutions to different problems that come with population growth. Land as well as urban and rural space is the most vital resource that humankind has (Falconer and Mitchell 2012). Managing the rural and urban spaces has various challenges and a number of contemporary urban concepts such as the smart city initiative can promote responsive and adaptive solutions to local challenges. Smart cities are digital cities or technologically-defined cities that use widespread broadband infrastructure to support e-Governance and a global environment for public transactions (Ferrer 2017). In sub-Saharan Africa, the smart city initiative has been the most talked about solution to various challenges faced in the region. However, there still exists a major gap between theory, practice and context. There is a need to fully understand the underlying frameworks for smart cities and their relevance to cities in the global South. Promotion of positive development and functional urban settlements comes from a complete understanding of the structures and systems of smart cities. This helps to maximise the opportunities that arise in implementing smart cities using frameworks that best fit the sub-Saharan Africa region.

New cities and new city plans across Africa differ in their spatial forms, locations, purposes/aims, marketing terms and relation to existing cities; these are connected to their local and national contexts (Joshi et al. 2016). The complexity of social systems in different areas of the continent increases the need to fully assess various frameworks for smart cities implored various countries and their relevance to southern Africa. Despite this, the smart city concept encompasses a wide array of loosely and broadly defined technological solutions to city challenges (Angelidou 2016). The ambiguities of these definitions represent a normative vision of the future which is concisely described and therefore still remains debatable amongst various actors in development. Despite the complex nature of smart cities, many countries in the developed world have contextualised and implemented the concepts in their cities and in their own ways to suit their social, economic and environmental areas. Many of the countries have developed smart cities as a way to minimise the use of energy in their urban centres. Therefore, the smart city initiative was an integrated and innovative solution in promoting the efficient use of the city's scarce resources and promoting the green concept. This shows that the development of smart city concepts is derived from the goals that cities wish to fulfil. These goals differ in their complexity across the globe as cities on their own are complex ecosystems harbouring different stakeholders with diverse interests. A Smart city is a harmonic blend of history with character, heritage, aesthetics, architecture, economy, environment and lifestyle.

The aim of this chapter is to assess and examine the various frameworks used by the developed world in the creation of existent smart city concepts. The chapter seeks to give a theoretical overview about the development of smart cities. It aims to answer the question: How well has the smart city concept travelled in various global contexts? Answering this question relies on an analysis of the effective aspects, limitations and methodologies used in the development of smart cities in various regions. The findings assist countries in the sub-Saharan region in coming up with appropriate frameworks for African cities and shed light on how the countries can effectively implement the frameworks in their own cities.

2.2 BACKGROUND AND HISTORY OF THE SMART CITY CONCEPT

This part of the chapter will clearly highlight how urban management shifted from traditional city management to the smart city.

2.2.1 TRADITIONAL CITY

Urban planning has been around for a long time and many of the urban planning principles date back to ancient Greece. The principles behind the theory of smart cities have evolved from the time of the earliest city planner Hippodamus of Miletus who was the inventor of the 'City Beautiful' concept that has shaped many cities in the world (Albino et al. 2015). Miletus is responsible for devising the city planning grid system. From the grid system, garden city and the development of skyscrapers, the issue of liveable cities has been the main goal of urban planners such as Sir Ebenezer Howard. He was convinced that for liveable and sustainable cities to take shape, planning and construction had to be a requirement (Hügel 2017). His views were influenced by his experience in the United States of America where he witnessed Chicago rebuilding itself according to its previous shape. A century later, the world has entered the age of the Smart City and the concept owes its existence to the garden city concept.

2.2.2 INFORMATION AGE

As was introduced in the earlier section, Table 2.1 lists the major milestones of the smart city concept.

Smart cities were developed as a way to move away from the traditional way of managing the urban environment. The road to smart cities dates back to the 1970s when Los Angeles created the first urban big data project 'A Cluster of Analysis of Los Angeles' in 1974. This is the time that cities started incorporating information and communication technologies (ICTs) into urban management. The term smart city was developed in the US inside the business environment of two corporations, IBM and CISCO (Cugurullo 2018). The term was developed to indicate an idealised city connected to topics of automation. The smart city concept that relates to building liveable cities has evolved over time due to technological advancement. This includes the existence of advanced infrastructure developments over the years for power usage, green building design, water management and management of greenhouse gases are some of the dynamics of today's cities (Crous et al. 2017). The main aim of smart cities is to improve economic growth and improve the quality of life of people by enabling local area development and harnessing technology that leads to smart outcomes. The first smart city was arguably Amsterdam with the creation of a virtual digital city in 1994. After that, many countries were incorporated in the smart city initiatives launched separately by CISCO and IBM.

Cities play a prime role in social and economic aspects worldwide and have a huge impact on the environment. Most resources are consumed in cities due to rapid population growth (Allen 2015). Thus, the ultimate goals of smart cities were optimum usage of space and resources along with an efficient and optimum distribution of resources. It also aims at increasing connectivity at various levels among citizens as well as between the administration and the population. Smart city definitions have evolved over time. In the 1990s, when the concept gained popularity, the concept was more linked to ICTs with regard to modern infrastructures within cities (Albino et al. 2015). It may have its origins in the Smart Growth movement of the late 1990s, which

TABLE 2.1
Major Milestones in the Smart City Concept

Year	Development
1974	Los Angeles creates the first big data project 'A cluster analysis of LA'
1994	Amsterdam creates a virtual digital city to promote internet usage
2005	CISCO put up US$25 million over five years for research into smart cities
2008	IBM Smarter Planet project investigated applying sensors, networks and analytics to urban issues.
2009	IBM gives US$50 million towards Smarter Cities campaign to help cities run more efficiently ARRA provides funding for US smart grid projects
2010	Japanese government names Yokohama as a smart city demonstrator project
2011	IBM names 24 cities as Smart Cities winners from 200 applicants Smart city expo world Congress held in Barcelona
2012	Barcelona deploys data-driven urban systems including public transit, parking and street lighting
2013	China announces first batch of pilot smart cities Smart London Board created by Mayor of London to shape the city's digital technology strategy
2014	China announces the second batch of 103 pilot smart cities
2015	China announces the third batch of 84 smart cities
2017	UK government launched the 5G test beds and trials programme
2019	US Federal Communications pick New York and Salt Lake City as 5G test beds
2020	Vietnam to start work on a new $4.2 billion smart city close to Hanoi with a completion target of 2028

Source: Adapted from Global Data Thematic Research (2020)

advocated new policies for urban planning. With time, various experts in cities have redefined the smart city concept and moved it from being technical to a more user-friendly concept that covers various aspects and dimensions that make up a city. It moved from being called an Information Society to Digital City or Smart City and today some call it an Internet of Things (IoTs) (Cugurullo 2018). These definitions are a brand, fashion, project, or a marketing concept but it is a way to define, under the same concept, the outcomes of the information revolution. In the urban planning field, the term smart city is described as strategic directions in policy-making (Datta 2015). Governments and public agencies at all levels are embracing the notion of smartness to distinguish their policies and programmes for targeting sustainable development, economic growth, better quality of life for their citizens and creating happiness (Tahir and Malek 2016). Thus, the smart city concept is no longer limited to diffusion of ICT, but it looks at people and community needs. Many cities in the world such as Barcelona have come to an understanding that Internet and new technologies are excellent opportunities to transform the city as well as reshape and rethink every aspect of the urban space; that is, energy, logistics, infrastructure, city management, housing, mobility, education, public space, healthcare, security and many more (Ferrer 2017). Therefore, this research assesses various aspects of the smart city initiative and its frameworks looking at their relevance in sub-Saharan Africa.

2.3 THEORETICAL UNDERPINNINGS

The research uses two main theories, which are the urban systems theory and the theory of smart cities. These concepts serve to highlight the complexity of the urban space and the need for assessing and examining different frameworks that work best in the context of sub-Saharan Africa.

2.3.1 URBAN SYSTEMS THEORY

Urban areas are complex spaces with lot of chaos and the need to unravel this chaos led to the introduction of the urban systems theory in the early nineteenth century (Hussein and Suttie 2016). A system is networks that consist of interconnected parts such as, housing, transport industry, commercial zones and institutions. An urban system is a bottom-up view of how the city works, based on the actions of the people who live there. The urban systems theory aspires to make the complexity of the city and intercity interactions understandable by isolating some of its constituent social processes and then relating these processes to others occurring both inside the city and between cities and the outside world (Jain and Courvisanos 2009). If one finds the interconnectivity and relationship between various flows of information, goods and money between cities, thick knots of inter-city interactions all of a sudden start to make sense. Therefore, to understand smart city frameworks for various cities, policy makers should first acquire the knowledge of their urban systems in their given contexts (Falconer and Mitchell 2012). We have a unique opportunity in the coming decades to shape the future of global society through innovation in urban systems.

2.3.2 A THEORY OF SMART CITIES

Smart city is an ultra-modern urban area that addresses the needs of businesses, institutions and especially citizens (de Falco et al. 2019). There is a need to understand the theoretical foundation of smart cities to develop the understanding of how technical methods can help achieve the pressing goals of existing and new cities. An epistemological view of smart city initiative emanates from understanding the main components that it holds and these include a smart economy, people, governance, mobility, environment and living (Falconer and Mitchell 2012). It is important to include phrases such as clean city, IoTs and friendly city in understanding the smart city concept, as they are also vital components in building sustainable or smart technologically led cities.

Smart economy relates mainly to the industry in a given city, public expenditure on GDP per head of city population and also the unemployment rate (Ferrer 2017). Smart people are the percentage of population with secondary-level education, foreign language skills, participation in life-long learning, individual level of computer skills and patent applications per inhabitant (Lim et al. 2018). Smart governance relates to the number of universities and research centres in the city, e-Government online availability, percentage of households with internet access at home and e-Government use by individuals. Smart mobility is sustainable, innovative and safe public transportation, pedestrian areas, cycle lanes, green areas, production of solid municipal waste, fuels, political strategies and perspectives, availability of ICT infrastructure and flexibility of labour market (Joshi et al. 2016; Sujataa et al. 2016). Smart environment is mainly concerned with ambitiousness of CO_2 emission reduction strategy, efficient use of electricity, efficient use of water, area in green space, greenhouse gas emission intensity of energy consumption, policies to contain urban sprawl and proportion of recycled waste. Lastly, smart living is the proportion of the area for recreational sports and leisure use, number of public libraries, total book loans and other media, museum visits, theatre and cinema attendance, innovation, transparent governance, sustainable resource management, education facilities and health conditions.

2.3.3 SMELTS SMART CITY FRAMEWORK

The SMELTS Framework (Figure 2.1) provides a detailed view of how smart cities should function and it also highlights different aspects that make up an urban system (Joshi et al. 2016). It has six significant pillars that are meant to give insights on the main aspects that define a smart city. These have been used to design a framework

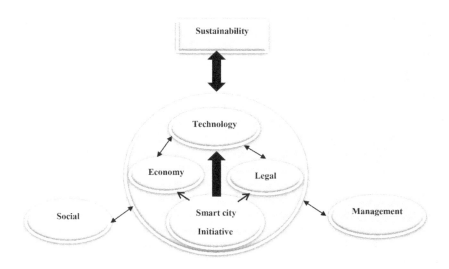

FIGURE 2.1 Showing SMELTS framework.

Source: Joshi et al. (2016)

that gives a more holistic view of the smart city initiative. The six pillars are social, management, economy, legal, technology and sustainability. Each of these will be elaborated upon in the following.

Social aspect of cities relates to the people, how they live and how they interact. This provides a new sense of possibility to the idea that smart cities are based on smart communities whose citizens can play an active part in their design (Meijer 2018). This is possible through the ability of citizens to actively communicate with each other as well as with groups and agencies that represent them. Therefore, the use of digital systems and ICT enables easier and faster communication amongst communities.

Management through governance of the urban space is a vital component in promoting development though the smart city initiative (Adamolekun 1989). In an effort to reduce limited transparency, leakage of resources, unequal city divisions and fragmented accountability digital or e-governance is an essential, effective and efficient administration tool of smart cities. The internet has become an open ground for citizens to be comprehensively involved in different aspects of decision making (International Federation for Information Processing (IFIP) 2018).

A good or bad economy can either drive or destroy smart city initiatives. Operational definition of a smart economy includes factors all around economic competitiveness as entrepreneurship, trademarks, innovation, productivity and flexibility of the labour market and integration in the national and global market (Malandrino et al. 2019). For a city to be called intelligent or smart, it should have the capabilities and capacities that allow innovation, capitalisation and efficiency in the use of its scarce resources. The smart city, like all models used for economic development, supports the maximisation of profits. The economic outcomes of the smart city initiatives are business creation, job creation, workforce development and improvement in productivity.

Evolution of smart cities cannot be successful without legitimate legal compliances. There is need for governments, policy makers and other actors involved in the smart city initiatives to come up with laws that assist and support the development of smart cities. The policies must conform and be compliant with the technical as well as non-technical requirements that are essential for urban growth. Therefore, before any decision can be made, rules and regulations pertaining smart cities must be kept in mind (Blackman 2003).

Technology is the backbone of the smart city initiative. Modern cities are getting smarter because of the rapid evolution of technology. Technology promotes smart sustainable cities as technology gives out data that helps map out problems that are likely to occur such as floods (Lim et al. 2018). This helps avoid, anticipate and mitigate the problem. Various devices and components must be connected with each other to facilitate real-time decision-making. Information plus technology therefore drives the smart city initiative in its quest to increase sustainability and improve the quality of life for citizens.

Sustainability is a way of economic and social development without disrupting the environment (Muchadenyika 2016). Smart city initiative should be implemented with the sustainability concept in mind. This aspect can be classified into social, economic and environmental sustainability. These aspects make up the major requirements of city environments comprising sustaining water, energy and food supplies, managing water and reducing greenhouse gas emission.

The SMELTS city framework has been utilised and adopted by many cities to achieve smart city standards and implement successful smart city projects. The elements that make up the SMELTS framework have laid a foundation on which many cities have set their own smart city frameworks. Another framework that has received increasing application in sub-Saharan Africa is the smart city transformation (Kumar et al. 2020).

2.3.4 SMART CITY TRANSFORMATION FRAMEWORK

The smart city transformation framework provides a summary of complex transformation processes that are needed for a city to be smart (Kumar et al. 2020). The framework contains four key areas, which are planning phase, physical infrastructure development, ICT infrastructure development and deploying solutions to transform a city into smartness.

The planning stage emphasises the importance of governance in building smart cities. Governance helps to enable coordination and integration across city departments to set up platforms for better service delivery (Muzondi 2014). In this framework, the public administrator is seen to be playing a lead role in the implementation of this smart city initiative. Urban planners, public administrators, civic groups and the citizens are key actors in the development of cities, their relationship, coordination and corporation is vital to sustainable growth in cities (Allen 2015). The greatest enemy to the progress of Zimbabwe's policies is the inconsistencies that are found in its policy-making system (Hahlani 2012). Beyond that, internal politics in development institutions and organisations also hinder development.

The physical infrastructure development phase signifies the greatest asset that a city has. This includes objects like roads, sewerage and energy networks. Thus, smart cities should focus mainly on core infrastructural development so as to build sustainable environments. The physical aspects that a city should focus on providing are adequate housing, water, energy supply, proper sanitation facility, waste recycle, efficient mobility, multi-modal transport, telecom connectivity, networked communication, safety of citizens especially for children, women and differently abled people, pollution-free environment and health and education facilities. These can be achieved through systems such as E-metres for power and water.

ICT infrastructure development is critical because efficient and effective information and communication software is important in building a smart city. Software components include application software, computer programs, data visualisation, data science, information system and mobile apps for various services. The implementation of smart technology can be advantageous if it can quickly detect or monitor various spatial or unusual events happening in and around the city. ICT, telecom and advanced network infrastructure deliver better e-services to its stakeholders.

When deploying smart solutions, 'smart' is the collective intelligence of software, hardware, cloud and sensing technologies to capture and communicate real-time sensor data of the physical world for advanced analytics and intelligent decision-making process. ICT systems help policy makers and city administrators to develop new digital solutions. For instance, in Harare, Zimbabwe there has been the establishment of smart E-metres for power and water to fix the problem of payments and credits.

2.4 METHODOLOGY

The thrust of this research is to highlight the frameworks used in the development of smart cities around the world. This was conducted through qualitative research that made use of desk study and documentary analysis. The theoretical framework for smart cities is used to analyse the progress in the adoption of the smart city concept in southern Africa. Theories tell precise and coherent stories about natural or human processes according to a set of laws that describe relationships between events or facts (Mutizwa-Mangiza 1986). The smart city frameworks discussed in this study describe the foundations on which smart cities are based and they reveal insights on how sub-Saharan Africa can increase the rate of adaption of the frameworks in cities. Secondary data sources that were used include journals, newspapers, books and policy documents. Triangulation which is the use of multiple sources of data was implored to validate and get access to different smart city frameworks. Documentary analysis also enhances the reliability of the research as it gives first-hand information (de Falco et al. 2019). Content analysis is the use of textual material in research and reducing it to more relevant, manageable bits of data. It is also a method of analysing the text of social investigation among the set of empirical methods (Kumar et al. 2020). This assisted the research in classifying information into different categories and sections to highlight frameworks used in smart city development.

2.5 FINDINGS AND DISCUSSION

2.5.1 IMPLEMENTATION OF SMART CITY FRAMEWORKS IN SOUTHERN AFRICAN COUNTRIES

Municipalities within Africa are faced with significant challenges related to the management of infrastructure. Most notably, Africa has seen rapid urbanisation over recent decades, and this trend is expected to continue (Crous et al. 2017; van Noorloos and Kloosterboer 2018). This has caused a lot of pressure on the continent's resources. Therefore, there is dire need for African cities to adapt and implement smart city initiatives to help deal with the urban challenges they are facing. Besides urbanisation, globalisation has also paved way for African cities to start implementing policies that are directed towards smart cities. The coming of the digital era has paved way for technology to assist in problem solving in cities. Smart cities require the deployment of innovative technologies including sensors, mobile technology and big data analytics across various infrastructure sectors including transport, energy, buildings, drinking water supply and wastewater treatment systems (Harrison and Donnelly 2011). There is need to critically assess the way smart cities have been implemented in the past as well as currently in other developed countries for relevance and application to African countries. Efforts to incorporate smart city initiatives through E-government services in southern Africa have been noted in countries such as Zimbabwe and South Africa (Green 2008; Muzondi 2014).

2.5.2 Zimbabwe

Zimbabwe has been adopting smart city initiatives in different aspects of its urban areas. Municipalities have introduced smart water metres in cities such as Kwekwe, Masvingo and Harare. The Zimbabwe Electricity Transmission and Distribution Company (ZETDC) has also implemented the use of smart electricity metres country wide. Kondongwe and Tanyanyiwa (2017) observe that 75% of the residents in Chitungwiza which is in the Harare metropolitan area are using prepaid electricity smart metres. The major problem that big cities such as Harare have faced over the years is traffic congestion. According to Mbara (2015), urban transport in Harare is in need of transformation and there is need to invest in mass transit systems that are efficient users of road space due to their high carrying capacities. This was implemented in 2019 when the government introduced Zimbabwe United Passenger Company (ZUPCO) buses for mass transit. In the transport sector, solar traffic lights have also been one of the successful smart energy projects implemented. In cities such as Harare and small towns like Zvishavane, the use of solar traffic lights has proved efficient in managing traffic, especially during power cuts and black outs. In 2018, the government of Zimbabwe ventured into urban renewal and regeneration projects in Harare, Bulawayo and Mutare. Some hostels in Mbare, Harare; Sakubva, Mutare; and Makokoba; and Bulawayo have gone through urban regeneration. However, the success of these smart city initiatives is measured on the basis of acceptance by citizens as well as revenue generation. The most accepted and successful smart city initiatives are the prepaid electricity smart metres. Kondongwe and Tanyanyiwa (2017) observe that Chitungwiza people have accepted smart metres as an efficient and effective way to monitor their expenses and usage of electricity. Smart city initiatives implemented in the country have generated revenue especially in the energy sector (Ropa 2015).

Due to various factors such as poor governance, economic instability and poor planning, service delivery is still poor in the country. Water service delivery is poor, especially in areas like Msasa Park, where some areas hardly receive piped water supplies (Gambe 2013). Building a sustainable framework for the implementation of smart city initiatives has been derailed by a number of structural and external problems in Zimbabwe (Rajah 2015). Zimbabwe has been beleaguered by economic, social and political turmoil in recent years, which has had a debilitating effect on its already-declining economy. Despite the country's dedicated national ICT policy adopted in 2005, there is no clear set framework and funds to fully tackle the smart city initiative. There is a disconnection between digital and physical environments and the lack of supporting regulations. This is supported by Muchadenyika (2016) when he states that the Harare City Council's failure to manage the urbanisation trends can be attributed to the economic and political crises as well as the lack of urban policies in Harare City and the institutional weaknesses of the legislation governing water and sanitation as well as other management systems.

2.5.3 South Africa

In South Africa, policies have been implemented that have assisted in moving cities towards smart city status. For instance, the country has incorporated an energy standard, SANS2004 (Green Building in SA), that aims at providing energy-saving practices as basic standards in the South African context. Cities such as Cape Town have implemented smart city projects such as free Wi-Fi in public places and cashless payment systems for public transport (Tshiani and Tanner 2018). E-government is one the main focal points of Cape Town's smart city strategies.

Cape Town has started to use more technology in its day-to-day management. In particular, smart metering is being used for electricity and water in 65% of the city's large administrative buildings. In the past, Cape Town also introduced smart city projects such as the SMART Cape Access. The aim of the SMART Cape Access Projects was to provide Cape Town citizens with free access to technology. More cities such as Johannesburg have recently introduced city management public Wi-Fi in many areas around the city and actively endeavoured to improve the city's broadband infrastructure to reduce the digital divide. This has been implemented as mobile phone usage in South Africa has increased dramatically over the years with the number of active mobile phones (over 66 million) now surpassing the number of people living in the country (51.8 million). Internet browsing through phones in South Africa is now teetering at 40% of the national population. About 34% of phone users make downloads from app stores, which is an indication of higher smartphone adoption (UCLG-A 2016). Therefore, Johannesburg boasts as the engine room of South Africa with the highest number of tech hubs in the continent. However, South Africa still faces privacy issues in implementation of smart city strategies. The country, as many other sub-Saharan countries, faces challenges in implementing successful smart city frameworks in some cities due to a number of factors. The factors include high costs, lack of proper knowledge, team integration within stakeholders, affordable technologies, proper communication strategies, no effective enforcement of laws by professionals and government and lack of stakeholders buy-in to the technology (Green 2008).

2.6 CONCLUSION

In conclusion, there are many smart city frameworks that can be implored by a city in implementing smart city initiatives. It can be seen that the SMELTS framework and the Kumar Framework are the most common frameworks used by African local governments. The smart city concept has evolved over the years. The dynamic development of innovative technologies has provided opportunities to build smart cities. The driving factors for smart cities have seen many cities expanding in their own complexities. This has given rise to different cities setting benchmarks, aims and foundations that underlie their smart city initiatives. It can be noted that implementation of successful smart city projects needs planning officials to undertake the programme using a holistic approach, that is, approaching the smart city initiative with a view of the social, economic, physical and technological aspects of a city. All the frameworks implored in this study make use of private and public partnerships and thus coordination, integration and good governance play a critical role in building smart cities. Due to political instability and corruption in some countries in southern Africa, it is still difficult to achieve sustainable and smart cities. Economic instability and civil unrest can cause smart city initiatives to fail. More countries like Zimbabwe do not have the capacity and technological knowhow to implement, install and run smart infrastructure systems. Smart city framework rests entirely on ICT as its bedrock. Thus, a country with limited access to these functions will limit its scope in the smart city initiative. The evaluation of smart city frameworks assists countries in sub-Saharan Africa to get a view of how best to implement their smart city projects. This assists them to help pick what is most appropriate and relevant to their given environments and problems as urban challenges across the world vary according to the place, environment and its people. There is need for policy makers to always develop a framework that helps shape their cities as planning and forecasting results in well-organised and sustainable cities.

REFERENCES

Adamolekun, Ladipo. 1989. *Issues in Development Management in Sub-Saharan Africa*. An EDI Policy Seminar Report, Number. 19. Washington, DC: World Bank.

Albino, Vito, Umberto Berardi, and Rosa Maria Dangelico. 2015. 'Smart Cities: Definitions, Dimensions, Performance, and Initiatives'. *Journal of Urban Technology* 22 (1): 3–21. Doi: 10.1080/10630732.2014.942092.

Allen, Adriana. 2015. 'Everyday Infrastructural Planning' in the Urban Global South: Urbanisation Without or Beyond Large Infrastructure? [OR] Access to Water and Service Coproduction in Peri-Urban Areas'. UCL Bartlett Development Planning Unit (DPU).

Angelidou, Margarita. 2016. 'Four European Smart City Strategies'. *International Journal of Social Science Studies* 4 (4): 18–30. Doi: 10.11114/ijsss.v4i4.1364.

Blackman, Tim. 2003. 'Urban Policy in Practice'. *Tailor and Francis E-Library*, 352.

Crous, P., G. Palmer, and R. Griffioen. 2017. 'Investigating the Current Status, Enablers and Restraints, and the Future of Smart City Adoption in Africa'. *IMESA*, 51–54.

Cugurullo, Federico. 2018. 'The Origin of the Smart City Imagery: From the Dawn of Modernity to the Eclipse of Reason'. In Lindner, C. and Meissner, M. (eds) *The Routledge Companion to Urban Imaginaries*. London: Routledge.

Datta, Ayona. 2015. 'A 100 Smart Cities, a 100 Utopias'. *Dialogues in Human Geography* 5 (1): 49–53. Doi: 10.1177/2043820614565750.

Falco, Stefano de, Margarita Angelidou, and Jean-Paul D Addie. 2019. 'From the "Smart City" to the "Smart Metropolis"? Building Resilience in the Urban Periphery'. *European Urban and Regional Studies* 26 (2): 205–223. Doi: 10.1177/0969776418783813.

Falconer, Gordon, and Shane Mitchell. 2012. 'Smart City Framework A Systematic Process for Enabling Smart+Connected Communities'. Cisco Internet Business Solutions Group (IBSG).

Ferrer, Josep-Ramon. 2017. 'Barcelona's Smart City Vision: An Opportunity for Transformation'. *Institut Veolia*, 70–75.

Gambe, Tazviona Richman. 2013. 'Stakeholder Involvement in Water Service Provision: Lessons from Msasa Park'. *International Journal of Politics and Good Governance*, Harare, Zimbabwe' 4 (4): 1–21.

Green, Lesley J. F. 2008. '"Indigenous Knowledge" and "Science": Reframing the Debate on Knowledge Diversity'. *Department of Social Anthropology, University of Cape Town* 4 (1): 144–163. Doi: 10.1007/s11759-008-9057-9.

Global Data Thematic Research. 2020. 'History of Smart Cities: Timeline'. *Internet of Things*. Available at https://verdict.www.vedict.co.uk

Hahlani, Crispen D. 2012. 'Bottlenecks to Integrated Rural Development Planning In Zimbabwe: A Focus the Midlands Province'. *The Dyke* 6 (2): 136–157.

Harrison, Colin, and Ian Abbott Donnelly. 2011. 'A Theory of Smart Cities'. *55th Annual Meeting of the International Society for the Systems Sciences*, 15.

Hügel, S. 2017. 'From the garden city to the smart city'. *Urban Planning* 2 (3): 1–4. Doi: 10.17645/up.v2i3.1072. Cogitatio Press. Lisbon, Portugal.

Hussein, Karim, and David Suttie. 2016. *Rural-Urban Linkages and Food Systems in Sub-Saharan Africa : The Rural Dimension*. 5. Rome, Italy: IFAD.

International Federation for Information Processing (IFIP). 2018. 'Digital Equity and Inclusion for ICT in Disaster Risk Reduction'. Available at www.itu.int/net4/wsis/forum/2018/Content/Uploads/DOC/f0908edbdf40412f84c164e626354c02/session-121_summary.pdf.

Jain, A., and J. Courvisanos. 2009. 'Urban Growth Centres on the Periphery: Ad Hoc Policy Vision and Research Neglect'. *Australasian Journal of Regional Studies* 15 (1): 3–24. Darling Heights, Australia.

Joshi, Sujata, Saksham Saxena, Tanvi Godbole, and Shreya. 2016. 'Developing Smart Cities: An Integrated Framework'. *Procedia Computer Science* 93: 902–909. Doi: 10.1016/j.procs.2016.07.258.

Kondongwe, Richard, and Vincent Itai Tanyanyiwa. 2017. 'Introduction of Pre-Paid Electricity in Zimbabwe: Insights from Chitungwiza'. *International Open & Distance Learning* 2 (4): 14.

Kumar, Harish, Manoj Kumar Singh, M. P. Gupta, and Jitendra Madaan. 2020. 'Moving towards Smart Cities: Solutions That Lead to the Smart City Transformation Framework'. *Technological Forecasting and Social Change* 153 (April): 119281. Doi: 10.1016/j.techfore.2018.04.024.

Lim, Chiehyeon, Kwang-Jae Kim, and Paul P. Maglio. 2018. 'Smart Cities with Big Data: Reference Models, Challenges, and Considerations'. *Cities* 82 (December): 86–99. Doi: 10.1016/j.cities.2018.04.011.

Malandrino, Ornella, Daniela Sica, and Stefania Supino. 2019. 'The Role of Public Administration in Sustainable Urban Development: Evidence from Italy'. *Smart Cities* 2 (1): 82–95. Doi: 10.3390/smartcities2010006.

Mbara, Tatenda Chenjerai. 2015. 'Achieving Sustainable Urban Transport in Harare, Zimbabwe: What Are the Requirements to Reach the Milestone?' *CODATU*, 14.

Meijer, Albert. 2018. 'Datapolis: A Public Governance Perspective on "Smart Cities"'. *Perspectives on Public Management and Governance* 1 (3): 195–206.

Muchadenyika, Davison. 2016. 'Sustainable Cities: Zimbabwe's Future Cities'. *The Space*. Available at www.thespace.co.zw.

Mutizwa-Mangiza, N. D. 1986. 'Formulating Research Proposals Urban and Regional Research', RUP Occasional Paper Number 1 (May): 1–15.

Muzondi, Loretta. 2014. 'Urbanization and Service Delivery Planning: Analysis of Water and Sanitation Management Systems in the City of Harare, Zimbabwe'. *MCSER Publishing, Rome-Italy* 5 (20): 2905–2915.

Noorloos, Femke van, and Marjan Kloosterboer. 2018. 'Africa's New Cities: The Contested Future of Urbanisation'. *Urban Studies* 55 (6): 1223–1241. Doi: 10.1177/0042098017 700574.

Rajah, Naome. 2015. 'E-Government in Zimbabwe: An Overview of Progress Made and Challenges Ahead'. *Journal of Global Research in Computer Science* 6 (12): 11–16.

Ropa, Darlington. 2015. 'A Study of the Impact of the Use of Prepaid Electricity Meters on Revenue Generation at Zimbabwe Electricity Transmission and Distribution Company (Zetdc)'. Master thesis? Harare, Zimbabwe: University of Zimbabwe.

Sujataa, Joshi, Saxena Sakshamb, Godbole Tanvic, and Shreyad. 2016. 'Developing Smart Cities: An Integrated Framework'. *Elsevier B.V.* doi: 10.1016/j.procs.2016.07.258.

Tahir, Zurinah, and Jalaluddin Abdul Malek. 2016. 'Main Criteria in the Development of Smart Cities Determined Using Analytical Method'. *Planning Malaysia Journal* 14 (5): 1–14. Doi: 10.21837/pmjournal.v14.i5.179.

Tshiani, Valerie, and Maureen Tanner. 2018. 'South Africa's Quest for Smart Cities: Privacy Concerns of Digital Natives of Cape Town, South Africa'. *Interdisciplinary Journal of E-Skills and Lifelong Learning* 14: 055–076. Doi: 10.28945/3992.

UCLG-A. 2016. Smart Cities—the African story. *Markets of Africa Ltd*, (5). ISSN: 7 00083 97207 1. Available at www.uclga.org.

3 'Smart City' Concept and Its Implications for Urban Planning Systems in African Cities

Ephraim Kabunda Munshifwa and Niraj Jain

CONTENTS

3.1 Introduction 25
3.2 Understanding the Smart City Concept: A Literature Review 26
    3.2.1 Definitions..26
    3.2.2 Ranking Frameworks 27
    3.2.3 Implications for African Cities 29
3.3 Zambia: Profile and the Smart Vision 30
    3.3.1 Smart Zambia Vision and the Smart City Concept 31
3.4 Door Step Conditions for African Cities 34
3.5 Conclusion 36
References..36

3.1 INTRODUCTION

African cities are facing myriads of challenges in urban planning and development emanating from rapid urbanisation and population growth. In a bid to find solutions, many are turning to new thinking such as the 'smart city' concept, similar to developments in Western countries (Harrison and Donnelly, 2011; Jasrotia and Gangotia, 2018; Boyle, 2019). However, the challenge is that many African countries are adopting this new thinking without fully investigating what this entails and what the doorstep conditions are to achieving that status. This chapter investigates the concept of a smart city and its link to urban planning. It asserts that urban planning provides the basis for the development of all towns and cities. It thus follows that in the movement from physical to smart cities, the availability of digital spatial data will be the starting point for the integration of geospatial information.

The chapter is arranged as follows: after this introductory section, the next section discusses in detail the concept of smart cities from its origin to current understanding. Section 3 focuses on Zambia and discusses the actions which the country has undertaken in its stride towards smart cities. From the literature reviewed and analysis of the case study, Section 4 outlines the doorstep conditions for African cities aspiring to attain the 'smart city' status while Section 5 concludes the chapter.

3.2 UNDERSTANDING THE SMART CITY CONCEPT: A LITERATURE REVIEW

3.2.1 Definitions

A number of studies (Harrison and Donnelly, 2011; Albino et al., 2015; Jasrotia and Gangotia, 2018; Boyle, 2019) start with the assertion that there is no specific definition for a smart city. Albino et al. (2015), for instance, provides a list of definitions of smart cities with most of these being post-2000. This chapter argues that these studies miss the fact that the concept is an evolving one which needs some form of categorisation to study it. This chapter categorises the definitions in periods of time, that is between 1998 and 2003, 2004 to 2007 and post-2008, simply referred to in this chapter as first-, second- and third-order definitions.

The 'first order' definitions of a smart city are derived from the smart growth movement in the US. For instance, Harrison and Donnelly (2011) and Jasrotia and Gangotia (2018) attribute the use of the word 'smart growth' to Bollier (1998) with its focus on how it could stop urban problems such city sprawl. The Southern Environmental Law Centre (SELC) also cited the same problems of poorly planned urban developments, traffic congestion and urban sprawl threatening the economy, communities and health and environment as necessitating new thinking in urban governance (SELC, 1999). The call at the time was to develop smarter growth policies to reverse these negative impacts of urban development. By 2000, the understanding of the smart growth concept was mainly found in urban development and planning literature with emphasis on urban spatial development, accessibility and transportation (Bennet and Biringer, 2000; Burchell et al., 2002; Southern Environmental Law Centre, 1999).

A review of the literature from 1998 to 2003 shows that the genesis of the smart city concept could be linked to attempts to solve urban problems relating to urbanisation and rapid population growth hence being referred to as the 'first order' smart city. By year 2000, a smart city was merely thought of as a 'vision'; a far-away, imaginary city (Hall et al. 2000).

A further perusal of the literature after 2010 (see Wygonik et al., 2015; Pan and Li, 2016; Chen et al., 2018) shows that the 'smart growth' movement has continued almost parallel to 'smart city' developments. Chen et al. (2018) defines this as the urban planning theory. Principles espoused by Smart Growth America have more to do with land use planning, city development and governance, without necessarily involving information technology (Smart Growth America, 2021). For instance, its emphasis is on mixed land uses and compact designs which signify intensification of development on already built-up land (Brownfields) and less conversion of green fields, requiring adjustment in thinking for urban planners.

The 'second order' smart city definitions relate to the involvement of technology companies. From about 2004, technology companies seemed to have gotten interested in the concept of urban growth which was now twisted to 'smart cities'. For instance, the concept gained prominence around year 2005 after the adoption of the term by technology companies such as Cisco, IBM and Siemens. The 2000s also saw the rapid growth of wireless internet communication (Wi-Fi) and smart phones with the capability to connect to the internet, boosting the quest for data-driven economies. Thus technology companies saw a business opportunity on smart cities to leverage on what Mattern and Floerkemeier (2010) calls the period of the 'internet of computers'. Information and communication technology (ICT), with the power of the internet, became the linking thread of all the actors involved in urban governance. From this perspective, use of ICT in urban governance is a recent inclusion despite its prominence in what now constitutes the concept of smart cities. In essence, smart cities have become the application of information technology to solve urban governance challenges related to traffic congestion, energy consumption and the environment.

The 'third order', post 2008, definition of a smart city is attributed to the 'birth' of the Internet of Things (IoTs); that is, the use of sensors, scanners, cameras, monitors, etc. for data capture at ultra-high speeds. Mattern and Floerkemeier (2010: 242) titled this 'from the internet of computers to the internet of things', signifying a shift in internet technology. Current definitions of smart cities emphasise the collection and analysis of real-time data for problem solving. Cisco Internet Business Solutions Group (IBSG) estimates that the IoTs was formally 'born' between 2008 and 2009 (Evans, 2011), marked as the point at which more devices were connected to the internet than the world population. Devices here include smartphones, tablets, sensors, cameras, etc. For instance, by 2010, an estimated 12.5 billion devices were connected to the internet while the world population was 6.8 billion, giving a ratio of 1.84 (Evans, 2011); that is each person had more than one device connected to the internet. Evans (2011) further calls this the next evolution of the internet and traces its roots to the Massachusetts Institute of Technology (MIT) at the Auto-ID Centre labs. Current definitions thus see smart cities as providing both virtual and physical platforms for different actors involved in urban development. Because of the prominence of ICT in these operations, some studies simply refer to the process as digitalisation of the city (Boyle, 2019).

3.2.2 RANKING FRAMEWORKS

As a step further, different parts of the world and organisations have formulated ranking or assessment frameworks to determine whether a city qualifies to be labelled smart or not; this section discusses some of these frameworks. In the process of developing these frameworks, some studies have argued that there is a difference between 'smart cities' and 'sustainable cities'. For instance, Ahvenniemi et al. (2017) argues that in the recent past there has been a shift in urban development from sustainability assessment to smart city goals. The study argues that there is a large gap between smart city and sustainable city frameworks in that the former lacks environmental indicators. The study proposes that assessments should focus not only on the use of technology but also on measures that contribute to the 'ultimate goals such as environmental, economic or social sustainability' (Ahvenniemi et al., 2017: 234). Thus, underpinning the achievement of these 'ultimate' goals is a functional urban planning system.

Recent developments in ranking frameworks or models for smart cities, coupled with review of the literature from as early as 1994, provide insight into the characteristics of smart cities. To understand the general characteristics of these cities, Ahvenniemi et al. (2017) analyses eight frameworks, namely: European Smart City ranking, Smart City Wheel, Bilbao Smart Cities study, Smart Cities Benchmarking in China, Triple Helix Model, Smart City Profiles and CityProtocal and CITYkeys. Based on this comparison, the study found that these frameworks emphasised economic and social impacts and less environmental impacts.

What is striking across the eight Smart City frameworks by Ahvenniemi et al. (2017) is the absence of urban planning components and more emphasis on economy, education, culture, science and innovation and well-being, health and safety. Even though urban planning is possibly captured under the 'built environment' sector, very few of these frameworks actually measure it. Reading between the lines, it can be deduced that urban planning is no longer one of the serious challenges for most Western cities, hence assumed as a given. In other words, urban planning systems have to first work as a basis for transformation into a smart city. The European Smart City ranking model[1] illustrates this point. This framework ranks cities based on six characteristics (Giffinger et al., 2007; Pinochet et al., 2019), namely:

1. **Smart mobility**: it relates to the provision of transport and ICT infrastructure to enable local, national and international accessibility. In addition, it emphasises the availability of ICT infrastructure and supports the integration between virtual and physical platforms.
2. **Smart economy**: it relates to the competitiveness of the city in relation to entrepreneurship, productivity, flexibility of labour, etc.
3. **Smart governance**: it emphasises participation in decision making by city residents, including provision of public and social services and transparent governance.
4. **Smart living**: it deals with the quality of life within the city, specifically it relates to cultural facilities, health conditions, safety of individuals, housing quality, education facilities, touristic attraction and social cohesion.
5. **Smart people**: it relates to social and human level of qualifications, creativity and participation in public activities. It further emphasises affinity to lifelong learning, flexibility, open-mindedness and social and ethnic pluralities.
6. **Smart environment**: it focuses on the natural resources available to the city as it deals with pollution, environmental protection and sustainable resource management.

Intrinsic in this European Smart City ranking framework is a functional urban planning system and reliable ICT infrastructure. Clearly, the current understanding of a

---

[1] A product of Vienna University of Technology (Department of Spatial Planning) and Partners

'smart city' goes beyond the digitalisation of processes to embracing the well-being of its inhabitants. Using this framework, ten top cities in the World have been classified as 'smart' starting with Singapore (Singapore), Seoul (South Korea), London (England), Barcelona (Spain), Helsinki (Finland), New York City (USA), Montreal (Canada), Shanghai (China), Vienna (Austria) and Amsterdam (Netherlands) (Smart City Governments, 2021). This framework is used later in this chapter to assess the level of development in Zambian cities.

3.2.3 Implications for African Cities

As seen in other parts of the World, attempts have been made to develop African 'smart' cities. For instance, Delusional Bubble (2021) lists 10 possible smart cities in Africa, namely Cape Town (South Africa), Konza Technopolis City (Kenya), Eko Atlantic City (Nigeria), Tatu city (Kenya), Hope City (Ghana), Waterfall City (South Africa), Vision City (Rwanda), Kigali (Rwanda), King City (Ghana) and Modderfontein (South Africa). A close look at this list reveals that some of these cities are still at the architectural design stage and under construction. Besides the traditional cities of Cape Town, Kigali and Modderfontein, the rest are newly planned largely privately funded mixed-use development projects.

Debates have thus ensued as to whether African cities can attain the level of development as demanded by the common understanding of a smart city or is it just a fantasy, dream or nightmare? For instance, Watson (2014) argues that African smart cities are largely a fantasy at the moment due to deep poverty and insufficient urban services; therefore, pursuing the smart city dreams without addressing these challenges will only result in further marginalisation and inequality in African cities. Thus, Watson (2014) further argues that urban development plans for most African cities are inappropriate in the context of 'smart' cities.

As a response to Watson (2014), Cain (2014) gives an example of a prestigious project in Angola which involved the development of Kilamba city in Luanda with 20,000 planned apartments using a public–private procurement approach and Chinese credit facilities. The study argues that without government's intervention through some subsidies, the apartments remained largely unsold rendering the project a potential flop. Cain (2014) questions whether financially the development of cities such as Kilamba city is feasible without the intervention of the State. Chiwanza (2021) presents similar arguments contending that there is nothing wrong in futuristic thinking; however, smart cities are not possible without solving the fundamental developmental problems of African societies. Chiwanza (2021) cites the planned Akon City to be located in Senegal on a US$6 billion budget. This futuristic metropolis will comprise luxury resorts, condos, offices, a hospital, a stadium and an artificial intelligence data centre on a 2,000-acre plot.

Other scholars, though, are more optimistic. For instance, Boyle (2019: 124) argues that African cities have an opportunity to 'leapfrog investments in technology'. Boyle (2019) further reports on the City of Cape Town's ambitious goal of becoming the first smart city in Africa. The study outlines the key characteristics of the digital city strategy of the city of Cape Town'. However, Boyle (2019) is not oblivious of the challenges in Africa and concludes that African governments need to ensure that local institutions aimed at addressing the challenges of cities such as urbanisation, housing deficits, informal settlements, etc. work. Simply investing in latest technology without understanding the deep-seated development problems of African cities is not a solution.

Studies also show that integration of diverse systems for attainment of the smart cities goal is also a challenge. For instance, Yu et al. (2019) argues that due to different systems running at the same time, integration as envisaged in smart cities can pose other challenges. For African countries, this is an old age problem where most of the technology is imported from other countries, posing serious integration challenges. Even when Boyle (2019) sees an opportunity for 'leapfrogging' investments in technology, caution should be exercised to ensure that: firstly, the technology is not too advanced for the level of development of a country, and secondly, that these technologies can 'speak' to each other.

Furthermore, other studies such as Porter et al. (2020) argue that 'smart' city planning often leaves out informal settlements in the loop, which already provides a challenge to African cities where the majority of its settlements are unplanned. The danger of wholesomely adopting the smart city concept is that African cities risk creating what de Soto (2000) calls 'bell jars', where a small part of the city is fully automated and moving to being smart while the greater part remains outside and undeveloped. Hence, Hammad et al. (2019) shows that within urban governance, smart cities can deal with issues relating to zoning, land-use allocation and location of facilities.

3.3 ZAMBIA: PROFILE AND THE SMART VISION

This chapter is largely based on in-depth review of the literature on smart cities, with particular interest on its link to urban planning. This section uses Zambia to highlight a number of points discussed in the literature. Zambia is classified as a lower middle-income country (LMIC) with a gross development product (GDP) of US$23.31 billion at 2019, a drop from US$26.31 billion in 2018 (World Bank https://data.worldbank.org/country/ZM). Like most African countries, Zambia faces myriads of challenges emanating from a weak economy, high rate of urbanisation and population growth resulting in increasing housing deficits, unplanned settlements and environmental degradation. World Bank statistics (see Table 3.1) further show that Zambia had a population of 17.86 million in 2019, an increase from 17.35 million in 2018. UN Habitat estimates that Zambia has a housing deficit of 2.8 million units, which is projected to increase to 5.6 million units in 2025 (Habitat for Humanity, 2020). Furthermore, statistics show that 70% of urban dwellers live in unplanned settlements. Income amongst Zambian citizens is also low, with a GNI per capita on US$1,440.00. Interestingly, citizens' adaption to information technology is very high with 89.2 per 100 having mobile cellular subscriptions. However, only 14.3% of the population use the internet. It is within this environment that 'smart' cities are being contemplated.

TABLE 3.1
Zambia Country Profile: Selected Variables

	1990	2000	2010	2018
World view				
Population, total (millions)	8.04	10.42	13.61	17.35
Surface area (sq. km) (thousands)	752.6	752.6	752.6	752.6
Population density (people per sq. km of land area)	10.8	14	18.3	23.3
Poverty headcount ratio at $1.90 a day (2011 PPP) (% of population)	55.2	52.1	65.8	58.7
People				
Income share held by lowest 20%	0.9	6.1	3.8	2.9
Environment				
Urban population growth (annual %)	2.6	1.4	4.2	4.2
Economy				
GDP (current US$) (billions)	3.29	3.6	20.27	26.31
GDP growth (annual %)	–0.5	3.9	10.3	4
Inflation, GDP deflator (annual %)	106.4	32.6	14	7.4
States and markets				
Mobile cellular subscriptions (per 100 people)	0	0.9	40	89.2
Individuals using the Internet (% of population)	0	0.2	10	14.3

Source: World Development Indicators database

3.3.1 SMART ZAMBIA VISION AND THE SMART CITY CONCEPT

One of the key components of smart cities is that governance systems, both national and local, need to work first to facilitate integration with the private sector (see Pinochet et al., 2019). Smart cities leverage on both public and private data to improve efficiency in service delivery for the city. Furthermore, the adoption of technology should be contextualised within the challenges of African cities.

At the national level, the Zambian government has embarked on a number of initiatives towards digitalising of various systems and processes. For instance, in 2015, the Smart Zambia vision was presented to the Parliament with its main objective being embracing a transformational digital culture for the country. Smart Zambia Institute (SZI), a Division under the Office of the President, was then established under Government Gazette Notice No. 836 of 2016 with a vision to digitise all government institutions. One of its first projects was the development of a National Data Centre and the ICT Talent Training. Phase I of this project was awarded to the Huawei Technologies Limited in 2012 (see Table 3.2) with a contract sum of US$65.5 million (GRZ/Huawei, 2012). This project involved the development of the building itself and supply of equipment for data storage, communication and training. It thus provides the nerve centre of a smart Zambia.

TABLE 3.2
Phase I Project—National Data Centre and the ICT Talent Training Project Costs

1	Facility	17,547,568.59
2	Data centre	14,161,353.17
3	Email system	2,498,592.00
4	ICT talent training	21,611,656.61
5	Telepresence system	4,483,924.75
6	Unified communication system	1,968,056.00
7	Assistant maintenance support	2,000,000.00
8	Freight and insurance	1,280,000.00
		$65,551,151.12

Source: GRZ/Huawei Technologies Limited, 2012[2]

In addition to the National Data Centre project, the government also implemented projects shown in Table 3.3 towards a smart Zambia.

In addition to these projects, the government also passed (or is in the process of passing) laws to facilitate the use of digital information in various areas: these laws include the Electronic Government Bill, N.A.B. 39 of 2021; The Cyber Security and Cyber Crime Act No. 2 of 2021; and The Data Protection Act No. 3 of 2021 (Tembo et al., 2018; GRZ, 2021a, 2021b, 2021c). Other initiatives towards a smart Zambia include the development of the National Spatial Data Infrastructure Programme and the National Land Titling Programme. The National Land Titling Programme was estimated to cost US$250 million, although the progress has been slow due to funding challenges (Tembo et al, 2018).

From the foregoing, it is clear that Zambia is spending massively towards implementation of ICT in government ministries and institutions. However, these projects are yet to directly address 'smarting' cities in the context discussed in the literature (Giffinger et al., 2007; Pinochet et al., 2019). For instance, it is true that the establishment of the SZI and the passing of laws such as the Data Protection Act, Electronic Communications and Transactions Act and Information and Communication Technologies open the country to the use of ICT both in public and private institutions. Furthermore, the National Spatial Data Infrastructure and the National Land Titling programmes, under the Ministry of Lands and Natural Resources, aim at improving the collection of data on land and geospatial information; however, progress on these two programmes has been slow due to a number of challenges relating to legal issues, manpower and equipment deficiencies and financial limitation (Tembo et al. 2018).

This chapter argues that despite these strides towards a smart Zambia at the national level, very few city initiatives have been reported on 'smarting' cities. For instance, World Construction Today (2021) reports on only one private project

---

[2] Also reported in Lusaka Times of January 5, 2015; www.lusakatimes.com/2015/01/05/pf-government-single-sources-huawei-us-65-5-million-scam-national-ict-project/

TABLE 3.3
Phase II Projects under Smart Zambia

Projects	Commencement of implementation	Objective	Status
Government National Data Centre	2012	Provides a hub for government wide net work	Tier I—Integrated Financial Management Information System (IFMIS) and Treasury Single Account (TSA).
Electronic services	2017–2018	e-payslips—provides for the automation of the production of payslips for public servants e-Cabinet—provides for electronic document and records management system e-pamodzi –	e-payslips—which in the initial stage captured 109,000 out of 169,000 civil servants on the electronic platform. Thus, by 2018, 190 civil servants were able to receive their payslips electronically
Government Wide Area Network	2014–2020	Provides for the connection of government ministries and institutions electronically.	Phase 1, implemented between 2014–15; it connected 20 government ministries and two institutions (Cabinet Office and Auditor General's office) Phase II, started implementation in 2016–2017; it connected Departments to ministries in nine provisional administration offices. Phase II ended in 2020, after connecting 152 Ministries and Departments.

Source: Smart Zambia Institute (SZI) (2021); www.szi.gov.zm/#

referred to as the 'Kalulushi Smart City' project in the Copperbelt Province, which is still at the architectural plan stage. The project is planned on a 67-hectare piece of land at an estimated cost of US$550 million. Among other developments, this 'city' is expected to build 200 residential units in addition to a shopping mall, hotel, office block, golf course, filling station and an adventure city (Construction Review, 2021). Compared to Singapore, Seoul, London or other planned cities in Africa (Senegal Nigeria, Kenya), this project dwarfs in comparison and can hardly qualify as a 'Smart city' using the European Smart City framework.

Overall, using the European Smart City framework to rank Zambian cities reveals clear deficiencies on this path. For instance, Zambia's ICT infrastructure lags behind the leading countries in sub-Saharan Africa with only 2,518 sites with fourth-generation (4G sites) internet access (World Bank, 2020), impacting the achievement of 'smart mobility'. In addition, Zambia's economic performance has been faltering in the past few years, recording a negative growth of 1.2% in 2020 (World Bank,

2021), inevitably affecting the quest for a 'smart economy'. Despite being a democracy, the participation of citizens in governance is low. For instance, despite participating in electing parliamentary representatives, interaction between these elected members and the citizens on governance matters is inadequate (Lolojih, 2010); many only return to their constituencies after five years to ask for re-election. Living conditions for at least 70% of the urban dwellers who live in informal settlements are poor, negating the quest for 'Smart living'. Because of lack of land (or poor) planning, recent studies show the allocation and development of property in environmentally weak areas of cities (Munshifwa et al., 2021). A combination of these factors explains why Zambian cities are far from qualifying as 'smart'.

As seen in the literature on the 'smart growth' movement, the precursor of smart cities, urban planning and related institutions need to work first (Bollier. 1998; Watson, 2014; Albino et al., 2015; Jasrotia and Gangotia, 2018; Boyle, 2019; Chiwanza, 2021). For instance, the *Urban and Regional Planning Act* of 2015 (replacing the *Town and Country Planning Act* of 1962 and the Housing Act of 1975) is the key legislation affecting urban planning in Zambia. In its introduction, it talks about ensuring sustainable urban and rural development by promoting environmental, social and economic sustainability in economic initiatives and control at all levels of urban and regional planning. The Act also recognises the need for a devolved system of urban governance to ensure multi-sectoral cooperation, coordination and involvement of various stakeholders. It is clear from the Act that the role of planning in urban areas is much wider than land use planning to include environmental, social and economic concerns. Thus, in the development of smart cities, this Act becomes a key coordinating instrument for local governments; at the moment, its link to the Smart Zambia initiative is unclear.

3.4 DOOR STEP CONDITIONS FOR AFRICAN CITIES

Globally, literature reveals that smart cities emerged from an attempt to address challenges facing cities such as rapid urbanisation, population growth, congestion, poor housing, urban sprawl, etc. (see Bollier, 1998; Wygonik et al., 2015; Pan and Li, 2016). In an attempt to solve these problems, information technology was seen as a means to help process information related to supporting integration of governance, energy and transportation in the implementation of development plans. However, it is undeniable that ICT companies are now driving the implementation of the smart city concept worldwide (Evans, 2011), at times overshadowing its original intent. It is therefore important at this point for African cities to pose and reflect on what an 'African smart city' entails and what the door step conditions are. This chapter isolates five key areas:

1. **Getting urban planning systems right**: It is clear that the genesis of the smart city concept is from urban spatial development. It is also clear that all modern 'smart cities' should have a workable development plan to guide it. A number of studies though show that most African cities are overwhelmed with rapid urbanisation and population growth challenges resulting in the

majority of its inhabitants living in informal settlements. Under 'Smart Living', smart cities aspire to provide quality life for its citizens through provision of health conditions and quality housing. Thus, the smart city dream cannot be realised if 70% of city dwellers are in informal unplanned settlements. The development of grandiose cities, such as Kilamba in Angola, Konza Technopolis in Kenya, Eko Atlantic in Nigeria, Hope City in Ghana and Akon City in Senegal, amidst poverty is seen as mere fantasising (see Cain, 2014; Watson, 2014; Chiwanza, 2021).

One of the demands for smart cities is mixed-use developments and more workable distances and integrated communities. For most African cities, the current town plans strictly separate commercial from residential uses, a product of traditional town planning.

2. **Ensure local public institutions work**: It is especially critical that local government structures work. The current state of most local government structures in Africa cannot support the demands of a smart city. This was also pointed out by Boyle (2019). Many of these local authorities are underfunded, lack autonomy, human resource constraints, etc. UN Habitat (2015) confirms the serious funding challenges developing countries face, resulting in failure for most of them to deliver on their mandates. For many of these cities, qualifying as smart cities is an impossible task. A number of studies testify to the failure in the delivery of services by local governments across African cities (Agba et al., 2013; Boris, 2015). For the smart city dream to be realised, these structures have to work.

3. **Develop appropriate ICT infrastructure**: Comparatively, Africa still lags behind other continents in ICT infrastructure. For instance, the *Infrastructure Consortium for Africa* (ICA) estimates that mobile phone penetration is at 63 per every 100 people for Africa, while Asia is at 89, Europe at 126 and North America at 109 (ICA, 2018). Similarly, internet users are at 25 per every 100 people for Africa, while Asia is at 42, Europe at 79 and The Americas at 66 (ICA, 2018). Households with internet access are at 15.4% for Africa while Asia is at 46.4%, Europe at 84% and North America at 64.4% (ICA, 2018). Thus, the *Infrastructure Consortium for Africa* argues that three quarters of Africa is offline, thus signifying a serious gap in the attainment of smart city status. In many cases, the connectivity is low, bandwidth is limited and coverage is minimal. Smart cities generate huge data, which require ultra-fast systems for data capture and processing.

4. **Develop effective collaborative structures**: Smart cities emphasise collaboration and information sharing between the public and private sectors. For urban planning purposes, the *Urban and Regional Planning Act* in Zambia also recognises the role stakeholders play in the development process. However, for smart cities with its many layers of actors, different ICT systems pose serious integration challenges (as noted in Anthopoulos and Vakali, 2012; Harrison and Donnelly, 2011). A number of African cities already have coordination challenges amongst the various central and local

government structures, which need to work for attainment of a smart city status.
5. **Complete the decentralisation process**: The autonomy of local governments is another important factor to ensure that cities work. In most African countries, city functions are closely tied to central governments, including the funding aspect. In many cases, only partial decentralisation has been achieved. As noted in the Zambian case, much of the activities towards achieving a smart Zambia are being done at the central government level with little or no activity at city level. A number of studies (see Ribot, 2002; Reddy, 2016; Kuenzi and Lambright, 2019; Arkorful et al., 2021) show that despite repeated attempts to decentralise, many of these processes are either incomplete, not performed to the satisfaction of citizens or politicised; the main reason being that by its nature, decentralisation involves transferring of power, which threatens most actors.

3.5 CONCLUSION

This chapter concludes that Africa has a lot of work to do towards smart cities. Based on the Smart City Governments assessment, no single African city qualifies as a smart city. For instance, using Zambia as an example, this chapter reveals that the country lags in ICT infrastructure, has a weak economy, inadequate citizen participation in urban governance and with at least 70% of its urban citizens living in informal unplanned settlements, negating the achievement of the basic tenets of smart cities.

Despite the opportunity to 'leapfrog' in terms of information technology, the chapter recommends that African cities should work on the basics, namely getting urban planning systems right; ensuring public institutions work; developing appropriate ICT infrastructure; developing effective collaborative structures and completing the decentralisation process. Furthermore, the chapter recommends more research to understand how an 'African smart city' should look; the use of other frameworks, such as the European Smart City framework, maybe a disservice to African cities. The chapter further contributes to the overall objective of the book by showing that achieving the development of African smart cities entitles first working on the basics of urban development.

REFERENCES

Agba, M.S., Akwara, A.F. and Idu, A., 2013. Local government and social service delivery in Nigeria: A content analysis. *Academic Journal of Interdisciplinary Studies*, 2(2), pp. 455–455.

Ahvenniemi, H., Huovila, A., Pinto-Seppä, I. and Airaksinen, M., 2017. What are the differences between sustainable and smart cities? *Cities*, 60, pp. 234–245.

Albino, V., Berardi, U. and Dangelico, R.M., 2015. Smart cities: Definitions, dimensions, performance, and initiatives. *Journal of Urban Technology*, 22(1), pp. 3–21.

Anthopoulos, L.G. and Vakali, A., 2012, May. Urban planning and smart cities: Interrelations and reciprocities. In *The Future Internet Assembly* (pp. 178–189). Springer, Berlin, Heidelberg.

Arkorful, V.E., Lugu, B.K., Hammond, A. and Basiru, I., 2021, January. Decentralization and citizens' participation in local governance: Does trust and transparency matter?—An empirical study. In *Forum for Development Studies* (pp. 1–25). Oslo, Norway: Routledge.

Bennet, H. and Biringer, J., 2000. *The Smart Growth—Climate Change Connection*. Conservation Law Foundation. [Online] Available from: www.clf.org/pubs/Smart_Growth_Climate_Change_Connection.htm [Accessed 18/06/2021].

Bollier, D., 1998. *How Smart Growth Can Stop Sprawl: A Fledgling Citizen Movement Expands: A Briefing Guide for Funders 1998*. Washington, DC: Essential Books.

Boris, O.H., 2015. Challenges confronting local government administration in efficient and effective social service delivery: The Nigerian Experience. *International Journal of Public Administration and Management Research*, 2(5), pp. 12–22.

Boyle, L., 2019. Finding Birnin Zana: In pursuit of an African Smart City. *Asset Magazine*, (76), pp. 125–128.

Burchell, R.W., Lowenstein, G., Dolphin, W.R., Galley, C.C., Downs, A., Seskin, S., Still, K.G. and Moore, T., 2002. *Costs of sprawl—2000* (No. Project H-10 FY'95).

Cain, A., 2014. African urban fantasies: Past lessons and emerging realities. *Environment and Urbanization*, 26(2), pp. 561–567.

Chen, X., Wei, L. and Zhang, H., 2018. Spatial and temporal pattern of urban smart development in China and its driving mechanism. *Chinese Geographical Science*, 28(4), pp. 584–599.

Chiwanza, T.H., 2021. The myth and deception of "smart cities" in Africa. *The African Exponent*, Monday April 19, 2021. [Online] Available from: www.africanexponent.com/post/8392-the-myth-and-deception-of-smart-cities-in-africa [Accessed 17/06/2021].

Construction Review, 2021. Kalulushi Smart City Project in Zambia nears construction phase. August 14. [Online] Available from: https://constructionreviewonline.com/news/kalulushi-smart-city-project-in-zambia-nears-the-construction-phase/ [Accessed 17/06/2021]

Delusional Bubble, 2021. Top 10 smartest cities in Africa this year. February 22, 2021. [Online] Available from: www.delusionalbubble.com [17/06/2021]

De Soto, H., 2000. *The Mystery of Capital: Why Capitalism Triumphs in the West and Fails Everywhere Else*. New York: Civitas Books.

Evans, D., 2011. *The Internet of Things: How the Next Evolution of the Internet is Changing Everything*. CISCO Systems. [Online] Available from: www.cisco.com/web/about/ac79/docs/innov/IoT_IBSG_0411FINAL.pdf [Accessed 17/06/2021].

Giffinger, R. and Pichler-Milanović, N., 2007. *Smart Cities: Ranking of European Medium-sized Cities*. Vienna, Austria: Centre of Regional Science, Vienna University of Technology.

GRZ, 2021a. *The Electronic Government Bill*. Lusaka: Government Printers.

GRZ, 2021b. *The Data Protection Act*. Lusaka: Government Printers.

GRZ, 2021c. *The Cyber Security and Cyber Crime Bill Act*. Lusaka: Government Printers.

GRZ/Huawei, 2012. Contract no. Huawei Ref. 00Y8941213000A—Smart Zambia phase I National ICT Development Project. [Online] Available from: www.lusakatimes.com/wp-content/uploads/2015/01/NationalDataCenterICTContract.pdf [Accessed 17/06/2021]

Habitat for Humanity, 2020. Country profile. [Online] Available from: www.habitat.org/sites/default/files/documents/Zambia%20Country.pdf [Accessed 17/06/2021].

Hall, R.E., Bowerman, B., Braverman, J., Taylor, J., Todosow, H. and Von Wimmersperg, U., 2000. *The Vision of a Smart City* (No. BNL-67902; 04042). Brookhaven National Lab., Upton, NY (US).

Hammad, A.W., Akbarnezhad, A., Haddad, A. and Vazquez, E.G., 2019. Sustainable zoning, land-use allocation and facility location optimisation in smart cities. *Energies*, *12*(7), pp. 1318–1340.

Harrison, C. and Donnelly, I.A., 2011, September. A theory of smart cities. In *Proceedings of the 55th Annual Meeting of the ISSS-2011*, Hull, UK.

Jasrotia, A. and Gangotia, A., 2018. Smart cities to smart tourism destinations: A review paper. *Journal of Tourism Intelligence and Smartness*, *1*(1), pp. 47–56.

Kuenzi, M.T. and Lambright, G.M., 2019. Decentralization, executive selection, and citizen views on the quality of local governance in African countries. *Publius: The Journal of Federalism*, *49*(2), pp. 221–249.

Lolojih, P.K., 2010. Zambian citizens, democracy and political participation. [Online] Available from: https://afrobarometer.org/publications/bp80-zambian-citizens-democracy-and-political-participation. Afrobarometer Briefing Paper No. 80.

Mattern, F. and Floerkemeier, C., 2010. From the Internet of Computers to the Internet of Things. In *From Active Data Management to Event-Based Systems and More* (pp. 242–259). Springer, Berlin: Heidelberg.

Munshifwa, E.K., Mwenya, C.M. and Mushinge, A., 2021. Urban development, land use changes and environmental impacts in Zambia's major cities: A case study of Ndola. In *Sustainable Real Estate in the Developing World*. Bingley: Emerald.

Pan, Q. and Li, W., eds., 2016. *Smart Growth and Sustainable Development: Selected Papers from the 9th International Association for China Planning Conference, Chongqing, China, June 19–21, 2015* (Vol. 122). Berlin, Germany: Springer.

Pinochet, L.H.C., Romani, G.F., de Souza, C.A. and Rodríguez-Abitia, G., 2019. Intention to live in a smart city based on its characteristics in the perception by the young public. *Revista de Gestão*, *26*(1), pp. 73–92.

Porter, G., Abane, A. and Lucas, K., 2020. *User Diversity and Mobility Practices in Sub-Saharan African Cities: Understanding the Needs of Vulnerable Populations. The State of Knowledge and Research*. Gothenburg, Sweden: Volvo Research and Educational Foundations.

Reddy, P.S., 2016. The politics of service delivery in South Africa: The local government sphere in context. *TD: The Journal for Transdisciplinary Research in Southern Africa*, *12*(1), pp. 1–8.

Ribot, J.C., 2002. *African Decentralization: Local Actors, Powers and Accountability*. Geneva: UNRISD.

SELC, 1999. Smart Growth in the Southeast: New approaches for guiding development. SELC. [Online] Available from: www.eli.org/sites/default/files/eli-pubs/d9.05.pdf [Accessed 18/06/2021]

Smart City Governments, 2021. Top 50 smart city governments [Online] Available from: www.smartcitygovt.com [Accessed 17/06/2021].

Smart Growth America, 2021. What is smart growth. [Online] Available from: https://smartgrowthamerica.org/our-vision/what-is-smart-growth/ [Accessed 17/06/2021].

Smart Zambia Institute, 2021. SMART Zambia institute: A smart and value centred public service. [Online] Available from: www.szi.gov.zm/ [Accessed 16/06/2021]

Southern Environmental Law Centre, 1999. Smart growth in the Southeast. [Online] Available from: www.southernenvironment.org/cases-and-projects/smart-growth-in-the-southeast [Accessed 17/06/2021].

Tembo, E., Minango, J. and Sommerville, M., 2018. Zambia's National Titling Programme: Challenges and opportunities. [Online] Available from: www.land-links.org/wp-content/uploads/2018/03/Session-06-05-Tembo-153_paper.pdf

The Infrastructure Consortium for Africa (ICA), 2018. Infrastructure financing trends in Africa. [Online] Available from: www.icafrica.org/en/fileadmin/documents/IFT_2018/ICA_Infrastructure_Financing_in_Africa_Report_2018_En.pdf [Accessed 17/06/2021].

UN Habitat, 2015. *The Challenge of Local Government Financing in Developing Countries*. Nairobi: UN Habitat.

Watson, V., 2014. African urban fantasies: Dreams or nightmares? *Environment and Urbanization*, 26(1), pp. 215–231.

World Bank, 2020. *Accelerating Digital Transformation in Zambia: Digital Economy Diagnostic Report*. Washington, DC: IBRD/World Bank.

World Bank, 2021. Zambia: Overview. [Online] Available from: www.worldbank.org/en/country/zambia/overview [Accessed 23/08/2021]

World Construction Today, 2021. Kalulushi Smart City Project in Zambia nears construction phase. Wednesday, August 25, 2021. [Online] Available from: www.worldconstructiontoday.com/news/kalulushi-smart-city-project-in-zambia-nears-construction-phase/ [Accessed 23/08/2021]

Wygonik, E., Bassok, A., Goodchild, A., McCormack, E. and Carlson, D., 2015. Smart growth and goods movement: Emerging research agendas. *Journal of Urbanism: International Research on Placemaking and Urban Sustainability*, 8(2), pp. 115–132.

Yu, J., Wen, Y., Jin, J. and Zhang, Y., 2019. Towards a service-dominant platform for public value co-creation in a smart city: Evidence from two metropolitan cities in China. *Technological Forecasting and Social Change*, *142*, pp. 168–182.

4 Transformation Pathways to Smart Villages
Lessons from Mulenzhe Village in Limpopo Province of South Africa

Emaculate Ingwani, Madilonga Trevor Mukwevho, Trynos Gumbo and Masala Thomas Makumule

CONTENTS

4.1 Introduction	41
4.2 Smart Concept	42
4.2.1 Smart Village Concept	43
4.2.2 Rural Landscape of South Africa	43
4.2.3 Regulatory Frameworks and Transformation Pathways to Smart Villages	44
4.3 Methods Operationalising the Study	45
4.4 Discussion and Analysis of Findings	46
4.4.1 Transformation Pathway to a Smart Village of Mulenzhe	46
4.4.2 Administration of Land Property Rights	47
4.4.3 Emergent Livelihood Opportunities	48
4.4.4 Opportunities and Challenges from the Transformation Pathway	49
4.5 Lessons Learnt from the Best Practice	49
4.5.1 Access to Basic Services	49
4.5.2 Entrepreneurship	50
4.5.3 Connectivity, Networks and the Internet of Things	50
4.5.4 Participatory Governance	50
4.6 Recommendations	51
4.7 Conclusion	51
References	52

4.1 INTRODUCTION

The smart concept is not a new phenomenon and is widely associated with advancements in technologies such as the Internet of Things (IoTs) used to promote smart cities (Zavratnik et al., 2018; Musakwa and Gumbo, 2017; Gumbo and Moyo, 2020).

According to Park et al. (2018:1), the IoTs refer to 'a set of technologies for accessing data collected by various devices through wireless and wired internet networks'. Smart city initiatives are slowly emerging in many African cities by applying the IoTs in spatial planning. Approximately, 50% of the world population lives in urban areas since 2017 and this proportion is expected to reach 66% by 2050 (Sharif, 2018). On the African continent, about 42% of the people live in rural areas and half of the population is expected to live in urban areas by 2030 (Institute for Security Studies, 2016). In South Africa, 33% of the population currently lives in rural areas (Macrotrends, 2021). Such developments call for concerted efforts by city planners to prioritise the IoTs in the development of human settlements that can meet the needs of current and future residents (Moffat et al., 2021).

This chapter presents a shift from an urban biased smart concept linked to cities to a rather localised rural perspective of the smart village concept reflecting on Mulenzhe Village located Limpopo Province of South Africa. The aim of this chapter is to reveal how the Mulenzhe Village transformed into a smart village through reorganisation of a social system. The objectives are: (1) characterise the pathway adopted to transform Mulenzhe Village into a smart village; (2) the opportunities and challenges from adopting the smart village pathway and (3) the lessons learnt from a best practice from the Limpopo Province of South Africa. After this introductory section, the key concepts from literature are interrogated followed by the methods section. The research findings are discussed following themes from the objectives and the chapter concludes by highlighting the recommendations.

4.2 SMART CONCEPT

Smart cities are characterised by industrial and communication technological functions (Park et al., 2018). However, there are layers of definitions and approaches to understanding 'smartness'. Not much has been documented about smart villages in Africa, which is largely regarded as a rural continent. Transformation pathways are more meaningful when interrogated contextually. Transformation pathways are simply choices by residents in collaboration with the local authorities in their desire to become 'smart' spaces.

There is no one-size-fits-all approach for defining smart spaces. These are complex and many. The smart concept is rooted in different circumstances of communities (Zavratnik et al., 2018). Smart concepts are associated with a plethora of advanced technologies used to plan cities such as infrastructure, health communication, transport and energy (Mohanty et al., 2016; Visvizi and Lytras, 2018). According to Qian et al. (2019), 'A city can be called "smart" when investments in human and social capital and traditional modern communication infrastructure fuel sustainable economic growth, high quality of life and wise management of natural resources through participatory governance'. Sutriadi (2018) aligns the smart concept with modern development discourses that involve electronic application, management and provisioning of services in urban areas as sustainable practices to improved well-being of citizens. The smart concept is linked to connectivity of infrastructure, technological, physical, business, social and information (Mohanty et al., 2016).

4.2.1 SMART VILLAGE CONCEPT

Urban functions that exhibit smart qualities in spatial planning advance the smart cities concept. On the other hand, the smart village concept represents a localised and micro-level replication of the smart concept from a rural perspective because village structures are absent in cities. It is therefore important to parallel smart cities and smart villages in spatial planning discourse (Zavratnik et al., 2018). According to Zavratnik et al. (2018:3), smart villages are autonomous and independent entities in the rural areas where inhabitants are enabled to make use of contemporary technological and social achievements, as well as infrastructure development in line with the provisions of sustainable development goals (SDGs). Sikora-Fernandesz and Stawaz (2016) also add that smart villages stimulate new approaches to empower local communities to meet the agenda of the SDGs by 2030. This shows that the 'smartness' of the villages is embedded in technological advancement and infrastructure development as well as soft socio-economic issues such as community empowerment enabled through participation in decision-making processes.

In this chapter, the smart village concept is conceptualised as an art, a concept, place and a science. As an art, a smart village is simply a community's way of expressing what they want to become and exhibits this through interaction in socially organised systems. As a concept, a smart village implies the transformation pathway adopted by rural areas to spaces characterised by the IoTs in terms of connectivity and linkages. As a place, a smart village is a human settlement that provides development opportunities to the inhabitants. As a science, a smart village is the interface of activities that are critical to enable the balance of opportunities for the rural residents through improved access to services such as education, health, housing, sanitation, transport, energy and good governance (Holmes, 2017). Just like a smart city, a smart village therefore seeks to achieve an integration of spatial planning functions in rural areas (see Park et al., 2018). For example, in Germany, the smart village concept focuses on improving the living conditions of rural population through enabled access to services and infrastructure (Zavratnik et al., 2018). Locational advantage of rural spaces to established urban centres is key to the development of smart spaces. However, the 'smartness' is never uniform as spaces are never homogenous.

4.2.2 RURAL LANDSCAPE OF SOUTH AFRICA

Rural villages of South Africa occupy the lowest rank on the hierarchy of cities and are mostly categorised as the former homelands. In South Africa, most rural settlements are located in the Limpopo, Eastern Cape and Mpumalanga provinces. Rural villages are sparsely populated and the majority of people depend on subsistence farming and other informal activities for their livelihood. John (2012) refers to rural villages of South Africa as hinterlands because they are located outside the urban agglomerations. In some cases, the villages of South Africa are urban and rural at the same time and thus are labelled as peri-urban areas.

According Ingwani and Gumbo (2016:3), peri-urban settlements are continuously evolving into affluent residential areas with low-cost and social housing estates and often emerge as informal settlements located on vacant spaces designated for other land uses. In the Limpopo Province, most peri-urban villages are located a few kilometres from urban centres and display a combination of urban and rural features that represent spaces with mixed rural and urban land uses fused together (Ingwani and Gumbo, 2016). Based on this definition, the Mulenzhe Village is an example of a peri-urban human settlement.

What makes peri-urban villages of Limpopo rural is the traditional authority system of land administration that is used to regulate land property rights in combination with spatial planning functions through the local municipality ordinances. A combined system of land administration that involves traditional authorities and municipal structures is absent in urban areas where the local municipalities are sorely in charge of land governance processes. As peri-urban villages are administered by both the local municipalities and the local traditional authorities, the local municipalities provide services and preside over the conventional spatial planning functions, while the traditional authorities administer common property land rights. Often, these overlaps in land administration between traditional authorities and the local municipalities lead to land use conflict(s) (Ingwani et al., 2019; Ingwani, 2021).

The rural villages of South Africa located in the hinterland are connected to the urban core by road networks and modes of transport such as buses and taxi minibuses. From Burgess's Concentric Zone Model (Burgess et al., 1925), the rural villages occupy the outer ring that defines the commuter zone and mark the edges of the peripheries of small rural towns (SRTs) that are service centres for the rural inhabitants (John, 2012). For example, the Mulenzhe Village is located on the hinterland of the Thohoyandou Town.

Many South African peri-urban zones of SRTs continue to witness rapid population growth as people look for spaces to settle (Hoogendoorn & Visser, 2015). Most of such human settlements located in the hinterland emerge as dormitory villages for people to stay while they obtain services and work in the SRTs (see Ingwani et al., 2019). In some cases, migrants use these peri-urban human settlements as their 'first stop' to migrate to larger towns and cities such as Johannesburg, Cape Town, Durban, or Pretoria. According to Budziewicz-Guzlecka (2019), peripheral regions usually do not have equal access to resources and markets as they are spatially distanced from the urban core. As SRTs slowly urbanise, the peri-urban villages in the hinterland positively respond to new development opportunities and gradually develop into 'new' centres of influence for the local residents. The transformation of villages in the hinterland of urban centres—the peri-urban—is therefore induced by a compound of external and internal processes and is therefore inevitable.

4.2.3 REGULATORY FRAMEWORKS AND TRANSFORMATION PATHWAYS TO SMART VILLAGES

The Constitution of the Republic of South Africa Act 108 of 1996, the Spatial Planning and Land Use Management Act (SPLUMA) 16 of 2013; the National Development Plan and local-level development plans are important regulatory tools that enable the transformation of spaces into smart villages. The Constitution of the Republic of South Africa provides for municipalities (local, district and metropolitan) to prioritise provisioning of basic services to all citizens without discrimination and to promote the local socio-economic development of the communities they serve. Chapter 2 of this Constitution spells out the Bill of Rights that clearly states the entitlements of the citizens in terms of access to services including housing, water, health, sanitation and food. It is therefore the role of the local municipalities as service providers to ensure that citizens under their jurisdiction access adequate services and enjoy improved standards of living and attain human wellbeing.

To achieve the provisions of the Constitution Act 108 of 1996, the government of the Republic of South Africa instituted the Spatial Planning and Land Use Management Act 16 of 2013 to guide the development agendas of municipalities and to harmonise the spatial planning effort. This spatial planning instrument sets parameters for the localised development plans including the Integrated Development Plans (IDPs) and the Spatial Development Plans (SDPs). An IDP is a municipal plan of action that sets development strategies informed by local-level community interests. Whereas an SDF guides the overall current and desired land uses within the municipal boundaries. Both IDP(s) and SDF(s) seek to promote sustainable development and the quality of life of community residents.

On the other hand, the SDGs and Agenda 2063 also (re)shape and (re)construct the local-level development strategies of this transformation in line with current experiences. Among the SDGs critical to the achievement of smart villages is Goal 11 that focuses on making 'cities and communities inclusive, safe, resilient and sustainable' by 2030 (Sharif, 2018). While SDG11 was singled out for the purpose of this discussion, it must be noted that all SDGs complementarily seek to achieve inclusive, safe, resilient and sustainable cities and communities in one way or the other.

4.3 METHODS OPERATIONALISING THE STUDY

To understand transformation pathway to smart village in Mulenzhe, this study adopted a qualitative phenomenological case study research design (Babbie, 2007). A phenomenological case study design revealed the motivations behind the transformation pathway adopted, as well as the challenges and opportunities encountered. The Mulenzhe Village was therefore purposively selected to provide empirical evidence on a real-life experience that typifies a transformation pathway towards becoming a smart village.

The Mulenzhe Village is located near the Nandoni Dam of the Thulamela Local Municipality in the Vhembe District of the Limpopo Province in South Africa. In local descriptions, the Mulenzhe Village falls under Mulenzhe area. The Mulenzhe area comprises of 402 villages of which the Mulenzhe Village is one of them.

Data were collected through direct observation of the phenomena during an excursion to the Mulenzhe Village in 2017 by the Department of Urban and Regional Planning (DURP) of the University of Venda. Additional qualitative data collection relied on conversations with experts from the local municipality and DURP. During the 2016 excursion, staff members and students of the DURP observed a real case of a smart village in a rural setting at the Mulenzhe Village. During the excursion, the chief as the local administrative authority briefed the staff members and students on the transformation pathway of the Mulenzhe Village towards achieving the smart village status. The observation of realities in the Mulenzhe Village provided an opportunity to triangulate evidence on the dynamics of a smart village concept with other sources to minimising biases (Bryman, 2012).

The debates on transformation into a smart village are mostly grounded on conversations with the chief in 2016 and the views of a spatial planning expert who resides in the Mulenzhe area. Views from these two key informants engender an understanding of the smart village concept from interpretations of individuals who are directly linked to the spatial planning practice. A review of the pertinent literature took centre stage to contrast the observed phenomena in the Mulenzhe Village with other communities undergoing similar circumstances around the world. The reviewed literature also shed light on conceptualising smart villages more generally.

Data analysis draws insights from a research conducted by Madilonga Trevor Mukwevho in 2017 that largely focused on the impact of the relocation process on household livelihoods in the Budeli Village located close to the Mulenzhe Village. The variables from the smart village framework by Zavratnik et al. (2018) helped us to explain the pathway adopted in the Mulenzhe Village and how the village continuously reimages itself into a smart human settlement. Data were thematically analysed with reference to these variables including access to basic services; entrepreneurship and sustainable livelihoods; connectivity, networks and IoTs; participation and governance; sustainable energy options and infrastructure provision.

These variables were derived from the SDG11 targets that seek to engender sustainable human settlements through 'elimination of slum-like conditions, provision of affordable and accessible transport systems; reduce urban sprawl, increase participation in urban governance, enhance cultural preservation; address urban resilience, climate challenges and pollution, enable sound urban management practices, and secure spaces' (Sharif, 2018:8). The underlying assumption is that achievement of these targets translates not only to sustainable human settlements but also smart spaces.

4.4 DISCUSSION AND ANALYSIS OF FINDINGS

The discussion and analysis of findings are organised around the themes from the objectives and are informed by the community residents' localised understanding of the 'smart' concept. The transformation pathway to a smart village is therefore processual.

4.4.1 Transformation Pathway to a Smart Village of Mulenzhe

Several strategies were adopted to transform the Mulenzhe Village into a smart village. The Mulenzhe area is a typical rural precinct located in one of South Africa's rural provinces—Limpopo. This Mulenzhe area falls under the Thulamela local municipality of the Vhembe district and the inhabitants are a mix of Venda and Xonga Indigenes. The Mulenzhe Village was originally located on the current site of the Nandoni Dam in the Vhembe district, 30 km from Thohoyandou town, the administrative hub of the Thulamela local municipality. The Mulenzhe Village is one of the 33 villages that were affected by the Nandoni Dam construction project in the Mulenzhe area. A total of 402 households from the Mulenzhe Village were relocated to a new site not far from their original homeland.

The history of the Mulenzhe Village dates to the apartheid era of South Africa before 1994 under the government of Venda homelands. Post-1994 saw the introduction and improvement of services to previously disadvantaged rural communities in South Africa. The construction of the Nandoni Dam in 1998 witnessed the relocation of 402 households from 33 villages located on the flood line of the Nandoni Dam that were considered a flood disaster risk zone in the Mulenzhe area. These villages were directly affected by the construction of the water reservoir and were relocated to a safer zone in line with the provisions of the Disaster Management Act of 1998 of South Africa. The affected households were compensated by the South African government for their losses on an asset-for-asset basis. From the affected households' perspective, the relocation meant loss of livelihoods and uncertainties—fear of the unknown.

The relocated households were compensated with new, modern, fenced, gated and well-demarcated houses with roof under tiles and security of tenure. The houses are endowed with services such as water and sanitation. The design considerations of the houses reflect the local culture which provides for a gazebo at each homestead to allow the residents to enjoy traditional homemade food prepared on open fire. The affected households were also compensated for loss of arable land and water use access rights to the Nandoni Dam. The design considerations of housing provision and reconfiguration of land property rights in the Mulenzhe Village engender the smart village concept.

4.4.2 Administration of Land Property Rights

Before relocation, land administration in the Mulenzhe Village was part of a wider communal and customary land tenure system, whereby the land rights were administered under the custody of the traditional authority on behalf of the state. Households owned individual arable and residential land parcels but grazing remained communal. Under the smart village concept, the land property rights of the community residents remained customary under the custodian of the chief and are provided based on the principles on which land can be or cannot be accessed by this community. The customary system of land tenure property rights in the Mulenzhe Village is regulated according to the local customs and traditions by individuals and as collectives and can be passed from generation to generation in perpetuity through inheritance.

At the same time, the land property rights in the Mulenzhe Village are autochthonous and are privately administered under the freehold land tenure system. Private freehold land tenure entails the terms and conditions on which land and other property rights can be accessed and owned privately by community residents (Bruce, 1998). Whereas a freehold land tenure system is 'held by an authority of a title deed such as a private individual, institution, or state' (Matondi and Dekker, 2011:1). Under this local arrangement of land administration structure in the Mulenzhe Village, all land belongs to the state and the chief holds the land property rights of the community residents as an estate in trust of the state. The chief remains the sole custodian of customary land tenure rights. This means that the households of the Mulenzhe Village own the houses that were allocated to them as personal property and are at liberty to use these dwelling units for any purpose they deem necessary as long as their actions are in conformity with the municipal zoning regulations for residential properties. In addition, the residents of the Mulenzhe Village individually own residential plots and farmland as private property. The categories of land uses in the precinct are institutional, commercial, residential, grazing, arable and forest as the land rights are privately administered under a trust. The land is titled with a deed and therefore has a market value.

From a smart village perspective, such land tenure arrangements enable the community residents to retain their cultural values as grazing land is owned communally. The role of the local municipality on behalf of the government under freehold land tenure is to provide services and infrastructure. This demonstrates a clear transformation to a smart village regulated by both traditional authority and municipal bylaws in a rural context.

Administration of land property rights under the freehold land tenure system enables community residents to tightly remain the custodians of land and other land property rights, thereby safeguarding them from any form of individualised land transactions such as land sales and land grabbing. Since land property rights in the Mulenzhe Village are only transferrable through inheritance, community identity and belonging are perpetuated. In addition, private land rights under the freehold land tenure system can be used as a form of collateral, which enables residents to obtain loans for investment through financial lending institutions such as banks. However, in some villages of the Mulenzhe area, land can be offered to migrants on compassionate grounds within the system of customary land tenure.

4.4.3 Emergent Livelihood Opportunities

Different forms of capital (natural, physical, social, human and financial) were observed in the Mulenzhe Village (see Scoones, 2009). There is not much change with regards to the practice of traditional and cultural activities on livelihoods options such as peasant farming, fishing, grazing and harvesting of forest products.

The houses that were part of the compensation package are important physical assets, which improve the quality of life of people. Since land property rights on individual residences are private, households can build houses to rent and for other activities to suit their individual needs. In addition, a road network provides the much-needed linkages within and outside the precinct. This enables the community residents to travel using different kinds of transport types such as buses, private cars and mini-bus taxis. In terms of economic capital, community residents of the Mulenzhe Village own cattle and other small livestock such as chickens and goats to supplement livelihood income by selling these animals and animal products. These animals have designated areas for grazing and penning—a practice that is usually relaxed and not enforced in many rural villages of South Africa.

Observations revealed that social forms of capital are represented in relationships between households in the Mulenzhe Village as residents network with each other

Transformation Pathways to Smart Villages

thereby reducing risks of insecurity. The smart village concept enables residents to pursue livelihood strategies under a guided model while at the same time adapt to new circumstances. Taking care of the local environment also is central to the livelihood options of the households in this village. For example, cutting down of trees and lighting of wood fuel inside the modern house are prohibited.

4.4.4 Opportunities and Challenges from the Transformation Pathway

The relocation process and the ultimate provisioning of new homesteads in the Mulenzhe Village marked the inception of the smart village concept as the community residents envisioned to become a best practice case different from other villages, which did not benefit from the relocation exercise under the same chief. From Harrison and Donnelly's perspective (2011:1), the pathway can be paralleled to experiential approaches to spatial planning embedded in local circumstances. Scaling up opportunities through sound spatial planning practices and deliberate municipal effort to develop a best practice in the precinct remains important. Predictable challenges can be reduced or twisted into opportunities to generate desirable future scenarios.

Enhanced rural–urban linkages emerge as one of the important opportunities from the smart village concept in the Mulenzhe Village as the residents are continuously straddling the rural–urban divide. This enables the residents to diversify their livelihood options and expand their social networks. Connectivity is an important element to achieving smart villages.

However, distorted development outcomes are clearly visible in the Mulenzhe area as the villages that benefited from the relocation exercise are far from being rural compared to those that never relocated. For example, in the boundaries of the Mulenzhe area, smart villages such as Mulenzhe are clearly bonded and demarcated. This means that the transformation of rural spaces to smart villages requires more effort through inclusive spatial planning strategies to spread the benefit incrementally and evenly across villages in the same area.

4.5 LESSONS LEARNT FROM THE BEST PRACTICE

Transformation of rural spaces into smart villages requires local-level visionary leadership; reorganisation of the social structure including land property rights and provisioning of adequate basic services and infrastructure to local communities by the local municipalities. The IoTs and connectivity are key for the transformation into smart villages. This requires integrated spatial planning that brings together infrastructure and supporting services.

4.5.1 Access to Basic Services

The Mulenzhe Village is endowed with basic services such as water, electricity, sanitation and road infrastructure. Infrastructure for services such as clinics and schools is prioritised and adapted to the needs of the residents.

4.5.2 ENTREPRENEURSHIP

The engagement of informal and formal businesses is vital to improve the quality of life. Most residents of the Mulenzhe Village run small businesses. Such entrepreneurial activities expand household income strings and livelihood options. The residents also sell farm produce at markets locally and in Thohoyandou Town. This enables the residents to pay for electricity, transport fares and other rateable municipal services. Similarly, in Czechia, small-family enterprises located in peri-urban smart villages specialising in agriculture and handcrafts expand household income streams and enhance household livelihoods (Vaisha and Stastna, 2019).

4.5.3 CONNECTIVITY, NETWORKS AND THE INTERNET OF THINGS

Connectivity plays a significant role in linking the Mulenzhe Village with the Thohoyandou town. A viable road network enables the residents of the Mulenzhe Village to obtain urban services through straddling the urban–rural divide. Connectivity is one of the key elements of a well-designed smart village.

In addition, the residents of the Mulenzhe Village have access to a reliable network for web browsing on computers and cell phones through service providers including Telkom, Vodacom, Cell C and MTN. Connecting to telecommunication networks is important in modern human settlements because connectivity advances strong social relationships of residents within and outside the precinct. According to Harrison and Donnelly (2011), the emerging role of technology needs to be recognised because micro-level interactions of residents and the resources that surround them are key to improved lifestyles. Yet, many rural communities of South Africa struggle for access to internet, although effort is currently made to expand this service. Budziewicz-Guzlecka (2019) also note similar concerns in Poland where 'saturation with internet access services is lower in rural areas than in cities'.

Many South Africans who live in rural areas do not have access to internet and most of these people could be staying in places that are inaccessible. While access to internet services is key to developing smart spaces (Park et al., 2018), contextualised smartness remains possible under these circumstances because smart villages are never an event but develop by harnessing local effort. Rural spaces can be smart too.

4.5.4 PARTICIPATORY GOVERNANCE

Participatory governance entails amplifying the voices of citizens and widening their choices on public platforms (Fischer, 2012). In the Mulenzhe Village, community residents use the Mulenzhe tribal council office for community meetings where communal concerns are deliberated. All community meetings (general), celebrations and resolution of conflict take place in the community hall. The chief and his advisors lead the community gatherings.

Participatory governance is also a critical enabler for accountability of public finance and building of strong and resilient institutions. Through participatory governance, community residents are empowered to express their interests and preferences in matters that affect their lives. In the Mulenzhe Village, participatory governance enables collaboration between the local municipality and the community residents. This enables sound service provision and improved lifestyles as the community residents are involved in the local-level decision-making processes.

4.6 RECOMMENDATIONS

There is need for deconstructing the narrative of associating smartness with urban spaces. It is therefore important to expand the definition of smartness beyond the rhetoric on technological advancement to capture fulfilment of the wellbeing of community residents, sustainable livelihoods and empowerment. This provides a more contextualised and holistic approach and advances an inclusive smart concept to include human settlements located in peri-urban spaces—the smart villages. This is possible when local municipalities incentivise rural areas that choose to transform into smart villages. This can be achieved through:

- Multi-stakeholder collaborative projects that engender smartness in villages.
- Strengthening local capacities to advance smartness by channelling municipal resources towards the contextualised needs of community residents.
- Strengthening the institutional capacity of the local communities to adhere to agreed reference points and set standards on the pathways to transformation to smart villages. Such pathways can be adopted by other villages elsewhere with ease.
- Awareness creation on the benefits of smart villages to residents, local government and traditional authorities. This enables the adoption and adaptation of existing municipal spatial planning regulations to meet the desired standards in smart villages.
- Public–private partnerships remain the key to pooling of resources to finance smart village projects.

4.7 CONCLUSION

Transformation pathways to smart villages are important to spatial planning functions in Africa. Lessons from a smart village—Mulenzhe—demonstrate differential outcomes from rural areas that transform to smart villages. Key spatial planning imperatives achieving successful transformation to smart villages include local-level visionary leadership; reorganisation of the social structure including land property rights and IoTs and connectivity and provisioning of adequate basic services and infrastructure. This requires integrated spatial planning, creativity and innovation through collaborative effort *between* the community residents, traditional authority and the local municipalities. A functional smart village concept is therefore an attribute inherent in the spaces themselves as well as in the livelihoods of community residents. Evidence towards redefining the smart concept from a rural African context is clearly apparent from the Mulenzhe Village case study where visionary leadership and community engagement were critical to transformation and adaptation to local needs. Smartness can be achieved in rural areas of Africa.

REFERENCES

Babbie, E. 2007. *The practice of social research*. 6th edition. Belmont: Wadsworth.
Bruce, J.W. 1998. Review of tenure terminology. *Tenure Brief*, (1): 1–8.
Bryman, A. 2012. *Social research methods*. 4th edition. Oxford: Oxford University Press.
Budziewicz-Guzlecka, A. 2019. Smart village as a direction for rural development. Proceedings of the 2019 International Conference on Economic Science for Rural Development, Jelgava, LLU, 9–10 May 2019, (52): 22–28.
Burgess, Ernest, Park, Robert E. and McKenzie, Roderick D. 1925. *The City*. Chicago: University of Chicago Press.
Fischer, F. 2012. *Participatory governance: From theory to practice*. Oxford: Oxford University Press.
Gumbo, T. and Moyo, T. 2020. Exploring the interoperability of public transport systems for sustainable mobility in developing cities: Lessons from Johannesburg Metropolitan City, South Africa. *Sustainability*, 12(5875): 1–16. Doi: 10.3390/su12155875
Harrison, C. and Donnelly, I.A. 2011. A theory of smart cities. Proceedings of the 55th Annual Meeting of the ISSS-2011, Hull, 17–22 July 2011.
Holmes, J. 2017. The smart villages initiatives: Interim review of findings. In *Cambridge Malaysian education and development trust*. Cambridge: Trinity College. 1–20.
Hoogendoorn, G. and Visser, G. 2015. The role of second homes in local economic development in five small South African towns. *Development Southern Africa*, 27(4): 547–562.
Ingwani, E. 2021. Struggles of women to access and hold landuse and other land property rights under the customary tenure system in peri-urban communal areas of Zimbabwe. *Land*, 10(6): 649. 1–14. Doi: 10.3390/land10060649
Ingwani, E. and Gumbo, T. 2016. Peri-urbanities as incubators of sustainable land use planning and development frameworks for the third space. 52nd ISOCARP Congress, 12–16 September 2016, Durban, South Africa. 775–786.
Ingwani, E., Musetha, R., Gumbo, T. and Moyo, T. 2019. The politics of digitized boundaries in Vhembe District Municipality of South Africa. REAL CORP 2019 Proceedings/Tagungsband, 2–4 April 2019. http://hdl.handle.net/10210/400813
Institute for Security Studies. 2016. Africa's future is urban. https://issafrica.org/iss
John, L. 2012. Secondary cities in South Africa: The start of a conversation. The Background Report. Pretoria: South African Cities Network. 6–19.
Macrotrends LLC. 2021. South Africa Rural Population 1960-2020 [online]. Available from: https://www.macrotrends.net/countries/ZAF/south-africa/rural-population [Accessed 26/12.021].
Matondi, P.B. and Dekker, M. 2011. Land rights and tenure security in Zimbabwe's post Fast Track Land Reform Programme. A synthesis report for LANDac. Harare: Ruzivo Trust. 1–47.
Moffat, F., Chakwizira, J., Ingwani, E. and Bikam, P. 2021. Policy directions for spatial transformation and sustainable development: A case study of Polokwane City, South Africa. *Town and Regional Planning*, 78: 46–64.
Mohanty, S.P., Choppali, U. and Kougianos, E. 2016. Everything you wanted to know about smart cities: The Internet of things is the backbone. *IEEE Consumer Electronics Magazine*, 5(3): 60–70.
Musakwa, W. and Gumbo, T. 2017. Impact of urban policy on public transportation in Gauteng, South Africa: Smart or dumb city systems is the question. In R. Álvarez Fernández, S. Zubelzu, and R. Martínez (Eds.), *Carbon footprint and the industrial life cycle: From urban planning to recycling*. Springer: Cham. 339–356.
Park, E., del Pobil, A. and Kwon, S.J. 2018. The role of internet of things (IoT) in smart cities: Technology roadmap-oriented approaches. *Sustainability*, 10(1388): 1–13.

Qian, Y., Wu, D., Bao, W. and Lorenzi, P. 2019. The internet of things for smart cities: Technologies and applications. *IEEE Network*, 33(2): 4–5.

Scoones, I. 2009. Livelihoods perspectives and rural development. *Journal of Peasant Studies*, 36(1): 1–26.

Sharif, M.M. 2018. Tracking progress towards inclusive, safe, resilient, and sustainable cities and human settlements. SDG11 Synthesis Report. United Nations.

Sikora-Fernandesz, D. and Stawaz, D. 2016. The concept of smart city in theory and practice of urban development. *Romanian Journal of Regional Science*, 10(1): 81–99.

Sutriadi, R. 2018. Defining smart city, smart region, smart village, and technopolis as an innovative concept in Indonesia's urban and regional development themes to reach sustainability. *IOP Conference Series: Earth and Environmental Science*, 202. Doi: 10.1088/1755-1315/202/1/012047

Vaisha, A. and Stastna, M. 2019. Smart village and sustainability. Southern Moravia case study. *European Countryside*, 11(4): 651–660.

Visvizi, A. and Lytras, M.D. 2018. Rescaling and refocusing smart cities: From mega cities to smart villages. *Journal of Science and Technology Policy Management*, 9(2): 134–145.

Zavratnik, V., Kos, A. and Duh, E.S. 2018. Smart villages: Comprehensive review of initiatives and practices. *Sustainability*, 10(2559): 1–14.

5 Smart Growth and New Urbanism, a Sustainable Approach towards Urban Redevelopment
Case of Chivhu

Monalissa Kaluwa, Chipo Mutonhodza and Leonard Chitongo

CONTENTS

5.1 Introduction 55
5.2 Theories of Urban Growth and Development... 57
    5.2.1 John de Groove's Smart Growth Concept 57
    5.2.2 Concept of New Urbanism... 57
5.3 Methodology.. 58
5.4 Data Presentation and Analysis... 59
    5.4.1 Chivhu Local Development Plan vis-à-vis Physical Development..... 59
    5.4.2 Challenges Faced by the Chivhu Local Authority in Trying to Avail More Land for Central Business District Expansion 61
    5.4.3 Smart Growth Concepts and New Urbanism as Remedies to Chivhu's Land Use Planning 62
5.5 Conclusion and Recommendations.. 65
References... 66

5.1 INTRODUCTION

Globally, lack of adequate space and infrastructure for further development within the central business districts (CBDs) is a manifestation of poor land use planning (Mwangi, 2012). Good governance, comprehensive land policies and sound land administration institutions are essential components for addressing the problems related to land use planning and management. According to Mwangi (2012), land being a scarce resource, there is a need to have in place sound land-use planning policies, which will ensure optimal utilisation with minimal conflict for sustainable economic and social wellbeing. In this regard, the smart growth and New Urbanism concept offer some guideline for identifying an adequate response to poor land use planning (World Bank, 2012). The world urban population will increase from 200 million in 2000 to 2 billion by the year 2050 (UN Habitat, 1996). This increase will definitely result in demand for urban land to cater for required developments for housing, infrastructure and other urban land uses. Thus, the adoption of smart growth and New Urbanism concepts can help in maximising urban economic growth through sustainable land use planning.

In America, the smart growth and New Urbanism initiative has led to compact, liveable and urban neighbourhoods, which attract more people and businesses (Ingrams et al., 2009). Smart growth and New Urbanism principles have been success stories in various parts of the world (John, 2013). Regionally, 'with high rates of political influence in urban planning and development in sub-Saharan Africa, conformity in land use development becomes a challenge in sustainable land use planning' (Mwangi, 2012). In the last decade, urban land use planning in Kenya was politicised. Politically and financially powerful members of the society acquired land in urban area without policies in national or local plans to guide development of the land they acquired (Burchell et al., 2000). These lands were then developed into urban property such as housing, commercial, industrial and institutional premises without conforming to relevant land use and development policies. This then leads to conflicting land use planning within urban areas; thus, smart growth and New Urbanism initiatives therefore can act as remedies towards urban redevelopment (Burchell et al., 2000).

Nationally, as with the case with most of Zimbabwe's local authorities, poor land use planning has led to non-conforming urban development and exhaustion of commercial land within urban centres leading to local planning authorities facing a dilemma on how to balance land use planning and local economic growth. Furthermore, given the rapid population increase within local urban areas, coupled with increased use of incremental planning, local authorities are advised to adopt trendy and sustainable planning paradigms so as to create sustainable cities nationwide. Chivhu is a case in point on the local context. The exhaustion of commercial land within urban CBD resulted in economic development being hampered as noted in Chikomba Rural District Council's report published in 'The Mirror' in 2017. The local authority thus decided to propose relocating the CBD to another site in order to avail the provisioning of commercial stands. This study thus proposes the assessment and cost–benefit analysis of urban redevelopment over relocation of the CBD to promote sustainable land use planning through the adoption of smart growth and New Urbanism initiatives so as to boost local economic growth of Chivhu. The chaotic and incompatible physical development of Chivhu CBD has negative implications on urban land use planning and can affect the rate and growth of Chivhu's local economic development. Coupled with a rocketing population increase and demand for commercial space in the CBD, the Chivhu local authority is faced with a dilemma on how to balance land use planning and economic growth. It is however important to acknowledge that though calls for the relocation of the Chivhu CBD have been made (though not yet approved), no research has been done to suggest the redevelopment of the existing town to achieve efficient and sustainable land use planning. Unless a strategy of promoting efficient land-use planning is considered, maximum growth and expansion of Chivhu CBD is not attainable.

Smart Growth and New Urbanism

The chapter focuses on assessing how the Chivhu town can adopt smart growth and New Urbanism concepts as sustainable strategies towards urban redevelopment. Objectively, the study assesses Chivhu's master plan vis-à-vis the physical development. This will assist in mainstreaming the smart growth and New Urbanism concepts in development planning of Chivhu town, ultimately leading to sustainable strategies that can promote the economic growth of the town. Significant amounts of literature on sustainable land use planning, urban redevelopment, smart growth and New Urbanism exist but detailed information within a Zimbabwean context is scarce. This study therefore seeks to fill in the existing knowledge gap by blending smart growth and New Urbanism with sustainable urban redevelopment in a bid to achieve economic revitalisation for the town of Chivhu. It weighs the benefits of urban redevelopment through smart growth and New Urbanism principles over relocating the Chivhu CBD as per the proposed plan by the Chivhu local authority.

5.2 THEORIES OF URBAN GROWTH AND DEVELOPMENT

The study adopted a dual theoretical framework by combining John de Groove's smart growth concept (1997) and the New Urbanism concept (pioneered by Carlthope, 1993).

5.2.1 JOHN DE GROOVE'S SMART GROWTH CONCEPT

De Groove (1997) is of the opinion that

> smart growth initiatives should emphasise on economic development with much focus exercised on policies to revitalise cities, reform local zoning to encourage compact development and infill, coordinate state agencies and their growth policies and overhaul capital investments to align with a sustainable agenda.

According to Ingrams et al. (2009), the smart growth concept calls for the 'Fix it First' programs to be adopted which is a move calling for investments to keep existing infrastructure in good repair before constructing new roadways. Kolbadi et al. (2015) argue that the smart growth concept aims at promoting compact cities. The rationale for the implementation of the smart growth concept relies heavily on a set of strategic benefits, which are conservation of countryside, less need to travel by car, reduced fuel emissions, better support for public transport, walking and cycling, better access to services and facilities, more efficient utility and infrastructure provision and revitalisation and regeneration of urban areas.

5.2.2 CONCEPT OF NEW URBANISM

New Urbanism is a movement in architecture and planning that advocates design-based strategies based on 'traditional' urban forms to help arrest suburban sprawl and urban-core decline and to build and rebuild neighbourhoods, towns and cities (Bohl, 2000). New Urbanism is an umbrella term encompassing the traditional neighbourhood development or 'neo-traditional town planning' of Andreas Duany and

Elizabeth Plater-Zyberk (Krieger and Lennerts, 1991), the pedestrian pocket and the transit-oriented design articulated by Carlthope (1993) and the 'quarters' approach of Krier (1998). The concept promotes the creation and restoration of diverse, compact, walkable, vibrant and mixed-use communities composed of the same components as conventional development but assembled in a more integrated fashion, in the form of complete communities. According to Gupta (2014), New Urbanism is one of the most important planning movements in this century and is about creating a better future for everyone.

New Urbanism is a model of sustainable urban development as evidenced by Toudeau's (2012) assertion that a number of sustainable development goals (SDGs) could be achieved through compact urban form, particularly where there is incorporation of mixed land use and housing type into the built environment. In addition, walking and social interaction further promote environmental and social sustainability which would resultantly enhance the economic sustainability of a region. Moreover, research studies on NU projects review how advocating for dense urbanism projects in or close to urban centres could promote compact development—a move that could lead to economic growth by unveiling more space for economic purposes; in this case, for urban redevelopment purposes (Toudeau, 2012).

For over two decades, urban designers have produced new model urban design codes and public policy recommendations for setting up intergovernmental planning institutions that foster New Urbanism. Furthermore, Carthope and Fulton (2001) state that since the late 1980s, states such as Florida and Oregon have amended existing or created new legislation that mandates or strongly encourage local governments to adopt plans and development regulations that require compact development patterns. This henceforth shows that Chivhu Local Authority has related cases to draw examples from in a bid to promote the redevelopment initiative through adopting or amending New Urbanism–related policies so as to promote urban liveability. New Urbanism is highly applicable to projects involving rehabilitation and retrofitting of existing urban designs and infrastructure (Bohl, 2000). In various situations, New Urbanism is presented as a genuine alternative to past urban renewal efforts (such as suburbanisation practices).

5.3 METHODOLOGY

Chivhu town is situated 146 km south of Harare in the Mashonaland East province of Zimbabwe along the Harare–Masvingo highway. The town developed as a result of colonial conquest by the white settler regime. The study adopted a mixed-methods approach where both qualitative and quantitative methods were applied. A case study approach was used to examine smart growth and New Urbanism concepts in Chivhu CBD. Data were collected through observation, questionnaires and key informant interviews. Fifteen respondents from the business community of 150 were randomly selected. Six key informants were purposively selected—three from the Chivhu Town Council and three representatives from the Department of Physical Planning (DPP). Secondary data sources included local newspapers, websites, municipal databases, layouts and municipal reports. Quantitative views regarding CBD rezoning vis-à-vis relocating were presented in the form of a graph. A thematic approach was used to present and analyse qualitative data.

5.4 DATA PRESENTATION AND ANALYSIS

5.4.1 Chivhu Local Development Plan vis-à-vis Physical Development

The local development plan (LDP) report of 2005 asserts that Chivhu has attempted to reconcile a number of isolated and disjointed incremental approaches to the development of the town centre since the 1890s. The expansion of the town centre and its development control has been largely without a substantive statutory plan save for layout plans, planning permission and the outdated 1960 Chivu outline plan. However, Abubakar (2008) argues that for a local authority to sustainably develop the area under its jurisdiction, it is important to revise master plans and LDPs so as to promote sustainable spatial planning and design. Moreover, revising LDPs also helps in linking up the economy and the spatial planning and design of the city (Dong, 2008). Therefore, having been without an updated statutory plan to guide its development, the existing dilemma Chivhu town is facing is henceforth an expected urban planning dilemma when it comes to development without updated LDPs.

5.4.1.1 Conformity versus Non-Conformity within the Chivhu Central Business District

With regard to the Chivhu LDP vis-à-vis physical development, the Chivhu CBD is characterised by conforming development. One respondent argues that formerly being a cowboy town with the current CBD sitting on what was formerly known as Enkeldoorn, a purely white settlement area, Chivhu's land use planning corresponds with the LDP. However, due to the increasing population and demand for land, Chivhu's current CBD can no longer accommodate the increasing demand. Thus, the Chivhu Town Council need to come up with an alternative remedy for the problem at hand.

According to one respondent, to address the existing problem the Chivhu LDP of 2005 clearly stipulates the need for the residents of former Enkeldoorn suburbs to incorporate their properties into the CBD zone through change of use. However, evidence shows that out of approximately 362 stands to be incorporated into the commercial zone, less than 20 properties abide to the proposal leaving at least 90% of the area operating as residential properties. Therefore, the statutory plan leaves most old residential properties out of the CBD. This explains why the Chivhu CBD is reported to have been sandwiched by residential stands leaving no room for further expansion and hence calls to relocate it.

However, one respondent argued that, 'a properly defined rezoned CBD would offer high economic returns to the local authorities through increased property values and investor friendly environment'. This is true with regard to stable but higher property values characterised by increased availability of commercial stands, coupled with lesser burden of service provisioning. Urban redevelopment through rezoning also helps in shaping strategies to deal with urban decline, decay and transformation

FIGURE 5.1 Non-conforming physical development within Chivhu town.

Source: Kaluwa (2018)

and thus it is important for the Chivhu Town Council to consider rezoning the CBD of Chivhu.

Regarding the issue of conformity of development, research findings also show that despite the CBD having conforming development, due to the increase in demand for commercial space, some residential property owners whose stands are located close to the CBD are subletting portions of their stands to small hardware operators without having followed the change of land use procedures (Figure 5.1).

Non-conforming development within residential stands is also enabled by lack of development control within the Chivhu local authority. In addition, non-compliance of the Chivhu residents to change of land use requirements is also leading to the rise of unregularised hardware shops.

5.4.1.2 Chivhu's Future Development Plans

According the LDP report of 2005,

> having been developed as a colonial administrative and trading centre rather than a centre for commerce, the centre is ill-equipped to support and sustain rapid and rampant population growth arising from the influx of people hoping to improve their lives in town.

With reference to the interviews conducted, one of the plans was to incorporate former white residential settlements into the commercial zone of Chivhu town centre through change of use. However, a set of challenges have led to the failure of this plan from becoming a reality, thus leading to the current land use planning chaos in Chivhu CBD.

5.4.2 CHALLENGES FACED BY THE CHIVHU LOCAL AUTHORITY IN TRYING TO AVAIL MORE LAND FOR CENTRAL BUSINESS DISTRICT EXPANSION

The research findings revealed that a number of challenges were being faced by the Chivhu Local Authority in their effort to acquire more commercial land through 'change of use'. The challenges include:

5.4.2.1 Poor Development Control

Evidence shows that, due to relaxed development control mechanisms within the local authority of the Chivhu town, the future development plans of the Chivhu CBD have not come into fruition. A case in point is the fact that properties situated within the former Enkeldoorn suburb have not exercised 'change of use' to enable commercial area expansion thus ending up with the current scenario of a 'sandwiched' CBD. In addition, owners of the residential properties are subletting portions of their properties to small hardware operators without consent from the Local Authority. Thomas (2001) argues that plans and policies provide a framework within which the development control process can take place and also help local authorities to achieve the vision they would have set themselves for the future development of the region. Henceforth, failure to exercise developmental control results in the failure to accomplish the set targets of the policy or plan in action.

5.4.2.2 Lack of Participatory Planning

Respondents mentioned that the Local Authority had not consulted the public during the initial stages of the preparation of the LDP of 2005. Hence, few people are giving in to the proposed plans for the expansion of Chivhu, thus leaving the Local Authority to come up with an alternative plan of relocating the CBD as they have failed to find a way to resolve the current situation. Hague and Jenkins (2004) connotes that in trying to deliver the best land use planning process, local authorities should adopt the concept of participatory planning rather public participation since 'the best land use planning process starts when the community requests it and . . . the council wants it, not when a government planner decides or finds it convenient to start'. Henceforth, public involvement from the initial planning stage to the implementation stage is crucial to build consensus about development proposals.

5.4.2.3 Lack of Public Cooperation

It is also argued that in 2015, the council engaged all property owners of dilapidated buildings so that they could demolish them and rebuild modern and upmarket structures. Despite it being a good call, three years down the line, little has been done to make this a success. This was a result of local authority having planned for the people not with the people, an ideology which is contrary to Hague and Jenkins (2004) who believe that the plan-making process should begin by putting issues of diversity, difference and even conflict at the centre of 'our' thinking and move from a planning process structured around a mentality of 'plan, consult, defend' to one that pre-negotiates, mediates unresolved differences and engages a plurality of participants in the implementation process.

5.4.2.4 Resource Constraints

It is also evident that due to the financial constraints of the Chivhu Town Council, several development projects are failing to come to fruition. An example is a quote from one interviewee, 'having anticipated that some residential stands in town will be converted into commercial stands through change of use, and having the desire to fully implement an urban rezoning plan, financial resources are the major drawback'. This is in line with what Hague and Jenkins (2004) point out that participatory approaches to planning need proper resourcing. Therefore, though it is significant for local authorities to exercise participatory planning, with limited resource base, the desired projects and plans are likely to fail.

5.4.3 SMART GROWTH CONCEPTS AND NEW URBANISM AS REMEDIES TO CHIVHU'S LAND USE PLANNING

Respondents showed mixed feelings regarding this subject. Most respondents from the category of business community proved not to know much Sustainable Urban Development concepts. Figure 5.2 shows respondents' understanding towards Smart Growth, New Urbanism and urban development.

5.4.3.1 Relocation vis-à-vis Redevelopment

From the research survey, many respondents argued that having to relocate the CBD did not seem to be the best idea to them as they feared for their business since they thought that the current Chivhu CBD could be overshadowed by the proposed one, thus leading to dwindling of business activities in the currently existing town. Figure 5.2 shows respondents' reaction towards redeveloping the CBD versus relocating it.

From the research findings illustrated in Figure 5.2, one can note that 14 of the respondents strongly agreed to the redevelopment of the CBD compared to relocation. In support of this position, one respondent went on to argue that in a bid to create a harmonious urban environment that is based on a well-balanced land use allocation and distribution system, the council should call for proper zoning of land for residential, commercial, institutional, recreational and other land uses. Thus, there is a need to rezone former Enkeldoorn residential stands to commercial stands. Smart growth strategies for sustainable urban development and revitalisation address a number of design concepts and principles such as balanced land use planning principles. Henceforth, the Chivhu Town Council should properly rezone the CBD in order to achieve balanced land use planning.

In addition, the Town Planning Officer highlighted that, of the 362 residential stands in the inner-city zone, which the Local Authority hoped to incorporate into the CBD through change of use, less than 20 stands were converted. The rest still remain as residential properties, a setback towards the Chivhu CBD's economic growth through shortage of commercial stands for economic investment.

Smart Growth and New Urbanism

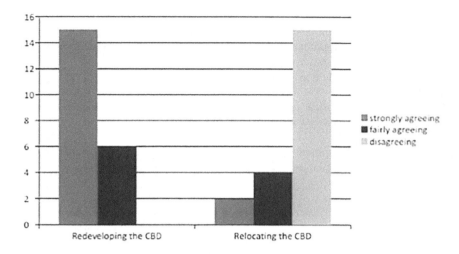

FIGURE 5.2 Respondents' views regarding CBD rezoning vis-à-vis relocating.
Source: Kaluwa (2018)

Moreover, research findings from the LDP report of Chivhu (2005) show that the second goal of the council was to optimise the use of land within the plan area. The council also wishes to encourage the delivery of land uses that will encourage a multi-functional and efficient urban settlement.

The town planner argued that the Smart Growth and New Urbanism development initiative is a sustainable move. However, poor resource endowment within the Local Authority and shortage of commercial space for development within the commercial zone has led to proposals of relocating the CBD to 2 km south-east of the existing town centre.

During the research survey, a respondent argued that a number of factors led the Council to come up with the relocation idea for the Chivhu CBD. These include:

- Need to avail more land for urban development
- Need to create more space for commercial stands
- Need to enhance economic development for the Chivhu town
- Need to attract investors, both local and foreign into Chivhu

This concurs with Millspaugh (2002) who noted that, for successful economic development of a region, a number of elements should be in place to promote an investor friendly environment. Relocating a business centre, despite it being a cumbersome task, which requires a large resource base, can successfully avail the region's economic potential.

Respondents from both the planning arena and the business community argued against relocation of the CBD. One respondent clearly stated that '*Relocating might not be the best way to address the current problem. It may lead to creation of a ghost town'*. This henceforth demonstrates that though relocation has the potential to avail land for economic development, with relation to the economic growth of the existing town, it would lead to negative economic impacts.

In addition, a respondent argued that:

> Relocation is likely to negatively affect existing development in that commerce will shift to the new site, the existing site will then deteriorate.

This shows that redevelopment is being viewed as the best remedy towards addressing the current land use planning situation of the Chivhu town. Another respondent clearly stated that:

> *I believe redevelopment is the best option though it has its own challenges; but in a bid to promote sustainability and reduce the possibility of creating ghost towns, it is the best.*

From this response, one can note that relocation has a potentially negative impact on the economic growth of the town. In addition, redevelopment seems to be taken and embraced as the solution to the existing land use planning problem in the area under study though people are aware of the associated challenges. Millspaugh (2002) asserts that economic motivation is an important factor to take into consideration when adopting the concept of city rebuilding. There is need to take into recognition the need for additional tax revenues to relieve the crisis of municipal finance; the desire to attract middle and upper-class families back to the city; the incentive of preserving old real estate investments and creating opportunities for new ones; and the urge, sometimes born in desperation, to create a *tour de force* of open space and architecture that will spark a 'revival' of the central city.

Another respondent also argues that:

> *Smart Growth and New Urbanism are good concepts to adopt when it comes to town planning especially on urban redevelopment considering that they can lead to creation of live-able cities which is good for economic development.*

The respondent believes that these concepts can offer a sustainable solution to the current situation in Chivhu since through redeveloping the inner-city and its surrounding area (rezoning), more stands can be unveiled though probably at a higher rate but good for the economic growth of the town.

Moreover, New Urbanism and Smart Growth have also been believed to offer a sustainable solution in Chivhu as there will be lower commuting costs and maximum utilisation of land and space. It also leads to the overall reduction of pollution and green-house gas emissions; thus, a sustainable strategy towards urban development. Due to a number of advantages associated with urban redevelopment, respondents tend to appreciate the adoption of sustainable urban development concepts rather than relocating the CBD as this would bring more harm than good to the economic and spatial development of the Chivhu town.

5.5 CONCLUSION AND RECOMMENDATIONS

The research findings show that Chivhu's physical development conforms to the LDP in use though it is outdated. However, with mounting pressure within the CBD and the increasing demand for commercial space, some owners of residential stands surrounding the Chivhu CBD are subletting portions of their lots to small hardware operators without undergoing proper change of use procedures as required thus leading to instances of non-conforming urban development within the Chivhu town.

The analysis of findings indicated that the current development of the Chivhu town is not controlled by any up-to-date statutory plan. Rather it has been largely influenced by disjointed incrementalism. However, though a properly defined statutory plan to guide the current and future development of the Chivhu CBD does not exist, the Chivhu Local Authority had hopes of availing more space for the expansion of the CBD through change of use of properties in the surrounding area. However, unavailability of resources within the coffers of Council and the lack of cooperation between the local authority and the property owners within that residential zone made it difficult for the plan to be a success. This henceforth left the council with only one option; that is, to relocate the CBD to avail more space for Chivhu's economic expansion and development.

The analysis of the research findings indicated that, though the majority of the business community had less understanding of the sustainable urban development concepts, redeveloping the CBD proved to be a better solution than relocating it. From the research findings, the major fear regarding the relocation of the CBD seemed to be the possibility of leading Chivhu to a ghost town. Thus, with a number of benefits associated with adopting sustainable urban development concepts such as Smart Growth and New Urbanism, redeveloping the CBD using these concepts seemed to be a better solution.

In addition, with the major aim of the study being to shape the land use planning of Chivhu so as to attract potential investors, the Smart Growth and New Urbanism concepts prove to be remedies to the existing land use situation of Chivhu. This is so since the concepts lead to creation of liveable and environmentally sustainable cities with mixed land use planning coupled with compact urban design. This would pave way for sustainable city growth since compact urban design within the rezoned area would accommodate more urban populace.

The conclusions of this study lead to the following specific recommendations that can be adopted to help come up with sustainable land use planning for Chivhu in order to avail more commercial space so as to revitalise and enhance economic growth and development of the town. Since adoption of the Smart Growth and New Urbanism concepts has proved to be a success story in many developing countries, it is high time local authorities forego old planning methods and embrace the sustainable urban development concepts. The concepts can be important tools for the revitalisation of towns and cities for enhanced economic growth and development.

Local authorities should embrace and adopt participatory planning in order to enhance community participation in local development planning so as to minimise public resistance and improve implementation of plans.

It is also recommended for the local authorities to adopt public educational and awareness campaigns on sustainable urban development concepts and initiatives for the development and growth of cities and towns. Local authorities should promote land use planning that is driven by the sustainable urban development concepts. Revision of LDPs should be done on a regular basis so as to enable the integration and adoption of modern sustainable urban development concepts in the designing and planning of cities and towns. The Ministry of Local Government should liaise with the national government on behalf of the local authorities for the government to come up with strategies to financially support local authorities in the implementation of future development plans, as in the case Chivhu.

Just like the change affecting local authorities, the Department of Physical Planning (DPP) should adopt and promote the use of sustainable urban development concepts in urban design and land use planning. It also has to promote revision of LDPs to meet the needs of the existing and future urban and regional populace. Moreover, the DPP should promote urban renewal and redevelopment initiatives to enhance the development of cities. It is also encouraged that the local communities embrace participatory planning and community involvement in urban planning and development as this enhances the sustainable growth of cities.

REFERENCES

Abubakar, J., 2008. *New American urbanism, reforming suburban metropolis.* New York: Skira Architecture Library.

Bohl, W., 2000. Analysing increasing process of urban open environments in process of urban development (Case study: Tehran metropolice), Utopia number 5.

Burchell, Robert W., Listokin, David and Galley, Catherine C., 2000. Smart growth: More than a ghost of urban policy past, less than a bold new horizon. *Housing Policy Debate*, 11(4), 821–879. Taylor & Francis. Abingdon-on-Thames, United Kingdom.

Carlthope, P., 1993. *The next American metropolis: Ecology, community and the American dream.* New York: Princeton Architectural Press.

Carthope, G. and Fulton, H., 2001. Sense of community and neighbourhood form: An assessment of the social doctrine of smart urban design. *Urban Studies*, 36, 1361–1379.

Chivhu Local Development Plan Report, 2005. Chikomba District Council.

De Groove, J., 1997. *Smart growth and the sustainable city concept.* London: Routledge.

Dong, E., 2008. *Cities in transformation.* Beverly Hills: Sage.

Gupta, P., 2014. Town planning in Delhi: A critique. *Economic and Political Weekly*, 4(40), 1591–1600.

Hague, C. and Jenkins, P., 2004. *Place identity, participation and planning.* London: Routledge. doi: 10.4324/9780203646755

Ingrams, G.K., Armando, C. and Carbonell, A., 2009. *Smart growth policies: An evaluation of programs and outcomes.* Cambridge: Kirkwood Printing.

John, T., 2013. *Smart cities and the global trends in urban planning.* Beverly Hills: Sage Publishers.

Kaluwa, M., 2018. *Smart growth and new urbanism: A sustainable approach towards urban redevelopment.* Bachelor of Science Honours Degree in Urban Planning and Development Dissertation. Great Zimbabwe University: Masvingo, Zimbabwe.

Kolbadi, N., Mohammadi, M. and Fahimeh, N., 2015. Smart growth theory as one of the main paradigms of sustainable city. *International Journal of Review in Life Sciences*, 5(9), 209–219.

Krieger, S. and Lennerts, R., 1991. Mixed land uses and commuting: Evidence from the American housing survey. *Transportation Research Part A*, 30(5), 361–377.

Krier, T., 1998. Retracting suburbia: Smart growth and the future of housing. *Housing Policy Debate,* 10, 513–540.

Millspaugh, Martin, 2002. Spatial analysis of urban smart growth indicators. *Research in Human Geography*, 77.

Mwangi, S., 2012. *How superstore sprawl can harm communities—and what citizens can do about it.* Washington, DC: National Trust for Historic Preservation. 120 p.

Thomas, P., 2001. The real cost of growth in Oregon. *Population and Environment*, 18(4), 373–388.

Toudeau, T., 2012. *Sprawl costs: Economic impacts of unchecked development.* Washington, DC: Island Press.

UN Habitat, 1996. *State of world's cities 206/07trends Sub-Saharan Africa urbanization and metropolitanization.* UN-Habitat SOWC 04/RB/4

World Bank, 2012. *Urbanization prospects: The 2011 revision highlights.* ESA/P/WP/224 Department of Economic and Social Affairs. Population Division: United Nations. New York.

6 Trans-Border Spatial Planning

Assessing the Musina–Beitbridge Twinning Agreement between South Africa and Zimbabwe

Shylet Nyamwanza, Peter Bikam and James Chakwizira

CONTENTS

6.1	Introduction	70
	6.1.1 Location of Study Area	71
6.2	Theoretical Framework	73
	6.2.1 Concept of Twinning Agreements	73
	6.2.2 Cities as Interdependent Networks	73
	6.2.3 Co-operational Learning as a Key Ingredient of Twinning Agreements	74
	6.2.4 Indicators of Twinning Agreements	74
	6.2.5 Twinning Linkage Success Factors	74
	6.2.6 Key Questions to Ask When Drawing Up Twinning Agreements	75
6.3	Methodology	76
	6.3.1 Administration of Questionnaires	76
	6.3.2 Key Informant Interviews	77
6.4	Results and Discussion	78
	6.4.1 The Terms of Reference of the Musina–Beitbridge Twinning Agreement	78
	6.4.2 The Implementation of the Agreement	79
	6.4.3 Challenges of the Twinning Implementation Action Plan	79
	6.4.4 Spatial Development Impacts of the Twinning Agreement	80
6.5	Recommendations	82
	6.5.1 Appointing Twinning Champions	83
	6.5.2 Implementing a Twinning Agreement Guide	83
	6.5.3 Establishing Reliable Funding Sources	83

DOI: 10.1201/9781003221791-8

    6.5.4   Establish a Working Joint Secretariat..83
    6.5.5   Harmonise Legislative Frameworks...84
6.6 Conclusion 84
References...85

6.1 INTRODUCTION

Twinning or sister city agreements are forms of legal and social memoranda of agreements between towns and regions to promote economic, cultural and commercial ties (Ahmad, 2001). However, the modern concept of town twinning has to do with fostering friendship and understanding to promote the development of cities through resource sharing and information exchange (Buxbaum, 2014). In recent times, twinning has increasingly been used to form strategic international business relationships between cities and countries (Beaverstock et al., 2000). Further to this, it was indicated that twinning with regard to city planning has undergone a paradigm shift and hence the use of terms such as global village, city state, community and neighbourhood development more than the term national or federal (UNDP, 2000). Twinning evolved over the years from being a historic instrument of building peace among European nations that have been previously at war into a worldwide phenomenon encompassing friendship, solidarity, culture, international understanding, humanitarian assistance, sustainable development and good governance (Hoetjes, 2009). De Villiers et al. (2008) indicate that twinning is a decentralised form of development; for example, building bridges of understanding between two cities. Furthermore, twinning agreements are instruments used for fostering international cooperation between adjacent cities to enable spatial development (Buxbaum, 2014). This suggests that there is genuine reciprocity of efforts and benefits for the two towns or cities involved. Discussions on twinning do not seem to capture the implementation component of twinning agreements, particularly the resolutions (Beaverstock et al., 2000). However, it was noted that what is more important is what has been achieved in a twinning agreement (Byukusenge, 2014). Similarly, there is also a disjuncture between the theory and practice of twinning relationships (Jayne et al., 2011).

In view of the opinions of the authors discussed earlier, twinning agreements can be derived for a specific local interest and it is important to note that twinning also responds to different motivations and desirabilities. Several countries in southern Africa entered into similar partnerships especially after the Southern African Development Community (SADC) Treaty of 1992, which acknowledged that, 'many people, land uses, and natural resources have always transcended national boundaries in the region'. In light of that, several Spatial Development Initiatives (SDIs) were established between and amongst countries in the SADC region. It was also because of the need to concentrate limited resources on the provision of hard infrastructure in areas with the highest economic potential and where leverage of private sector investment is most likely to be achieved (Phillips, 2002). Some of the SDIs in southern Africa include the Trans-Limpopo SDI, the Maputo Development Corridor, the Lubombo SDI, the Limpopo Valley SDI and the Okavango Upper Zambezi Tourism Initiative (OUZIT) (Phillips, 2002).

The establishment of our case study, the Musina–Beitbridge twinning agreement of 2004, can be traced back to the Trans-Limpopo SDI of 2001 that was signed between the Limpopo Province of South Africa and the Matabeleland South Province of Zimbabwe in 2001. This SDI was planned as a spatial development and economic strategy in the development corridor that stretched from Victoria Falls in Zimbabwe, through Beitbridge into South Africa's Limpopo Province through Musina and up to Polokwane.

6.1.1 Location of Study Area

Musina Local Municipality is one of the four local municipalities within the Vhembe District Municipality situated in the Northern part of the Limpopo Province in South Africa (Musina IDP, 2016). Musina town was established in 1929. According to Stats SA (2012), Musina Local Municipality had a population of 68,359 in 2011 and a growth rate of 5.53% from 2001 to 2011. The Beitbridge border links South Africa, Zimbabwe and African countries to the north. Musina and Beitbridge share a border post through which approximately 9,000 people pass every day, that is, 270,000 per month and 3,240,000 per year (Netsianda, 2011). During peak periods such as Easter and Christmas holidays, the numbers can soar to about 25,000 per day. It is a major trade route where most goods imported from overseas for countries such as the Democratic Republic of Congo, Malawi and Tanzania pass through. This shows the importance of the two towns as transit towns. It was in view of the importance of the two towns that a twinning agreement was signed in 2004 under the auspices of the Trans-Limpopo SDI of 2001. (Byukusenge, 2014).

Figure 6.1 shows the location proximity between Beitbridge and Musina. The proximity of the two border towns which enhanced the signing of the twinning agreement is also shown. Musina town is 15.6 km away from the border post whilst Beitbridge town is 177 metres away from the border post. An international road traverses the two towns—from the South African side it is called N1 and from the Zimbabwean side the name of the road changes to A6. In addition to this, what also separates the two towns is the Limpopo River as shown in Figure 6.1.

The aim of the twinning agreement was to make Musina and Beitbridge reach a development agreement for the benefit of the two towns. This is because the spatial development cooperation of 2004 was aimed at signing a collaboration that would address issues of infrastructure development thus easing the movement of people, sharing information and expertise as well as promoting foreign direct investments through the establishment of Special Economic Zones (SEZs) (Twinning MOU, 2004). However, the study revealed that not much was achieved. The implementation strategies to achieve the objectives were inadequate and not clear. The aim of the study was to unpack the strategies of the twinning agreement between Musina local municipality in South Africa and Beitbridge municipality in Zimbabwe. The Musina and Beitbridge twinning agreement was used as a case study to first examine the impact of the approach on spatial development and in the second place to determine the implementation of the agreement with respect to spatial planning and the harmonisation of the twinning agreement between the two towns from two countries. To put the study into perspective, the terms of reference of the twinning agreement

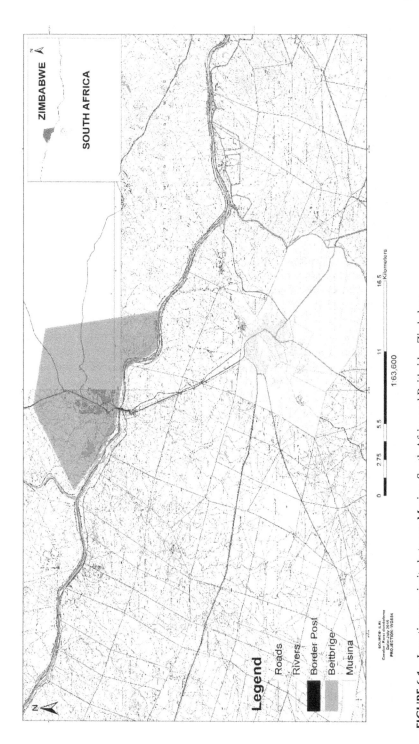

FIGURE 6.1 Location proximity between Musina, South Africa and Beitbridge Zimbabwe.

Source: Musina SDF (2018), UNIVEN GIS Section (2016) and Beitbridge TC report 2018

were unpacked to determine how it was implemented before discussing the impact and the spatial development component.

6.2 THEORETICAL FRAMEWORK

The purpose of this study was to have a deeper understanding of the implementation of spatial development twinning agreement between two towns from different countries because there is limited literature on the impact of such agreements. The review of literature was centred on the theory of twinning agreements between two towns and spatial development to determine the extent to which the objectives of the agreement were achieved. The first step was to determine if the terms of reference of the agreement are linked to the spatial impact of the objectives of the agreement. The snowball sampling method was used to determine the sampling size of the respondents. Secondary data sources in relation to the impact of the agreement on spatial development and the residents of both towns were utilised.

6.2.1 Concept of Twinning Agreements

The first recorded twinning agreement was undertaken in 1920 between Keighley, West Yorkshire and Poix du Nord in France (LGIB, 2001). However, after the end of the Second World War, the concept spread rapidly, with British cities forging links with European cities which were devastated by the war conflicts. Twinning was then seen as an effective tool in promoting peace and reconciliation. European twinning originally was essentially motivated by peace consolidation within Europe (SCI, 2004). The concept of sister-city relationships, a similar philosophy of 'citizen diplomacy', followed but took a global perspective. In the United States of America, the concept of sister cities programs traces its roots to 1956 when President Eisenhower proposed a people-to-people citizen diplomacy initiative and the organisation 'Sister Cities International' was formed (SCI, 2004). It was originally part of the National League of Cities, but became an independent, non-profit organisation in 1967 due to its tremendous growth and popularity (LGIB, 2001).

Over the years, the concept of twinning evolved from its origins of confidence-building between European towns into a global phenomenon encompassing friendship, solidarity, culture, awareness-building, international understanding, humanitarian assistance, sustainable development and good governance (UNDP, 2000). The participants have also changed. The United Nations Development Programme recalls that when city-to-city cooperation began links were usually between town halls. They were led by political representatives consisting largely of high-level visits between the twinned towns, supplemented by cultural and sporting exchanges (UNDP, 2000).

6.2.2 Cities as Interdependent Networks

The increasing interdependency and networks between cities are explored in 'world city network analysis'. The concept evolved from world-city and global-city theories. This analysis examines urban connections from a global perspective, a departure from city studies that had previously been specifically concerned with national urban systems (Brown et al., 2010). While the world-city network analysis sees cities as 'highly concentrated command points in the organisation of the world economy', these points are connected in an 'interlocking network' in the world economy (Brown et al., 2010). It is argued that globalisation can be understood in terms of world city network analysis and global commodity chain analysis, that cities have spatial relations, which have transnational functions materially within their spheres of influence (Beaverstock et al., 2000). This suggests that cities are no longer isolated but intertwined in a world of flows, linkages, connections and relations.

6.2.3 Co-operational Learning as a Key Ingredient of Twinning Agreements

The learning theory was proposed as an approach to municipal twinning (Hoetjes, 2009). The learning theory falls under the class of collective action theories of development planning. Learning theory revolves mainly around the questions, 'are twinning partners learning from each other?' and 'do they communicate effectively?' The theory focuses on the co-operation aspects of the partners such as planning, inputs, implementation, outputs and targets achievement. The key ingredient is cooperation between twin towns and what they want to achieve beneficially. The basis of this theory is that twinning is a win-win situation where all partners must benefit. The cooperation aspects are crucial for the realisation of the anticipated effects (Hoetjes, 2009).

6.2.4 Indicators of Twinning Agreements

De Villiers et al. (2008) undertook a ground-breaking study linked to the discipline of spatial planning. He proposed a management and planning model for municipalities to invoke municipal twinning agreements. This model consists of fundamental success factors, including strategy formulation, identification of potential partners, evaluation and availability of resources, strong leadership, communication and public participation (De Villiers et al., 2008). He further suggested some traits such as trust, cultural sensitivity and flexibility to complement these success factors. Table 6.1 lists possible success indicators for successful twinning agreement implementation with the corresponding implementation indicators.

Table 6.1 outlines which factors contribute to a successful twinning agreement between two parties. The twinning factors have corresponding indicators. The list is by no means exhaustive but these are just the principal issues to take note of (De Villiers et al., 2008).

6.2.5 Twinning Linkage Success Factors

In view of the indicators of twinning implementation outlined in Table 6.1, greater importance should be given to the success factors, which impact the potential success of such arrangements between or amongst cities. These success factors were first discussed by Tjandradewi and Berse (2011). Figure 6.2 is the diagrammatic illustration of the twinning linkage success factors.

TABLE 6.1
Twinning Agreement Factors and the Corresponding Implementation Indicators

Twinning indicators	Implementation factors
Strategic formulation of twinning agreements	Strategies with respect to effectiveness and efficiency
Identification of willing alliance partners	Political and technical leadership willing to partner
Evaluation of potential partners to agree on key terms	Trust; Reciprocity; Commitment; Understanding; Cultural sensitivity; Risk; Flexibility
Negotiations of both parties	Negotiation champions that are qualified and experienced.
Community participation	Effective involvement of community leaders and groups
Availability of resources	Budget, Donor-funded or not, Structural arrangements and staff complements.
Support twinning agreements on both sides	Identification of champions on both sides to run the agreement.
Similar spatial development perspectives	Programs and project to achieve spatial development.

Source: Derived from De Villiers et al. (2008) Concept of twinning agreement success factors and indicators

Figure 6.2 shows seven key success factors of city-to-city relationships suggested by Tjandradewi and Berse (2011). Participation in projects between cities should be at the civil society levels of the cities. Reciprocity should be demonstrated in mutual trust and respect by both parties. Tangible outcomes related to the objectives of the agreement should be monitored in the partnership. Recognition from the central government is essential to ensure the sustainable impacts of exchanges through networking and institutionalisation of their actions (Tjandradewi and Berse, 2011). The participation of senior level officials and decision makers is critical in paving the way for cooperation, mobilisation of resources and support from different departments and other institutions. Intensive publicity in local media on the cooperation will increase citizen awareness on the benefits of the cooperation (Tjandradewi and Berse, 2011).

6.2.6 Key Questions to Ask When Drawing Up Twinning Agreements

Key questions are crucial when negotiating and drawing up twinning agreements (Connell et al., 1996). The questions enable the planners and implementers of the agreement to determine the outcomes or deliverables in advance to know whether they are feasible or not. Figure 6.3 is the concept of some of the key questions to ask when drawing up twinning agreements between cities.

These six questions can be asked before signing any twinning agreement (Connell et al., 1996). The questions can be used to guide the implementation strategy of the specific terms of reference of the twinning agreement. At the end of the implementation period for a particular question, the outcome will depend on whether the questions were answered or not and what should be done to deliver during implementation (Connell et al., 1996).

FIGURE 6.2 Twinning linkage factors.

Source: Adapted from Tjandradewi and Berse (2011)

6.3 METHODOLOGY

A trans-border twinning agreement performance impact survey method was used to explore the implementation of a twinning agreement using the concepts of trans-border twinning agreements. The process of assessment involved evaluating the achievements of cooperation objectives against predetermined criteria. Trans-border twinning survey falls within the ambit of an independent assessment of ongoing or completed agreement involving two towns which may or may not be within the same political boundary but more likely to be in two different countries (Hsu, 2003). This means that the process of assessment involves evaluating the achievements of cooperation objectives against predetermined criteria. The study used key informant interviews, questionnaires and direct field observations to determine the impact of the implementation of the concept on the residents of the two border municipalities.

6.3.1 ADMINISTRATION OF QUESTIONNAIRES

Three hundred and eighty-five questionnaires were administered to the residents of Musina and Beitbridge. The use of both open-ended and closed questions was

Trans-Border Spatial Planning

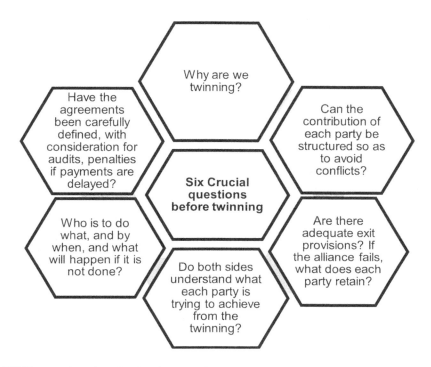

FIGURE 6.3 Questions for drawing up twinning agreements.

Source: Adapted from Connell et al. (1996)

appropriate to this study to assist the researcher to have an insight into all the factors surrounding the implications of the implementation of the twinning agreement and its impact on the residents in Musina and Beitbridge border towns.

6.3.2 KEY INFORMANT INTERVIEWS

Fourteen key informants participated in the interviews. Data collected assisted the analysis to understand which factors the municipal departments considered and which resources were put aside during the planning, implementation and management of the Musina–Beitbridge twinning arrangement. The authors assumed that the key informants possess the knowledge essential in determining the effectiveness of the terms and conditions of the twinning agreement. Data collected from key informants helped to derive strategies on how to improve trans-border spatial planning through twinning. Based on the organisational structures of Musina and Beitbridge, local municipalities, key informants from the following departments of both Musina and Beitbridge were interviewed: the mayor's office; the chief executive officer's office; the technical division director's office; the department of immigration, the Beitbridge border post and the spatial planning department.

6.4 RESULTS AND DISCUSSION

This section discusses the terms of reference of the twinning agreement, the implementation of the agreement and the challenges and the impact on the local residents in both towns.

6.4.1 THE TERMS OF REFERENCE OF THE MUSINA–BEITBRIDGE TWINNING AGREEMENT

At the beginning of the study, the twinning Terms of Reference were examined to explore the tangible benefits which were achieved since the agreement was signed in 2004. The main features of the terms of reference of the twinning agreement between Musina and Beitbridge were two-fold, namely: to strengthen the multilateral development of economic, trade, scientific, cultural and human relations and to assure that firms and friendly relations in the two municipalities strengthen development programmes between the two parties. The terms of reference and the memorandum of agreement (MOA) opened the gate for strengthening of friendly cooperation through the exchange of development ideas under eight themes as shown in Table 6.2.

The eight twinning agreement themes and the implementation details were among the deliverables to be achieved. However, our focus was on the theme of 'economic administrative prospects' because they relate to spatial planning and the implications of the impacts on the local residents from both towns.

TABLE 6.2
Themes of the Musina–Beitbridge Twinning Agreement

Twinning agreement themes	Implementation details
Tourism and conservation	Joint tourism development initiatives. Management of mobile resources like hunting
Environment and engineering	Solid waste management. Water quality control
Women and children	Matters of child abuse and disabled children
Education, arts, sports and culture	Partnership in education, arts, drama, poetry and culture. Sport. Traditional dancing and music
Health	Administer health-related issues such as HIV/AIDS and reduction in teenage pregnancies
Transport and disaster management	Transport coordination and disaster management as well as Information and resources sharing
Safety and security	Joint community-based crime prevention strategy
Economic and administrative prospects	Administration: Exchange of information, knowledge and expertise between the two municipalities; easing movement of local people; Employment creation; infrastructure development and foreign direct investments

Source: Extract from the Musina–Beitbridge twinning agreement (Twinning MOU, 2004)

6.4.2 THE IMPLEMENTATION OF THE AGREEMENT

We used a trans-border twinning agreement performance impact survey method to assess the achievements of the cooperation by analysing the outcomes of the twinning agreement between Musina and Beitbridge using the concepts of trans-border twinning agreements. The goal of assessing the implementation of the agreement was to determine the extent to which the objectives of the twinning agreement have been achieved and consequently determine the spatial development impact. The study provided results on what was achieved and what was not achieved. For instance, infrastructure development was one of the key planning challenges that needed to be addressed in 2004 according to the twinning MOA. Although there was evidence of joint meetings soon after the twinning MOA was signed, there is no evidence of joint implementation of what constituted spatial development planning by the two towns.

6.4.3 CHALLENGES OF THE TWINNING IMPLEMENTATION ACTION PLAN

The key informants from both Musina local municipality and Beitbridge Town Council were asked to indicate the twinning agreement implementation challenges, which were hindering them to achieve the spatial development goals of the twinning agreement in both municipalities. The challenges indicated are illustrated in Figure 6.4.

The challenges indicated were manifested by the pressing problems of spatial development such as lack of bus ranks, decaying infrastructure, inadequate ablution

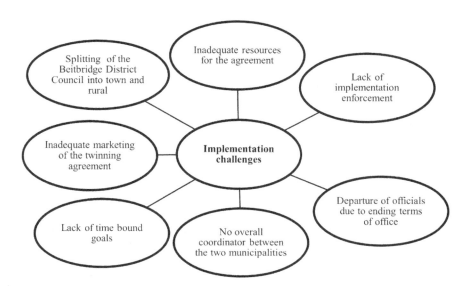

FIGURE 6.4 Implementation challenges that hindered the achievement of spatial development goals.

Source: Nyamwanza (2017)

blocks, inadequate pavements for pedestrians, inadequate tarred roads and traffic congestion in the central business districts (CBDs) of both the towns. By 2016, the two local municipalities did not have enough funds to develop their CBDs. There is overdependence on government grants. One of the aims of the twinning agreement was to share ideas on projects, which could generate income for the development of Musina and Beitbridge towns, but this was inadequately addressed. Furthermore, the local residents expressed dissatisfaction and outlined a number of issues that they felt should have been addressed by both parties. They indicated that they were not properly engaged in the twinning agreement and because of that they were not aware of what was happening on the ground. The key informants admitted that although they had a keen interest in the twinning agreement, they did not have any strategy or responsible person to champion it. The Musina Integrated Development Plan of 2016 indicated that the joint implementation committees from Beitbridge and Musina needed to be resuscitated for the twinning to work. The twinning agreement between the Musina local municipality and Beitbridge Town Council faced a number of challenges that resulted in a lower impact on trans-border spatial development due to the slow implementation progress caused by the lack of funds.

6.4.4 Spatial Development Impacts of the Twinning Agreement

In an attempt to have an overview on whether the spatial development planning goals were achieved, the key informants were asked to score the spatial development goal achievements out of 10 with 1 being the lowest score and 10 being the highest score. Table 6.4 shows their responses.

Overall, the highest average score for all goals was 3.6 out of 10, which is 36%. This means the key informants felt that the spatial planning goals of the Musina–Beitbridge twinning agreement had achieved only 36% of its goals, which is not close to the mid-mark. However, it is important to note that in Musina, the scores for SEZ projects were very high, up to 70%. In Beitbridge, SEZ projects were still far behind, with most key informants scoring a zero and only two scoring 1 and 2, respectively. In terms of SEZs, Musina is at a more advanced stage, whereby they have already demarcated their proposed SEZ sites and are working on site reports. However, the Beitbridge Town Council is still working on choosing their sites.

On average, the key informants indicated that 34% had been achieved in terms of addressing infrastructure challenges. In terms of tarring roads, the Musina local municipality still has a backlog of about 20 km of gravel roads that have to be tarred and 25 km backlog of tar roads that have to be upgraded/resurfaced (Musina IDP, 2016). Both towns still experience high levels of congestion because there are inadequate pavements for pedestrians. There are inadequate ablution blocks in both towns considering the high transit population travelling between the two towns. In terms of bus ranks, Musina contains one formal rank that is inadequate for buses. As a result, buses have to park at Musina filling stations because there is no space for them to park in the rank. Furthermore, the Beitbridge bus rank is still an open gravel space with no infrastructure. This open space floods during rainy seasons because there are no drainage facilities. Most of the roads in Beitbridge are still in gravel form. There are also inadequate pavements in Beitbridge.

TABLE 6.4
Key Informants' Overview on Spatial Planning Goals Achievement

Spatial planning goal	Musina local municipality's key informants' rating (Out of 10)							Beitbridge Town Council's key informants' rating (Out of 10)					Border post (SA and Zim)			Mean
	K1	K2	K3	K4	K5	K6	K7	K8	K9	K10	K11	K12	K13	K14	
1. Addressing inadequate and decaying infrastructure in Musina and Beitbridge	5	6	7	6	4	4	3	5	0	4	3				34%
2. Sharing of information and expertise sharing with respect to spatial development planning.	5	3	1	0	0	2	0	6	0	3	3				16%
3. Assembling inputs to implement the terms of reference in the twinning agreement.	2	2	1	0	0	1	0	5	3	0	0				7%
4. The progress of establishing Special Economic Zones (SEZ) on both sides.	6	7	5	6	5	0	0	1	2	0	0				23%
5. Easing the movement of people	0	0	0	0	0	0	0	0	0	0	0	n/a	n/a	n/a	0
Average score	3.6	3.6	2.8	2.4	1.8	1.4	0.6	3.4	1	1.4	1.2				16

Source: Nyamwanza (2017) (Sample size of key informants: 14)

Information sharing was achieved up to a level of 16%. The general feeling from all the respondents was that there was not a direct implementation plan on how information and expertise was supposed to be shared. Twinning officials were supposed to meet four times per year. However, after 2006 there was no evidence of such a joint meeting being conducted. Inputs that had been set aside to implement the terms of reference were only 7% of what was needed. The progress for SEZs was only 23%. Easing the movement of people had not been achieved at all. This also corresponds with what the local citizens indicated that their movement across the border had not been eased at all. Furthermore, the key informants indicated that there were no strategies to integrate spatial planning legislations from different countries.

One key informant indicated that the goals were partially achieved during the early years of the twinning agreement. They managed to conduct meetings, field visits, address Beitbridge sanitation problems and solid waste disposal and the unhygienic effluent that was affecting the Limpopo River around 2006. All the other key informants indicated that the goals were not achieved. The key informants from the Beitbridge Town Council indicated that the separation of Beitbridge town and Beitbridge rural in 2007 left the twinning responsibilities hanging because till date they are not sure whether it is the responsibility of the town council or the rural council. Eighty percent of the key informants also indicated that there was no one to champion the twinning agreement.

On the contrary, the key informants from the Musina Town Council indicated that the major reason why the goals were not achieved was because of the economic recession that has affected Zimbabwe since 2005, which led to the suspension of many trans-border development projects in Zimbabwe as a whole. Key informants explained that soon after the signing of the MOA between 2004 and 2007, meetings were conducted regularly and the goals were identified. Plans were put in place to implement them. However, from 2009 onwards, the plans started falling apart. These included the economic meltdown (Parkins, 2011), establishment of two local municipalities in Beitbridge (Netsianda, 2011) and lack of a stand-alone budget for the twinning agreement. These contributed to the reduction in the prioritisation of the twinning goals. Most respondents also explained that they needed an overseeing champion to lead the twinning projects. This is supported by Buxbaum (2014) who unpacked Johannesburg's city-to-city partnerships and concluded that the absence of a twinning champion can contribute the failure to achieve the goals and objectives of the twinning agreement.

6.5 RECOMMENDATIONS

This section advances recommendations on issues that should be considered to ensure the success of twinning agreements. Although these recommendations are in response to the spatial planning component of the twinning agreement between Musina in South Africa and Beitbridge in Zimbabwe, they can apply to other southern African countries who are in similar partnerships. These countries include Mozambique, Zambia, Angola, Botswana and Namibia amongst others who are members of the SDIs in southern Africa listed earlier in Section 6.1.

Trans-Border Spatial Planning 83

6.5.1 Appointing Twinning Champions

The current implementation committee of the Musina local municipality and Beitbridge Town Council should appoint twinning champions. These professionals should not only be involved from the start in initiating interest in the twinning but also actively and continuously work at making the twinning work in line with the terms of reference. Twinning champions should be alert all the time and continuously scan the environment for the continuous improvement and well-being of the twinning agreement. They should also have excellent communication skills and the authority to delegate whenever there is need. Twinning champions can also be referred to as 'link persons' because they are the officials who continue to link the two municipalities. Key informants alluded to the fact that they did not know the person who was supposed to press the button to revive the Musina–Beitbridge twinning. Therefore, the twinning champions should address this void because they will be overseers from both sides. It is also very important to appoint town planners in the spatial planning task team as they play an important role in the area of spatial development cooperation. Each task team should have a professional who will automatically be the champion.

6.5.2 Implementing a Twinning Agreement Guide

A twinning agreement implementation guide should be established for twin cities. This guide should contain a solid implementation plan for the twinning agreement to work. The guide should contain all the guidelines on how to implement twinning agreements. Guidelines on how to formulate twinning objectives, assemble inputs, evaluate progress, division of labour, engage the public and marketing strategies should be part of the twinning guide. There should also be guidelines on how to determine the life span of the twinning agreement. Specific objectives should be clearly articulated with linked projects and milestone trackers. Roles and responsibilities should be clearly outlined so that if something is not done properly, they can easily track the responsible persons. Twinning marketing strategies to attract investors and to engage the public should be outlined in the twinning guide. For instance, marketing through media, radio stations, internet and billboards should be clearly outlined. A twinning agreement implementation guide is essential so that all stakeholders are well informed of their roles and responsibilities.

6.5.3 Establishing Reliable Funding Sources

Twinning agreements cannot function without funding. Hence, there is a need to establish funding models for twinning agreements. Generally, funding mechanisms depend on the type of project to be funded. Therefore, the twinning funding models should be based on the type of twinning agreement, the stakeholders involved as well as getting support from the national government of the twin municipalities involved.

6.5.4 Establish a Working Joint Secretariat

The study revealed that the absence of a place to make joint decisions with respect to the determination of information was a major obstacle in realising the expected outcomes of the twinning agreement. As a result of this, implementation of the twinning agreement was not run effectively from a central secretariat where each party will have instant access to hard and soft copies of the minutes of meetings from previous years. The prospect of a joint twinning agreement provides the proper workstation for both parties and decisions will not be a hear-say because it will serve as the centre for the twinning champions for the period of time that the agreement will last.

6.5.5 HARMONISE LEGISLATIVE FRAMEWORKS

A legislative framework should combine principles, standards and morals on how twinning stakeholders are supposed to act and conduct themselves. As indicated by the key informants, twinning stakeholders needed a framework which they could abide by. The way the twinning had been structured shows that there were no principles that bind the stakeholders. Even if someone stopped performing their duties, no legal framework could be used to ensure that they performed the tasks assigned to them. Therefore, for such kinds of partnerships to work, there is a need for some form of rules and regulations that should be in line with the national policy framework. This will encourage the stakeholders to meet the deadlines because they will be bound by the legislative framework (De Man et al., 2001).

6.6 CONCLUSION

The importance of an agreement framework guide is important to achieve the objectives of twinning agreements between two parties. This is particularly true in trans-border agreements between towns and cities in different countries. Similar partnerships in Africa can draw lessons from our case study; for instance, partnerships that were established based on other Southern Africa SDIs such as the Maputo Development Corridor SDI, the Lubombo SDI, the Limpopo Valley SDI and the Okavango Upper Zambezi Tourism Initiative (OUZIT). This study has shown that several success indicators were not considered to achieve the spatial planning objectives of the Musina–Beitbridge twinning agreement. In addition, very little was achieved with respect to the provision of basic services like water and electricity, tarring of roads and the establishment of SEZs. This study has shown that several twinning implementation success indicators were left out. For instance, considering the key questions that should be asked to guide the implementation of the agreement, employing spatial planning twinning champions to guide and give direction as well as defining every step of the agreement through public participation to get every sector of the public involved. Hence, the respondents indicated that the impact on them was very minimal. In addition, the agreement did not have a central secretariat for the coordination of the implementation programs and projects. The results of this study hopefully provide guidance to explore deeper the concept of twinning agreements, especially between two towns located on each side of international borders, not only on spatial planning objectives but also for other city twinning objectives. The results of this assessment of these two towns could be considered as a basic starting point to build on the concept of a twinning framework guide to ensure that the goals and objectives of twinning agreements are achieved effectively.

REFERENCES

Ahmad, G.A. 2001. *Innovative Twinning of Cities.* Paper presented at the First International Convention and Exposition-Twin Cities 2001, Twinning Cities for Smart partnership and Mutual Prosperity, 16–17 April 2001, Melaka.

Beaverstock, J.V., Smith, R.G. and Taylor, P.J. 2000. World City Network: A New Metageography. *Annals of the Association of American Geographers*, 90 (1), 123–134.

Beitbridge Town Council Annual Report. 2018. [Online]. Available from: www.beitbridge.com [Accessed 18 January 2016].

Brown, E., Derudder, B., Parnreiter, C., Pelupessy, W., Taylor, P.J. and Witlox, F. 2010. World City Networks and Global Commodity Chains: Towards a World-systems' Integration. *Global Networks*, 10 (1), 12–34.

Buxbaum, G. 2014. *Unpacking Johannesburg's International City-To-City Partnerships.* Johannesburg: University of Witwatersrand.

Byukusenge, J. 2014. Findings of 3 Case Studies: Cross-border Local Economic Development (LED) in EAC and SADC EALGA. March 2014. Arusha, Tanzania. pp. 2–14.

Connell, D., LaPlace, P. and Wexler, K. 1996. *The Partnership and Alliances Audit: A Company Self-assessment.* Zurich: Strategic Direction Publishers.

De Man, A., Duysters, G. and Vasudevan, A. (eds.). 2001. *The Allianced Enterprise: Global Strategies for Corporate Collaboration.* London: Imperial College Press.

De Villiers, J.C., de Coning, T.J. and Smit, E. 2008. *Twinning and Winning.* Cape Town, South Africa: University of Stellenbosch Business School (USB) Leader's Lab.

Hoetjes, B.J.S. 2009. Trends and Issues in Municipal Twinnings from the Netherlands. *Habitat International*, 33, 157–164.

Hsu, Y. 2003. *Montreal's Twinning with Shanghai—A Case Study of Urban Diplomacy in the Global Economy.* Doctoral dissertation in the department of Communication Studies. Montreal, Quebec, Canada. Concordia University.

Jayne, M., Hubbard, P. and Bell, D. 2011. Worlding a City: Twinning and Urban Theory. *City: Analysis of Urban Trends, Culture, Theory, Policy, Action*, 15, 25–41.

LGIB (Local Government International Bureau). 2001. The Links Effect: A Good Practice Guide to Transnational Partnerships and Twinning of Local Authorities. *LGIB International Report Number 3*, October.

Musina IDP (Integrated Development Plan) Full Report 2015/16 Review 2016. [Online]. Available from: musina-idp-2015_16.pdf (cer.org.za) [Accessed 28 February 2018].

Musina SDF (Spatial Development Framework) Full Report 2014/15 Review. 2018. [Online]. Available from: www.musina.gov.za/official-documents/spatial-development-framework/ [Accessed 21 January 2018].

Netsianda, M. 2011. Municipalities Want to Improve Service Delivery. Beitbridge News Correspondent. *News 24 Archives*. [Online]. 27 April 2007. Available from: www.news24.com/Africa/Zimbabwe/SA-scraps-Zim-border-passes-20071204 [Accessed 10 November 2015].

Nyamwanza, S.A. 2017. *Outcomes of Trans-border Spatial Development Cooperation: Insights from Musina and Beitbridge Twinning Agreement.* Masters' dissertation in the Department of Urban and Regional Planning. University of Venda. South Africa.

Parkins, N.C. 2011. Push and Pull Factors of Migration. *American Review of Political Economy*, 8 (2), 6.

Phillips, A.J. 2002, June. Transboundary Areas in Southern Africa: Meeting the Needs of Conservation or Development. In *Ninth Conference of the International Association for the Study of Common Property, Victoria Falls, Zimbabwe* (Vol. 1721, p. 2002).

SCI (Sister Cities International). 2004. An Introduction. [Online] Available from: www.sistercities.org/sci/ [Accessed 5 April 2015].

Statistics South Africa (Stats SA) 2012. Census 2011: Statistical Release (revised). Available from: www.statssa.gov.za/publications/P03014/P030142011.pdf [Accessed 20 March 2019].

Tjandradewi, B.I. and Berse, K. 2011. Building Local Government Resilience through Cityto-City Cooperation. In Rajib Shaw and Anshu Sharma (eds.) *Climate and Disaster Resilience in Cities (Community, Environment and Disaster Risk Management* (Vol. 6, pp. 203–224). Bradford: Emerald Group Publishing Limited.

Twinning MOU. 2004. *Musina Local Municipality and Beitbridge Rural District Council Memorandum of Twinning Agreement.* Musina Local Municipality, South Africa.

UNDP (United Nations Development Programme). 2000. The Challenges of Linking. Bureau for Development Policy. [Online]. Available from: http://magnet.undp.org/Docs [Accessed 4 June 2019].

UNIVEN GIS Section. 2016. The Proximity between Musina and Beitbridge [Map]. Scale 1:63,600. Thohoyandou, South Africa: Geographical and Information Systems Department, University of Venda.

Section III

Context and External Drivers of Sustainable and Smart Spatial Planning

7 Transnational Land Governance for Sustainable Development
A Comparative Study of Africa and Latin America

Pamela Durán-Díaz

CONTENTS

7.1 Introduction 89
7.2 Background: The Role of Culture for Sustainable Land Development 90
7.3 Methodology 93
7.4 Land-Related International Governance Indicators 94
    7.4.1 The World Bank's World Governance Indicators 94
    7.4.2 Transparency International's Corruption Perceptions Index 97
7.5 The Impact of Weak Governance in Latin American Sustainable Development 101
7.6 Transnational Land Governance towards Sustainable Development in Africa 102
7.7 Conclusions 104
References..105

7.1 INTRODUCTION

The United Nations defines sustainable development as development that meets the needs of the present without compromising the ability for future generations to meet their own needs, achievable by harmonising the so-called three pillars of sustainability: economic growth, social inclusion and environmental protection (also referred to as economy, ecology and society). To do so, the 2030 Agenda for Sustainable Development (UN General Assembly, 2015) demarks a set of 17 Goals and 247 indicators, a number of which relate to land. Likewise, TerrAfrica partnership defines sustainable land management as 'the adoption of land use systems that, through appropriate management practices, enables land users to maximise the economic and social benefits from the land while maintaining or enhancing the ecological support functions of the land resources' (TerrAfrica, 2009).

In addition to the harmonisation of the three pillars, targeted policies, institutional support, multilevel multi-stakeholder and partnerships are needed to attain sustainable land management. Hence, sustainable development could only be achieved when grounded on good governance structures. Land governance involves 'a procedure, policies, processes and institutions by which land, property and other natural resources are managed. [. . .] In every society, sound land governance is the key toward the achievement of sustainable development' (Essien, 2015), and thus plays a crucial role in setting and safeguarding an appropriate framework for sustainable land management, multisectoral collaboration and co-production of sustainable spaces between the State, citizens and the private sector.

According to Espinoza et al., 'in countries like Zimbabwe, South Africa or Central American countries, the fight for gaining access to land has left a trace of violence and political unrest' (2016). Hence, there is need for a study to correlate stakeholder's perceptions on sustainable development and governance in Africa and Latin America, two regions in the Global South.

This chapter presents a comparative study on the similarities and differences between African and Latin American land governance and its impact on sustainable development, based on international governance indicators and stakeholders' perceptions. The research approach taken was a mixed-method research to correlate the quantitative data from international governance indicators from the World Bank and Transparency International and the qualitative data (narratives and perceptions) from interviews, debates and focus group discussions (FGDs) with regional stakeholders from Africa and Latin America. The discussions revealed that, despite their apparent dissimilarities, Africa and Latin America have common challenges in the implementation of sustainable development strategies, with weak governance and conflict being the most urgent issues to address.

With the aim of learning from the regional differences to trace a feasible path towards sustainable development, this chapter begins with this introduction after which a background section about the role of culture for sustainable land development. Thirdly, it presents the methodology followed for data collection and analysis. Then, it presents the analysis of the land-related international governance indicators from the World Bank and Transparency International in both regions. After that, it explores in the implications of weak land governance in Latin America based on the experience of the Amazon rainforest, before showcasing the Great Green Wall for Sahara and the Sahel Initiative as an example of transnational land governance in Africa. Finally, it presents the conclusions and recommendations for land policy development.

7.2 BACKGROUND: THE ROLE OF CULTURE FOR SUSTAINABLE LAND DEVELOPMENT

The social component of land use should be taken into consideration when developing and implementing land management instruments. This can, for example, be done through a purposeful, that is, conscious documentation of how land should be owned and used and how the rights should be exercised in line with the values of communities and socio-historical contexts. Such a land policy should additionally incorporate its intentions on how to secure both formal and informal land tenure, being the actual relationships between people and land in terms of uses, access, rights, ownership, custodianship and cultural practices. This is because insecure land tenure and poor land policies lead to environmental risks and faltering economy, which can compromise the sustainability of our livelihood in both urban and rural settings across the world.

In such contexts, urban and regional planning focuses on developing programs for the use of land to anticipate population growth patterns and provide basic services, infrastructures and housing, while strengthening the sense of belonging through community engagement.

The most basic definition of sustainability lies on meeting the present needs of the community without compromising the resources of future generations (Burndtland, 1987); however, this concept is not new.

> In Akyem Abuakwa as in Africa, the management of land is rooted in religious and sacred beliefs. Most tribes share a common reverence for land as the foundation of community existence. The earth was regarded as a sanctuary of the souls of the departed ancestors who commanded the living to use the land wisely. As Nana Ofori Atta, the paramount chief of Akyem Abuakwa said 'land belongs to a vast family of whom many are dead, a few are living and countless are still unborn'. Land was the sanctuary of the departed souls of the ancestors and they commanded the living to use the land wisely.
>
> (Amankawah, 1989)

Such an approach to land reveals a sense of time conceived as a complex circular process, rather than as an incisive linear arrow. This way of regarding time weaves it, makes it cyclic and loads it with responsibilities in relation to past and future generations.

In Mexico, the Mayan, Nahua, Zapoteca and Mixteca peoples, sculpted their cosmovision and community life around the bond between death, earth and life. More than six million people from about 40 different Indigenous groups hold rituals associated with the celebration of the dead, which takes place every year at the time of the corn harvest. This transition between a time of scarcity and a period of relative abundance is dedicated to sharing the first crops with the ancestors. The principles of reciprocity that rule between the community and their lineages make the celebration of death a symbolic retribution, since the agricultural cycle of corn would be inconceivable without the intervention of the ancestors (CONACULTA, 2006).

Thus, both African and Mesoamerican pre-Colonial cultures share a sense of intergenerational equity deeply rooted in the conception and management of land, around which the cosmovision, social dynamics and use of resources are built. Ergo, they should be fertile lands for sustainable development. However, this is unfortunately not the current reality.

For instance, the Bandiagara Escarpment in the Dogon country of Mali is a geological fault, a cliff of sandstone that extends for over 150 km and rises 500 m. When standing in the lower sandy flats of the South, the archaeological and ethnological structures inserted in the site for 5000 years emerge: houses, granaries, altars, caves, tunnels, graves, sanctuaries and vertical connections. Some homes can only be reached by climbing and abseiling, hanging from ropes made out of baobab fibres. The upper part of the cliff is irrigated by the Niger River, which overflows during the rainy season and flows through the cracks of the escarpment, forming waterfalls that also infiltrate water into the rock and is stored in underground reservoirs. This water harvesting method allows the Dogons to irrigate the few fertile fields that exist and to model the clay to continue modifying their landscape.

As water and mud are the determinant factors for the location of different villages throughout the cliff, the Great Mosque of Djenné is in fact the largest sacred adobe building in the world. It has been exposed to weather conditions since its construction in the thirteenth century and, built of mud, the mosque has eroded over time. This fact implies a continuous investment in terms of resources and work force to preserve it. To do so, a local festivity to restore the Mosque takes place every year. In 2009, however, after a series of heavy rains, one part of the Mosque collapsed. As it has been recognised as a World Heritage site by UNESCO since 1989, it received a large injection of funds to be restored. They respected the original architecture of the façade and the layout, but the construction technique was modernised with materials significantly more durable than mud. The project engaged the community and led to a temporary large economic mobilisation concentrated in the area. In the long run, the impacts of such intervention are unclear. The restoration of the mosque with local materials, such as clay, straw and natural binders, involved an annual pilgrimage to restore the buildings, which ensured the activation of local economy through the exchange and collaboration of several villages. On the contrary, the restoration with imported materials such as clay mixed with cement as binder meant the loss of the annual festival to rebuild the mosque (Durán-Díaz, 2016). Intangible values were also altered because the ancestral construction methods for sacred places were not respected, and therefore, the locals have nearly stopped using the Mosque as the religious place it was, turning it into a scenery for tourists.

To give the mosque back to the locals, enormous efforts had to be made to revive the restoration festival with local materials as in the past. Thus, although the restoration of the Djenné mosque was successful from the architectural point of view and in terms of efficient resource management, it did not have a positive impact on the quality of life and economy of one of the poorest countries in the world. It is noteworthy that, in the World Bank's list of the 50 poorest countries in the world according to their gross domestic product based on purchasing power parity (GDP-PPP), 32 are in Africa. Mali is in position 21 (World Bank, 2020a).

Across the world, in 2017, three major earthquakes made Mexico shake in horror, two of which were devastating. The first one, a magnitude 8.1 earthquake happened on September 7 with the epicentre near Chiapas in the Pacific Coast. It caused more than 90 deaths and injuries to over 200 people. The second one, a magnitude 7.1 earthquake, hit the centre of the country on the 19th of September causing 355 fatalities and 6,011 injured (Senado de la República, 2017). It happened precisely on the 32nd anniversary of the 1985 earthquake that buried Mexico City under the rubble, with an estimated death toll of between 6,000 and 40,000 people (World Bank, 2012)

(Sistema Sismológico Nacional, 2008). There has been an undeniable improvement since then. However, this progress was not due to a quick response from the official institutions but rather because of a simulacrum held in the morning to commemorate the 32 years of the earthquake. Hence, the population was prepared. What was the response of the official institutions and the civil population? There was a 24/7 media coverage of the actions taken by the different rescue teams. The Admiral of the Navy José Luis Vergara gave detailed information about the operation to rescue from a collapsed school, a 12-year-old girl named Frida Sofía. After a couple of days of full coverage, the lie was uncovered: Frida Sofía never existed (BBC Mundo, 2017). There were also discrepancies in information, such as the Embassy of Canada in Mexico twitting the shipment of 1500 tents for shelters (Embassy of Canada in Mexico, 2017), while the Mexican Tax Agency expressed their gratitude for the 750 tents received (SAT Mexico, 2017). The Secretary of Foreign Affairs subsequently clarified the inconsistent information (Secretaría de Relaciones Exteriores, 2017), but the trust on the official institutions amid the crucial moments of crisis was already broken. From fake heroic operations to international aid that was sent but not received nor distributed, thousands of building permits that were issued without complying with the regulations came to light. Each news had its downside, except for the work of rescue dogs who located more than 70 victims and the multitude of volunteers who mobilised *en masse* to offer their help.

In both cases, the implementation of strategies of any kind, namely built heritage restoration projects without previous consultation with the local community, or non-transparent emergency responses, evidences a lack of sensitivity and a weak governance structure. The examples of Mali and Mexico extend over most African and Latin American countries.

7.3 METHODOLOGY

This research relied on two methodologies:

1. A comparative descriptive analysis of how, where and when which sort of indicators on land governance are constructed. The international governance indicators taken into account for the purpose of this research are the World Bank's (2020b) Worldwide Governance Indicators and Transparency International's (2020) Corruption Perception Index. The datasets of both sources are interactive, open access and updated on a yearly basis.
2. A discursive qualitative analysis of debates and interviews with 109 international stakeholders and FGDs with 9 African and 16 Latin American participants. The FGDs took place in June 2021 as two regional workshops of three hours each, in which each of the participants analysed a case study of their choice. The series of debates and webinars related to sustainability of intangible and built heritage were conducted during the course of July 2021, within the framework of the High-level Political Forum of the United Nations and the Our World Heritage Sustainability Debate Month with international experts on world heritage and sustainability. Such debates and webinars are recorded and publicly available (Our World Heritage, 2021).

7.4 LAND-RELATED INTERNATIONAL GOVERNANCE INDICATORS

A number of sustainable development goals (SDGs) and their indicators relate directly to land management, such as SDG1 No Poverty, SDG2 Zero Hunger, SDG5 Gender Equality, SDG11 Sustainable Cities and Communities and SDG15 Life on Land. The Land Portal has undergone an effort to harmonise land-related monitoring initiatives that tracks, maps and makes publicly available land datasets. To do so, Meggiolaro et al. (2018) provides a descriptive overview of the reach of international initiatives such as the Global Land Indicators Initiative (GLII) by the Global Land Tool Network, the Global Property Rights Index (PRIndex), the Global Donor Working Group on Land Platform led by USAID, the Monitoring and Evaluation of Land in Africa (MELA) and the International Land Coalition (ILC) Dashboard.

As an input to previous efforts, this chapter contributes to the analysis on sustainable land management from the perspective of local stakeholders' perceptions in confrontation to international governance indicators such as the World Governance Indicators by the World Bank and the Corruption Perceptions Index by Transparency International.

7.4.1 THE WORLD BANK'S WORLD GOVERNANCE INDICATORS

The aforementioned examples of Mali and Mexico display the same root cause for major problems that hinder sustainable development in Africa and Latin America: a power structure whose processes and institutions are not trustworthy. A government that does not understand the dynamics of vernacular identity, that sells its resources to the highest bidder, that turns its face to the tourist and its back to the locals, that submerges its inhabitants under water and is as inefficient as one that allows its people to crush under the rubble for a lack of an action plan.

The root cause is, thus, a failure in governance. To understand this issue, we should define separately government and governance. The government is the group of people who rule the State, while governance is the relationship between the government and the citizens they lead. Thus, governance is the verdict of the actions of a government.

On the one hand, the government focuses on legislation and decisions. Governance, on the other hand, consists of processes and measures of how to put different decisions in the agenda. The bequest of governance is making and implementing decisions, ideally based on handling requests and complaints effectively and efficiently. Therefore, while the government is responsible for the formal implementation of policies through the law, governance is responsible for the formal, informal and traditional forms of government.

According to the World Bank,

> governance consists of the traditions and institutions by which authority in a country is exercised. This includes the process by which governments are selected, monitored and replaced; the capacity of the government to effectively formulate and implement sound policies; and the respect of citizens and the state for the institutions that govern economic and social interactions among them.

(World Bank, 2020b)

Transnational Land Governance

To determine whether governance is effective or weak, one must analyse the processes and results in terms of measurable parameters. To do so, the European Governance: a White Paper (European Commission, 2001) presents five interconnected principles:

1. **Openness**: institutions must act with transparency.
2. **Participation**: stakeholders must be involved.
3. **Accountability**: the distribution of roles for legislation and implementation must be responsive and go beyond the transparent use of financial resources.
4. **Effectiveness**: goals must be clear and constantly monitored.
5. **Coherence**: policies and actions must be consistent and easy to follow, even to solve complex challenges.

For these principles to work properly, local European governments trace national action programs with clear steps and measurable parameters for the local authorities to adopt and commit to improving their performance.

In the international context, the World Bank (World Bank, 2020b) has measured the global state of governance in 200 countries since 1996 making use of the World Governance Indicators, according to six dimensions:

1. Voice and accountability that captures the perceptions of the extent to which a country's citizens are able to participate in selecting their government as well as freedom of expression, freedom of association and free media.
2. Political stability and absence of violence that captures the perceptions of likelihood of political instability and/or politically-motivated violence, including terrorism.
3. Government effectiveness that is understood as the perceptions of the quality of public services, the quality of policy formulation and implementation and the credibility of the government's commitment to such policies.
4. Regulatory quality that refers to the ability of the government to formulate and implement sound policies and regulations that permit and promote private sector development.
5. Rule of law that captures perceptions of the extent to which agents have confidence in and abide by the rules of society and, in particular, the quality of contract enforcement, property rights, the police and the courts as well as the likelihood of crime and violence.
6. Control of corruption that captures the perceptions of the extent to which public power is exercised for private gain, including both petty and grand forms of corruption as well as the 'capture' of the state by elites and private interests.

Figure 7.1 presents the interactive data of all governance indicators aforementioned in Latin America and the Caribbean and Sub-Saharan Africa in 1996, 2010, 2015 and 2020 in graph view. The data presents no significant variation in 24 years in both regions for voice and accountability and Rule of Law. However, the slight improvement in political stability and absence of violence in Latin America and the Caribbean is noteworthy albeit yearly fluctuations. This is not the case for Africa, where conflict

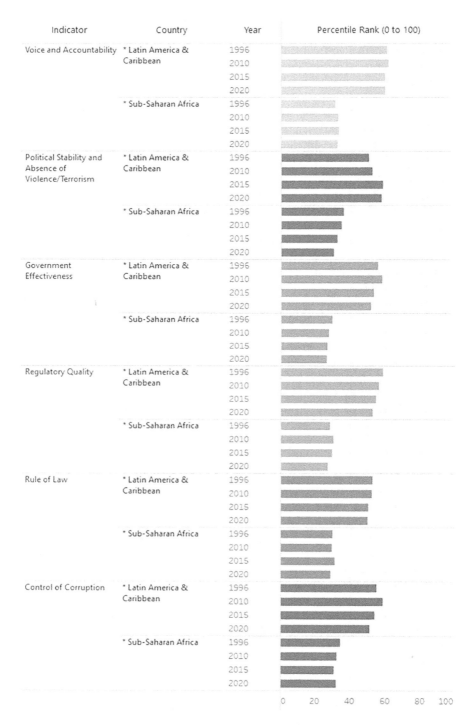

FIGURE 7.1 Worldwide governance indicators of Latin America and the Caribbean and sub-Saharan Africa from 1996 to 2020.

Source: World Bank WGI (2020) available at: https://info.worldbank.org/governance/wgi/Home/Reports (accessed on July 15, 2021)

Transnational Land Governance 97

is still one of the biggest challenges with at least 15[1] countries with ongoing conflicts, regardless of the African Union's Master Roadmap of Practical Steps to Silence the Guns in Africa by the Year 2020 (African Union, 2016). However, although only six Latin American countries[2] seem to have ongoing conflicts, it is remarkable that the Mexican Drug War actually had the highest death toll in the world in 2020, with over 30,000 deaths according to the Armed Conflict Location & Event Data (ACLED, 2020).

The levels of government effectiveness and regulatory quality are declining in both regions. This issue goes hand-in-hand with a consistent lack of capacities in participatory approaches for policy formulation, implementation and monitoring, a failure to consult and engage stakeholders in decision-making, as revealed during the FGDs.

We want to propose a governance committee, because we need to strengthen all the capacities of the decision makers at the local and state level. . . . Along with this governance committee, we should include a component of capacity building to learn how to make decisions, how to interpret information, how we know which path to follow.

(FGD, 2021)

It would be very interesting to be able to develop the capacities of the local governments of the community leaders, for us to realise how our lives would change if sustainable development is not addressed. I feel that seeing it as a matter of public policy is important. That is, we have to work very punctually so that we can begin to demonstrate the tangible, real, medium term, short term impacts, in order to be able to begin to generate knowledge.

(FGD, 2021)

I think that when you start to generate those capacities at the local level, you can ensure that the municipalities will generate a change in the way they look for solutions

(FGD, 2021)

The control of corruption is somewhat improving in Africa, but frantically declining in Latin America. Corruption was a recurrent issue in the FGDs and therefore requires a deeper consideration.

7.4.2 Transparency International's Corruption Perceptions Index

To dive deeply into the control of corruption, the Corruption Perceptions Index by Transparency International (Transparency International, 2020), which assesses the general perception of the public sector by the citizens in a percentile rank, through the aggregation of data from a number of different sources. The countries in other regions in the world that score above 90 points tend to have higher levels of expression

---

[1] South Sudan, Sudan, Somalia, Democratic Republic of Congo, Burundi, Chad, Nigeria, Mozambique, Central African Republic, Mali, Kenya, Niger, Ethiopia and Burkina Faso.
[2] Nicaragua, Guatemala, Mexico, El Salvador, Peru and Colombia.

and press freedom, access to information concerning public expenses, standards of integrity for public officers and independent judicial systems.

Figure 7.2 presents the gradient map and scores of the Corruption Perceptions Index for Latin America and the Caribbean in 2020. It is notable that 20 out of 23 countries score below 50. Moreover, the levels of perceived corruption in 14 of 23 countries have worsened since 2012. Uruguay and Chile hold the top scores with 71 and 67 points, respectively, and Nicaragua (22), Haiti (18) and Venezuela (17) hold the bottom scores. Argentina, Ecuador and Guyana are the significant improvers, while Chile, Guatemala, Nicaragua and Venezuela are the significant decliners based on their new maximum or minimum score since 2012.

Except for Chile, Costa Rica and Uruguay, the low scores of Latin America and the Caribbean reflect an overall distrust in public institutions, which are generally dysfunctional and unreliable due to poor implementation and weak enforcement. The Latin American experience evidences that governance tends to weaken when the law is complex, inconsistent or outdated; access to justice, courts and dispute mechanisms are confusing, expensive or lengthy; institutions are not clear, are overlapped and are not coordinated; decision-making processes are not transparent; public servants are unmotivated, poorly paid and poorly trained and civil society is weak.

Figure 7.3 presents the gradient map and scores of the Corruption Perceptions Index for the African Union in 2020. It is noteworthy that 44 out of 49 countries score below 50. On the other hand, it is remarkable that the levels of perceived corruption in 21 out of 49 countries have improved since 2012. Seychelles, Botswana and Cape Verde hold the top scores with 66, 60 and 58 points, respectively, and Equatorial Guinea (16), Sudan (16), South Sudan (12) and Somalia (12) hold the bottom scores. Angola, Côte d'Ivoire, Ethiopia, Senegal and Tanzania are the significant improvers, while Congo, Malawi, Mozambique, Madagascar, Liberia and Zambia are the significant decliners based on their new maximum or minimum score since 2012.

It is important to note that although Somalia presents the lowest score not only in the African Union but also in the whole world (position 179/180 together with South Sudan), it nonetheless presents its highest score since 2012. The qualitative information gathered in the FGDs, interviews and debates provide a new perspective on how to interpret these indicators. For instance, Rwanda is almost at the same level as Costa Rica, which is particularly impressive considering the recent history of Rwanda. In 1994, the hegemonic Hutu government perpetrated a genocide in which between 500,000 and 1 million people were killed during the country's civil war in a cruel attempt to exterminate the Tutsi ethnicity. After 25 years, Rwanda recovers its social structure partaking of inclusive initiatives, such as involving citizens in cleaning public spaces monthly. As a result, Kigali is the cleanest capital in Africa and a role model to the world. Their policy implementation to prohibit the manufacture, import, sale and use of single-use plastic objects is worth replicating.

Burkina Faso reached 40 points, almost as much as Argentina. This is interesting, considering the 2015 *coup d'etat* in Burkina Faso, which failed due to the timely intervention of the African Union and the United Nations, who restored Michel Kafando as president and supported the democratic transition of the country.

Tanzania scored almost the same points as Colombia, and both are on the rise. Colombia is achieving it with participatory planning and law enforcements to end

Transnational Land Governance

FIGURE 7.2 Corruption perceptions index map and scores for Latin America and the Caribbean in 2020.

Source: Transparency International (www.transparency.org/en/cpi/2020/index/twn) consulted on 03.10.2021

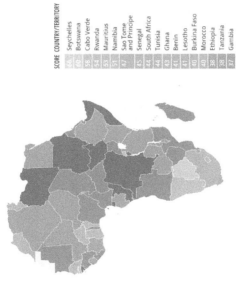

FIGURE 7.3 Corruption perceptions index map and scores for the African Union in 2020.

Source: Transparency International (www.transparency.org/en/cpi/2020/index/twn) consulted on 03.10.2021

impunity, while Tanzania in implementing climate-change mitigation policies, such as the ban on single-use plastics.

Ethiopia scored almost as well as Ecuador, both being significant improvers in their region. In the African context, Nobel Peace Prize Winner (2019) and Prime Minister Abiy Ahmed has done immense efforts to achieve peace and international collaboration, initiating the dialogue to resolve the conflict with Eritrea. This reform project is facing threats due to the widespread political tensions within the ruling party.

The main takeaway of this new perspective is the affirmation that it is challenging to ensure that development measures are truly sustainable and inclusive when grounded on weak governance structures. In the absence of participatory approaches, there are impositions and policies that extend over time and/or are truncated before being implemented. Even minor improvements in the governance principles influence the advancement towards sustainable development.

7.5 THE IMPACT OF WEAK GOVERNANCE IN LATIN AMERICAN SUSTAINABLE DEVELOPMENT

The current environmental threats to the Amazon rainforest due to corruption and lack of collaboration was a recurrent concern mentioned by Latin American participants in our FGDs. The Amazon rainforest is a tropical forest that extends over an area of 7 million km^2 and spreads over nine different countries in Latin America: Brazil (with 60% of the surface), Peru (13%), Colombia (10%), Venezuela, Bolivia, Ecuador, French Guyana and Suriname. Being one of the most diverse ecoregions in the world, it houses 20% of the world's bird species, 50% of the world's plant species and 390 billion trees. It has been acknowledged as one of the seven natural wonders of the world and the lung of the planet. Culturally, it is also a diverse region.

Nevertheless, the total area of the Amazon rainforest has reduced by 20% since 1960 to make way for livestock, agriculture, the timber and rubber industry and mining. Uncontrolled and progressive deforestation has increased fire risks every year. For instance, in August 2019, more than 900,000 hectares were burnt in the 40,000 fires caused by climate change and the illegal practice of slash and burn to deforest for agricultural purposes. In addition to the immediate impact on the global climate, we must consider the carbon dioxide and monoxide emissions from burnt trees, the irreversible damage to the biodiversity and the threats to Indigenous tribes living in the rainforest.

It is therefore important to see the protection measures taken by the different countries. The contradictory policies of Ecuador have exalted Indigenous grassroots uprisings. On the one hand, the Fossil Fuel Law from 1971 (Consejo Supremo de Gobierno del Ecuador, 1971) enables the exploration and exploitation of all hydrocarbon deposits within the Ecuadorian territory, including seas and forests. On the other hand, the Organic Law for the Comprehensive Planning of the Amazon Special Territorial Circumscription (Consejo Supermo de Gobierno del Ecuador, 2018) aims at enhancing a socioeconomic model rooted in the sustainable management of natural resources. The ambiguity raises conflicts blurred by lengthy bureaucratic procedures that enable the immediate exploitation of the Amazonian resources. In response, the female Waorani leader Nemonte Nenquimo led in 2020 a successful Indigenous campaign to protect 500,000 hectares of the Amazon rainforest against oil extraction, now endorsed by the Ecuadorian Constitutional Court (Blasco, 2020).

Within the framework of the 2019 Amazonian Summit, the leaders of Colombia, Brazil, Peru, Ecuador, Bolivia, Surinam and Guyana signed the 'Leticia Pact' (Ministerio del Medio Ambiente y Desarrollo Sustentable de Colombia, 2019). The agreement aims at addressing many of the causes for deforestation, insecurity, forest fires and other environmental risks, through an action plan to boost collaboration, community development, female and Indigenous empowerment, capacity building, transparency and accountability. However, 60% of the Amazon rainforest is in Brazilian territory and the current government states that the only way to protect the Amazon is by opening it to economic development through the enhancement of private investments. Data from the National Institute for Space Research (INPE) reveals that the Brazilian Amazon has suffered its worst deforestation since 2008, impacted by the policies implemented by President Jair Bolsonaro. Such policies have defended the industries that drive the destruction of the rainforest, at the expense of regional environmental protection system and undermining indigenous rights. According to INPE, between August 2019 and July 2020, the total loss of forest vegetation cover amounts 11,088 km^2, 9.5% higher compared to the same period on the previous year (Instituto Nacional de Pesquisas Espaciais, 2020).

Despite the existence of bilateral policies such as the 'Leticia Pact' and international non-binding agreements such as the Paris Agreement, the weak governance structure across the region threatens the Amazonian rainforest. To date, the only feasible solution is to rely on the aid and the pressure from international governments from the Global North to safeguard the Amazon.

7.6 TRANSNATIONAL LAND GOVERNANCE TOWARDS SUSTAINABLE DEVELOPMENT IN AFRICA

The Great Green Wall for Sahara and the Sahel Initiative (GGWSSI) was a recurrent example of good governance practices mentioned by African participants in our FGDs. The GGWSSI is an initiative led by the African Union since 2007 to combat desertification and requires the collaboration of, so far, 21 sub-Saharan countries from Senegal to Djibouti, going through Mauritania, Mali, Burkina Faso, Niger, Nigeria, Chad, Sudan, Eritrea and Ethiopia. The project consists of the reforestation of an area of 50 km wide by 8,000 km long with resilient and drought-resistant vegetation, a cross-section of the sub-Saharan desert, covering the Sahel and the Horn of Africa. In such a transnational collaboration, 100 million hectares of degraded land will be restored, absorbing 250 million tons of CO_2 and creating 10 million jobs respectful with the environment.

Africa is building a mosaic of green infrastructures to restore eroded land, on which more than 70% of African economy lies. According to the report by the United Nations Convention to Combat Desertification (UNCCD, 2020), the GGWSSI has achieved since 2007 the following:

- Ethiopia has produced over 5 billion plants and seeds, trained 62,759 people on food and energy security as well as maintenance of biodiversity and restored 1 million hectares of land.
- Senegal has planted over 18 million trees, restored over 850,000 hectares of degraded land for communities, trained 2,120 people on food and energy security as well as maintenance of biodiversity and positively affected 322,221 inhabitants with this project.
- Nigeria has produced 7.6 million plants and seedlings, trained 1,205 people on food and energy security as well as maintenance of biodiversity, reforested 2,801 hectares of land and created 1,396 jobs for the inhabitants.
- Sudan has produced 1.9 million plants and seedlings, trained 1,716 people on food and energy security as well as maintenance of biodiversity and restored 85,000 hectares of land.
- Burkina Faso has produced 16.6 million plants and seedlings, trained 26,869 people on food and energy security as well as maintenance of biodiversity, improved 51,633 households and restored 29,602 hectares of land. In addition, they recovered the traditional Zaï practice of the Indigenous communities, consisting of digging pits in the earth of 20–30 cm deep to concentrate compost and collect rainwater. Compost with manure attracts termites that dig tunnels, which causes the soil to break down better and thus the cultivated plants grow stronger roots. Since this ancient agricultural system was recovered, sorghum and millet crops have improved by up to 500%.
- Mali has produced 135,472 plants and seedlings, trained 1000 people on food and energy security as well as maintenance of biodiversity and reforested 6,297 hectares of land.
- Eritrea has produced and planted 128.8 million trees, terraced and afforested 52,930 hectares of degraded land and terraced 65,231 degraded farmlands.
- Niger has produced 146 million plants and seedlings, trained 1,200 people on food and energy security as well as maintenance of biodiversity, created 21,487 jobs and restored 363,928 hectares of land.

These are the immediate effects, directly related to land and rural development. Furthermore, it contributes to 16 of the 17 SDGs, helping the environment, combating desertification, generating employment, providing food security, ensuring access to water, resolution of territorial conflicts, securing land tenure and many more.

In environmental matters, monkeys are stopping to steal the crops and return to the forest. Ethiopia is recovering forest agriculture for coffee growing, avoiding conflicts over natural and agricultural land use. Blockchain technology is being used in transactions to ensure fair trade in coffee. Indigenous wisdom is combined with technological innovations.

In social matters, now that water is more easily found in the subsoil due to the humidity retained by the trees, not only can the land be cultivated but also girls can go to school. Since water harvesting is a traditionally feminine task for sub-Saharan people, girls do not have to travel miles a day to find water in the desert. This creates an environment of inclusion and equity.

Moreover, the African Union acts as a neutral mediator on the emerging governance issues, a vital role to manage a project that involves 21 different countries and millions of stakeholders with competing interests and power. This sets an example of transnational land governance to the rest of the world.

7.7 CONCLUSIONS

From this comparative study of stakeholders' perceptions versus international governance indicators in Latin America and Africa, two regions in the Global South, we can learn that:

1. Land governance is the foundation of the three pillars of sustainable development (ecology, society and economy). However, there is an urgent need to widely recognise a fourth pillar to this, namely culture, because cultural values influence the degree to which people execute laws, policies and informal habits towards land. The interview excerpts confirm and underline that strategies, projects, policy formulation and implementation to advance towards sustainability are void when done disregarding public participation, the local culture and governance principles, as such strategies do not respond to the needs of all stakeholders and puts contestants' interests in conflict, within and across regions and generations.
2. Global datasets on international governance indicators are better understood when contextualised. Although the research datasets on world governance summarise the results on the views on the quality of governance and corruption perception unbiasedly, the insights of local stakeholders enrich significantly the comprehension of regional challenges that could be scaled up to a global approach. For instance, Africa scored nearly half of the points of Latin America in the percentile rank of the World Governance Indicators (WGI). However, the slight advances in Africa in political stability, voice and accountability and control of corruption denoted a meaningful progress towards sustainable development. The decay in Latin America in the same indicators entailed severe crises and environmental risks, despite scoring higher than Africa.
3. Land governance is most effective towards sustainability when executed in a transnational framework. Land governance, understood as the rules, processes and structures through which decisions are made about access to land and its use, the way in which the decisions are implemented and enforced, and in which competing land interests are managed, works consistently when addressed at a transnational level. The national and local initiatives and policies may intend to secure land rights, manage environmental risks and develop land in a sustainable way. However, these efforts are partial when a shared ecosystem is handled differently across administrative borders. We could overcome these challenges through the enhancement of international collaboration and the application of international frameworks such as the FAO's Voluntary Guidelines on the Responsible Governance of

Tenure and the Paris Climate Agreement as well as regional political frameworks such as the Leticia Pact in South America and the Global Mechanism FLEUVE project for the GGWSSI in Africa.

The similarities and differences between Africa and Latin America indicate that the means to strengthen land governance are:

- International support and cooperation for conflict resolution
- Consistent political dialogue
- Awareness raising and capacity building on regional challenges and tools to overcome them
- Technical, political and financial support at the municipal, provincial, national and transnational levels
- Participatory and inclusive approaches in stakeholder engagement.

In conclusion, it is imperative to strengthen research-policy linkages to put science at the service of land governance. Sustainable development is not an aim but a path that requires awareness, capacity building and scientific advice for policy-making and decision-making. Human activities cause biodiversity loss, floods, droughts, forest fires, water shortages and compromise our food security. Conflicts over resources are forcing the most vulnerable groups to emigrate. Young people, women and Indigenous communities are uprising to demand immediate action.

Collaborative research making use of land-related datasets and international governance indicators could contribute to land management matters, in terms of assessing the environmental quality, response management, understanding of land dynamics and conflicts, respect for local culture and the use of new technologies, among others. The academic community plays an important role in raising awareness, educating the community, advancing in technology innovation and disseminating scientific knowledge in which decisions should be based.

REFERENCES

ACLED, 2020. *Disaggregated Data Collection, Analysis & Crisis Mapping*, s.l.: The Armed Conflict Location & Event Data Project.

African Union, 2016. *Master Roadmap of Practical Steps to Silence the Guns in Africa by Year 2020*, Lusaka: s.n.

Amankawah, H. A., 1989. *The Legal Regime of Land Use in West Africa: Ghana and Nigeria*, s.l.: Pacific Law Press.

BBC Mundo, 2017. El caso de 'Frida Sofía' en la escuela derrumbada por el terremoto en México revive la historia de 'Monchito', el niño del sismo de 1985 que tampoco existió. *BBC News*.

Blasco, L., 2020. Nemonte Nenquimo: "No esperen que sólo los pueblos indígenas defendamos la Amazonía, es una lucha de todos". *BBC News*, 1 December.

Burndtland, G. H., 1987. *Report of the World Commission on Environment and Development: Our Common Future*, Oxford: Oxford University Press.

CONACULTA, 2006. *La festividad indígena dedicada a los muertos en México*, Mexico: Consejo Nacional para la Cultura y las Artes.

Consejo Supremo de Gobierno del Ecuador, 1971. *Ley de Hidrocarburos No 1459, Decreto Supremo No 2967*, s.l.: s.n.
Consejo Supermo de Gobierno del Ecuador, 2018. *Ley Orgánica para la Planificación Integral de la Circunscripción Territorial Especial Amazónica*, s.l.: s.n.
Durán-Díaz, P., 2016. Extreme Cultural Landscapes in the Context of Globalisation: Mali, China and Mexico. In: *The Future of Professional Fields. Views from the Mexican Diaspora in Europe,* Barcelona: CreateSpace Independent Publishing Platform, pp. 258–280.
Embassy of Canada in Mexico, 2017. *1500 Tents that Will Serve as Selters for the Victims of Mexico's Recent Earthquakes Arrived Today from Canada*, s.l.: Twitter.
Espinoza, J., Kirk, M. & Graefen, C., 2016. Good Land Governance: Between Hope and Reality. *zfv—Zeitschrift für Geodäsie, Geoinformation und Landmanagement*, 2.
Essien, E., 2015. Exploring the Food Security Strategy and Scarcity Arguments in Land Grabbing in Africa: Its Ethical Implications. In: E. Osabuohien, ed. *Handbook of Research on In-Country Determinants and Implications of Foreign Land Acquisitions*, s.l.: IGI Global, p. 28.
European Commission, 2001. *European Governance A White Paper*. [Online] Available at: https://ec.europa.eu/commission/presscorner/detail/en/DOC_01_10 [Accessed 15 July 2021].
FGD, 2021. *Focus Group Discussion on Stakeholders Perceptions* [Interview] (21 May 2021).
Instituto Nacional de Pesquisas Espaciais, 2020. *Ministério da Ciência, Tecnologia e Inovaçoes*. [Online] Available at: www.gov.br/inpe/pt-br [Accessed 23 January 2021].
Meggiolaro, L., Sato, R. & Sorensen, N., 2018. *Land Governance and the Sustaianble Development Goals (SDGs): Consolidating and Harmonizing Monitoring Initiatives*. Washington, DC: World Bank.
Ministerio del Medio Ambiente y Desarrollo Sustentable de Colombia, 2019. *Pacto de Leticia por la Amazonía*, Leticia: s.n.
Our World Heritage, 2021. *Our World Heritage—Sustainability*. [Online] Available at: https://ourworldheritage.org/sustainability/ [Accessed 1 August 2021].
SAT Mexico, 2017. *Hoy México recibió 750 casas de campaña de Canadá para apoyar en las labores de rescate*, s.l.: Twitter.
Secretaría de Relaciones Exteriores, 2017. *El gobierno de Canadá realizó un envío de ayuda para damnificados, consistente en 1500 casas de campaña, gracias por su solidaridad*, s.l.: Twitter.
Senado de la República, 2017. *Recuento de los daños 7S y 19S: a un mes de la tragedia*, Mexico: Instituto Belisario Domínguez. Senado de la República.
Sistema Sismológico Nacional, 2008. *El sismo de 1985 en cifras*, Mexico: Universidad Nacional Autónoma de México.
TerrAfrica, 2009. *Using Sustainable Land Management Practices to Adapt and Mitigate Climate Change in Sub-Saharan Africa*, s.l.: s.n.
Transparency International, 2020. *Corruption Perceptions Index*. [Online] Available at: www.transparency.org/en/cpi/2020/index/twn [Accessed 23 July 2021].
UNCCD, 2020. *The Great Green Wall Initiative*, s.l.: s.n.
UN General Assembly, 2015. *Transforming Our World: The 2030 Agenda for Sustainable Development*, s.l.: UN General Assembly.
World Bank, 2012. *FONDEN Mexico's Natural Disaster Fund—A Review*, Washington, DC: The International Bank for Reconstruction and Development.
World Bank, 2020a. *Open Data GDP per capita, PPP*, s.l.: The World Bank Group.
World Bank, 2020b. *Worldwide Governance Indicators*. [Online] Available at: http://info.worldbank.org/governance/wgi/ [Accessed 15 July 2021].

8 Are We There Yet? Prospects and Barriers to Implementing Smart City Initiatives in Harare, Zimbabwe

Abraham R. Matamanda

CONTENTS

8.1 Introduction 107
8.2 Conceptualising the Smart City 108
8.3 Prospects and Limitations to Smart City Initiatives in Harare 112
    8.3.1 Characterising Harare in the Broader Socio-Spatial Planning Context of Zimbabwe 112
    8.3.2 The Institutional and Regulatory Framework for Smart City Activities and Processes 112
    8.3.3 The Twinning of Harare as Leverage for Smart City Development 113
    8.3.4 Data Price as a Penalty for the Urban Poor and Stifle to Internet Connectivity 114
    8.3.5 The Vibrant Civil Society as an Enabler to Citizen Participation 115
    8.3.6 The Tertiary Institutions as Living Labs for Smart Ideas in Harare 115
    8.3.7 Budgetary Complacency and Institutional Constraints 116
8.4 Conclusion 117
References 118

8.1 INTRODUCTION

Like elsewhere around the globe, governments in Africa have been formulating plans to implement smart initiatives aimed at spurring sustainable urban development in the rapidly urbanising African cities (Watson, 2015; Siba and Sow, 2017; Smart Africa, 2017). The smart city Rwanda master plan has been conceptualised to implement smart initiatives in Kigali and other cities in Rwanda (Rich, Westerberg and Torner, 2019). Similar cases are Konza Techno City, Kenya, Eko Atlantic in Nigeria and Hope City in Ghana (Ajibade, 2017). Huet (2016) opines that smart cities are the key to Africa's third revolution because they enhance cities' sustainability and resilience.

Regardless of the increasing focus and attention on smart initiatives, governments in the Global South, especially in Africa, are often criticised for adopting urban plans and visions from the Global North. The smart city concept is one such plan. Watson (2014) argues that these envisaged plans are 'urban fantasies' that do not conform to the local realities of the African cities characterised by poverty, civil wars and financial woes. Therefore, the smart city initiatives seem to be far-fetched fantasies that may even exacerbate the existing spatial injustice, considering how the plans sideline the socio-economic realities of the communities (Watson, 2015). Consequently, the smart cities have resulted in the production of white elephants or abandoned projects as citizens fail to relate with the ultimate products.

Zimbabwe is not an exception, as its cities have been grappling with multiple problems. The major challenges are experienced in emerging settlements that characterise settlements on the margins of formality and informality (Chigudu, 2019). The emerging settlements primarily accommodate the urban poor, which exacerbates spatial justice and adequate human well-being. It is thus critical to find ways to enhance the liveability of the emerging settlements as various proponents indicate the need for identifying alternative urban transformation strategies for the future of Zimbabwe cities (Muchadenyika, 2020). Although smart city development in Zimbabwe may help reduce the existing challenges in human settlements, there is a need to begin by understanding the local realities and analysing how the best practices are adopted to formulate and implement smart initiatives.

This chapter describes the smart city initiatives and then evaluates which (internal and external) factors hinder and potentially support the development. The chapter focuses on emerging human settlements, using Harare, the capital city of Zimbabwe, as a case study. Methodologically, the data interpretation is qualitative, using secondary and primary data sources. Secondary data sources that include policy documents, pieces of legislation, including the Regional Town and Country Planning Act (Chapter 27:12) of 1976, Urban Councils Act and the Public Health Act, have been reviewed. Key informant interviews were conducted with purposively selected professionals working in private and public sectors. The data were subsequently analysed using thematic and content analyses.

The subsequent sections start with a literature review on the definitions and conceptualisations of smart cities and the prospects and limitations of achieving such cities in an African context. Next, a brief background of Harare reflects on the socioeconomic development trajectory of the city. The prospects and limits to smart city initiatives in Harare are presented followed by the implications for practical smart city planning projects and a concluding section that wraps up the chapter.

8.2 CONCEPTUALISING THE SMART CITY

The literature review undertaken by Moura and Silva (2019) showing more than 27 definitions indicates that 'smart city' is an indistinctive concept that is not always consistently used. Some scholars (Inkinen, Yigitcanlar and Wilson, 2019; Moura and Silva, 2019) consider smart cities as intelligent or digital cities, emphasising the notion of technology in defining smart cities. Yigitcanlar, Han and Kamruzzaman (2019) opine that a holistic approach is required that recognises the utility of other components beyond technology in understanding smart cities.

Albino, Berardi and Dangelico (2015) assert that people are clever and not passive because they bring creativity, diversity and knowledge critical in understanding urban problems. In this vein, Kummitha and Crutzen (2017) postulate the two approaches to understanding smart cities: technology-driven and human-driven. The former recognises technology as the ultimate solution for urban problems, while the latter identifies citizens as significant players in improving urban life through their ingenuity and intellect. Therefore, effective smart cities integrate technology with human and intellectual capital, social capital, entrepreneurial capital and relational and environmental capital to develop a holistic approach to formulating solutions to resolve problems overwhelming cities. Just like any model or concept, there are always problems associated with it. Therefore, a typical smart city consists of technology (in its different forms) and human and social capitals. Smart cities must also ensure that the citizens' social, environmental and economic needs are realised without compromising each other.

Urbanisation in many African cities is still in its infancy stages and thus their prospects of smart city initiatives are multifold. According to Senkhane (2016), many African cities are increasingly becoming technological hubs critical in smart city development. Caragliu, Bo and Nijkamp (2011) argue that smart cities emphasise the crucial role of high-tech and creative industries in long-run urban growth. Cities such as Johannesburg in South Africa, Cairo in Egypt, Kigali in Rwanda and Nairobi in Kenya have invested in and modernised the IT ecosystem. A growing patchwork of entrepreneurs, tech ventures and innovation enables the viability of this IT ecosystem (Senkhane, 2016). These innovation hubs are vital grounds where novel and transformative ideas are formulated and subsequently implemented in different sectors such as water metering, public transport and energy harnessing from solar or solid waste. An example is the Silicon Savannah in Nairobi, Kenya, a technological hub that offers various digital opportunities that include affordable 3G internet connections that facilitate mobile payment services booming in the city (Moime, 2016).

However, the limited investment in baseline infrastructure for technology remains a major obstacle to information and communication technology (ICT) advancement in most African cities. Although the limited infrastructure may be an obstacle to smart city development in Africa, mainly due to the high costs of installing such infrastructure, the absence of such infrastructure may be a blessing in disguise. Deloitte and Touche (2014) explain that the emerging settlements make it possible to install the latest available ICT technology, thus removing the costs of upgrading or removing existing infrastructure.

The majority of the population in Egypt under 30 years is incredibly tech-savvy, which positions the country as one of the most digitally smart countries in the world (Senkhane, 2016). Moreover, the entrepreneurial spirit among the youth is an excellent ingredient for innovation and education, which is critical in breeding new ideas that are often divergent from the traditional ways of service delivery in cities which have often been costly. Young people are more likely to adopt technology, making it feasible to implement smart city projects oriented on technology (Siba and Sow, 2017).

The increasing use of technology and social media by the youth is beneficial in smart city initiatives. Social media is a platform for participation in civil matters relating to the social component of smart cities, citizen e-participation. The Smart

Africa (2017) report highlights that 'social media allows a boundless, continuous conversation that provides leaders, citizens, and private organisations with crucial data and insights about the life of the city'. Similarly, Vázquez and Vicente (2018) established a positive correlation between e-participation in smart cities and age and educational attainment.

The smart city initiatives tend to be national projects which often become the responsibilities of national governments and not local governments. An example is the Rwanda smart city master plan, a national development framework that aims to spearhead smart initiatives in developing and planning cities in Rwanda. The smart city initiatives of Rwanda have been lauded on the premise that it is an exemplar for African cities (Manirakiza et al., 2019). However, although there is some merit in the smart initiatives being introduced and spearheaded by the country's President Paul Kagame who is at the forefront of the smart city initiatives. Consequently, the initiatives tend to be politically oriented as the president seems to advance a particular agenda which, in some instances, may be the brainchild of specific external organisations. Baffoe, Ahmad and Bhandar (2020) raised a concern that the developments in Kigali are deeply rooted in capitalist and neoliberal ideologies that are inclined mainly towards wealth accumulation by a few elites.

The smart developments seek to support and sustain private sector development through market mechanisms that ultimately compromise governance and social justice that are equally critical elements in the making of smart cities. The challenges include developing cities beyond the reach of the poor because planning tends to become for the elites, only resulting in the poor being left behind. As Datta and Odendaal (2019) put across, smart cities are nested in power dynamics. The few elites advance a particular agenda that benefits them, usually at the expense of the marginal communities. Hence, in most instances, the initiatives tend to benefit a few corporates while grossly disadvantaging the poor.

The foregoing also raises a critical issue relating to smart cities: modernisation and adopting what Watson (2014) termed the 'urban fantasies'. Many governments try to mimic foreign designs that do not relate to the local realities in developing smart cities. This ideology of modernism had been a significant constraint of urban development projects in many African countries where the fetish for planning 'order' and aesthetics usually takes precedence over social justice matters. Examples of demolitions, evictions and displacements of marginal communities when 'big developments' come beg the question of the smartness of the projects when the social aspects are ignored (Kaika, 2017; Shamsuddin and Srinivasan, 2020).

On the contrary, the notion of modernism is not bad. It poses an opportunity associated with the visions, dreams and aspirations of a better world through copying from the best practices. These initiatives that are copied may have to be localised to suit the local conditions. This has been an opportunity in Africa through globalisation, allowing collaboration and cooperation among national and local governments. City twinning allows for training exchanges for human capital, information and technical exchanges and collaborations on technical initiatives (World Health Organisation, 2001). The twinning of cities in the North and South and South and South facilitates engagement, learning and sharing resources and expertise.

Vázquez and Vicente (2018) highlighted that e-participation among citizens in smart city initiatives is best explained through citizens' political interest and external political efficacy. The main challenge thus becomes the centralisation of the programmes, which may be politicised and thus fail to integrate public concerns while also being irrational. Yet, successful smart city initiatives must be based on technology and, most importantly, the needs of the citizens (Weber, 2019).

Universities and technical colleges act as incubators that educate vibrant learners who then develop innovative solutions to spur smart city initiatives. The evidence from the various ways universities in South Africa partner with the corporate world and develop innovative technologies contributing to smart city development. For example, rainwater harvesting in informal settlements of Cape Town, brick making using ashes from coal mining in University of Zululand and initiatives from Western Cape University (Grobbelaar, Tijssen and Dijksterhuis, 2017). The role of universities as living labs for innovation thus takes a centre stage and considering the university student massification in Africa, these tertiary institutions thus offer critical, innovative solutions in developing a smart city.

Smart city initiatives tend to be cost-effective and require enormous capital outlays to enable the projects to kick start and be sustained. For example, approximately US$15.5 billion was required to develop the Konza Techno City in Kenya (Ajibade, 2017). These costs are beyond most local governments' reach, which ultimately makes funding a significant constraint to smart city development.

Lastly, poor governance is the major obstacle to the success of such programmes in Africa. The authoritarian leadership characteristic of most African cities where the central government mingles in managing the affairs at the local scale jeopardises the success of many smart city projects. Democratic decision-making is next to non-existent in many African countries. The case of Rwanda shows an authoritarian government. Tanzania and Uganda exemplify how transparency is absent in decision-making coupled with the interference of central government in local government affairs in Harare and how the persistent corruption and embezzlement of public funds are meant to be channelled for urban development. The mismanagement of the US$144 million from China to the City of Harare for upgrading the water infrastructure in the city is an indication of how poor governance and management retards significant development (Piesse, 2019). Consequently, such practices repel investors and potential partners intending to invest in PPPs, thereby stifling the smart city development programmes.

Key lessons can be drawn from the African examples of smart city development presented. First, the youthful population offers an opportunity as this is the innovative and technologically competent age group. Second, the university massification in African countries over the past decade contributes to establishing living labs that become innovation incubators. Third, the limited development in many cities means the cost of installing new infrastructure tends to be less than when retrofitting where existing infrastructure exists. Misgovernance, corruption, limited financial resources and the adoption of plans that do not match local African realities and aspirations stifle the success of developing smart city projects.

8.3 PROSPECTS AND LIMITATIONS TO SMART CITY INITIATIVES IN HARARE

8.3.1 CHARACTERISING HARARE IN THE BROADER SOCIO-SPATIAL PLANNING CONTEXT OF ZIMBABWE

Harare has perpetuated spatial injustice from the onset, which contradicts the smart city development. Since the early 2000s, Movement for Democratic Change (MDC)[1] mayors and councillors have been in control of local government affairs in Harare. This situation has caused significant tensions with the central government, which has remained under the Zimbabwe African National Unity-Patriotic Front (ZANU-PF) (Ndawana, 2020).

The collapsing economy coupled with the poor governance resulted in much disgruntlement among the urbanites. Civil unrest characterised by protests in major cities, especially in Harare, ensured. These protests were orchestrated and communicated mainly through social platforms. The resentment towards the civil society in Harare linked to MDC has disrupted the activities of these organisations, which in some instances have been agents of social change through their advocacy work in vying for improved services among the citizens (Matamanda, 2021).

The indigenisation and empowerment of the locals have been a development agenda spearheaded by the government since the mid-2000s to empower the youth and women to engage in different development projects. Individuals and groups received start-up capital to fund their projects, mainly supporting innovation and industrialisation beginning at the grassroots (Kamete, 2004). Albeit being politicised, the idea was noble as this provided a platform where the youth would foster innovative ideas contributing to the country's socio-economic development.

8.3.2 THE INSTITUTIONAL AND REGULATORY FRAMEWORK FOR SMART CITY ACTIVITIES AND PROCESSES

Harare has committed to becoming a smart city, evident in the city's regulatory and development plans and strategies that provide insights into where the city is heading, how it seeks to get there and the obstacles that come along the way. First, the Harare Strategic Plan 2012–2025 (HSP) envisages achieving a world-class city status by 2025 through several initiatives to become a smart city (City of Harare, 2012). Through the HSP, the city of Harare has committed to provide first-class service delivery and promote investment through the following initiatives:

- Upholding citizen participation in all matters relating to social services. Evidence is from the e-platform, where timely information is frequently disseminated.
- Service standards through which residents can participate in service-level benchmarking and determine the level and quality of council services they receive.

---

[1] The main opposition political party that was established in the late 1990s by trade unions and middle-income citizens following the disgruntlements with the Mugabe government at the time.

- A 'One-Stop' service centre permitting online access to services and appraising the existence and details so that citizens can access services.
- Value for money by providing social services in an economical and efficient way.

Statutory instruments guide the development of settlements in Zimbabwe and the master plans are blueprints that continue to guide urban development in Harare. The Harare Combination Master Plan (HCMP) of 1992 guides the future development of Harare by prescribing a southward expansion of the city that would result in spatial concentration in this area previously zoned for farming (City of Harare, 1992). The HCMP envisages compact development through densification and transit-oriented development to mitigate urban sprawl.

The legislation is somehow too conservative, rigid and archaic because it remains a legacy of the colonial government and fails to cope with contemporary issues (Chigudu and Chavunduka, 2020). At times, the statutes do not accommodate local realities resulting in planning failures. First is the RTCPA—described as the planning bible for urban development in Zimbabwe. This legislation was inherited from the colonial government and developed based on modernism, emphasising development control as the pinnacle of urban planning and development. Modernism is not bad but how it tends to be interpreted and implemented in Harare becomes a problem. Second are the demolitions, evictions and displacements that have been rife in the recent past, where informal settlements and informal activities in Harare have been demolished on the presumption of cleaning up the city (Matamanda, 2021). The RTCPA supports these demolitions by mandating local authorities to 'control' development and engage in demolitions or evictions when the development does not conform to the plans. But, how can a city be smart when the rights of certain individuals are undermined? Moreover, are there no other means of integrating such land uses and activities into the 'smart' ideologies rather than making other citizens worse off, thus widening the inequality gap between the haves and the have nots?

Specifically, the Harare Strategic Plan 2012–2025 (HSP) seeks to make Harare a world-class city by 2025. This also resonates with the national development agenda that aspires to make Zimbabwe a middle-income country by 2030. The HSP identifies different urban development strategies that need to be implemented to make Harare a smart city by 2025. Among these strategies are enhancing service delivery, combating climate change and other environmental problems, enhancing leadership through good governance and curbing ill-practices such as corruption, supporting livelihoods for the poor while upgrading and rehabilitating the public transport services in Harare (City of Harare, 2012).

8.3.3 The Twinning of Harare as Leverage for Smart City Development

City twinning initiatives must be used as a platform to spur smart developments. This is evident from the city of Harare, which has an active twinning arrangement with several cities. Of utmost importance are the active bilateral relations that Harare has with three global cities, Nottingham in the United Kingdom, Cincinnati in Ohio State in the United States of America and Munich in Germany. The twinning arrangement with Munich has been in place since 1996. Several exchange programmes have been undertaken over the years, including human capital training, especially in IT training, meant to assist Harare in setting up ICT and internet connectivity. Considering the encroachments that occur undetected, and thus creating a mess in land use tenure and governance as evident in Hopley, the twinning of cities thus allows the employees to be capacitated with the requisite tools and skills critical in advancing Harare's smart city agenda.

Moreover, the twinning arrangement with Munich has led to the implementation of enormous capacity-building programmes. These programmes resulted in improvements in water billing and revenue collection in Harare (Chifera, 2015). Still, the success of such an arrangement usually depends on the foreign relations that the central government has with the global community. This brings to attention the argument that cities may not have the autonomy to implement their development agendas. Instead, the central government tends to scrutinise everything, and at times, diplomatic decisions may be made, which seemingly frustrate the smart city initiatives.

8.3.4 DATA PRICE AS A PENALTY FOR THE URBAN POOR AND STIFLE TO INTERNET CONNECTIVITY

The existence of social media platforms among the residents and their ability to use such platforms for communication and mobile banking is a positive stride that may be a basis for smart city development in Hopley. The majority of the residents have smartphones that allow them to make online mobile transactions. The connectivity of internet services is generally good in urban areas where the services provided by Econet Wireless Zimbabwe are accessible even in settlements such as Hopley Farm.

The major challenges tend to be the disruptions in connectivity resulting from power outages. Moreover, the cost of data is very high and hence a hindrance to connectivity to the internet among many. Some reports claim that the country has the highest data prices in the world. For example, a monthly package of Wi-Fi 8 giga data from Econet Wireless Zimbabwe cost US$13 while the same package costs US$24 in NetOne. In this context, the cost of data becomes exorbitant, especially for the residents in Hopley who are living in poverty and have daily incomes of less than US$1 (Matamanda, 2020). Ndoro (2020) has bemoaned how such expensive tariffs hinder citizens' access to information, thus compromising their ability to participate in matters concerning them.

The obtaining situation in Harare shows that little has been done with regards to investments in affordable internet connectivity as has been done in cities such as Nairobi in Kenya. As pointed out by Moime (2016), technological hubs offering digital opportunities have been established in the Silicon Savanah in Nairobi. This has not been the case in Harare. While other countries are creating enabling environments that allow the private sector to provide such facilities and infrastructure, as evident in Johannesburg and Kigali, Rwanda (Senkhane, 2016), the government in Zimbabwe seems to be stifling the growth and development of such initiatives. Moreover, the digital technology industry seems to be monopolised by Econet and NetOne, who enjoy the monopoly and yet sabotage the development of smart cities.

8.3.5 The Vibrant Civil Society as an Enabler to Citizen Participation

Civic engagement is increasingly being undertaken on social media platforms through Facebook, WhatsApp and Twitter. There are several good practices in the social media platform, for example, the initiatives by the Combined Harare Residents' Association (CHRA) in Harare, which coordinates and assists citizens in representing their interest to the elected Council of Harare and the Executive Branch of Government. In this way, the CHRA has been instrumental in its advocacy programmes that challenge the local authorities on different matters concerning the city's socio-economic development. As a result, the CHRA has provided a platform through which the citizens' views are brought to the attention of the authorities which is critical in smart cities. However, it can be noted that such e-participation is politically motivated to some extent and involves certain individuals and groups advancing their political mandates and not the concerns of the citizens.

8.3.6 The Tertiary Institutions as Living Labs for Smart Ideas in Harare

The presence of institutions of higher learning that have a thrust on technological advancement and smart city initiatives is a major enabling factor that can be harnessed to spur smart city initiatives in Harare. Harare is home to the University of Zimbabwe, Harare Polytechnic College and Harare Institute of Technology, major tertiary institutions providing training in STEM subjects necessary for achieving smart cities. These universities engage in innovative programmes such as harnessing solar energy, biogas for waste management, designing systems for monitoring land uses, traffic movement and billing systems. The University of Zimbabwe has conducted feasibility studies and designed projects that include clean energy options identified as those contributing to the development of smart cities. This project has been undertaken by the University of Zimbabwe in partnership with the United Nations International Children Education Fund (UNICEF). The Harare Institute of Technology has undertaken similar projects, offering degrees in solar energy and renewable energy. The establishment of the Zimbabwe Manpower Development Fund (ZIMDEF) meant to fund the development of critical human capital also contributes to tertiary students' technological competence and literacy as the thrust of the funding in recent years has been on STEM degrees. In 2016, the government also engaged with the Scientific, Industrial Research and Development Centre (SIRDC) to develop a road map of opportunities and pathways for STEM careers that are significant in industrialising and modernising Zimbabwe (Chakanyuka and Musiiwa, 2016).

The foregoing section illustrates the attention the government has been paying in creating tertiary institutions that train graduates who are innovative and contribute to the modernisation and industrialisation of the country. This is a critical step considering the argument raised by Grobbelaar, Tijssen and Dijksterhuis (2017), who have stressed the role of universities in supporting sustained growth and development based on technology. The support from the government in trying to inculcate an entrepreneurial and innovative spirit in the human capital has also been commendable and is a necessary step as Siba and Sow (2017) highlighted the merits of such an investment in human capital. The former Minister of Higher and Tertiary Education,

Professor Jonathan Moyo, reiterated the need for fostering human capital development when he mentioned that:

> There is no politician who can create jobs and there is no politician who has ever created jobs. Politicians only come up with necessary policies and laws that create the environment. This is what we are good at creating, the environment. Entrepreneurs and investors create jobs. Investors invest in innovative ideas, and these ideas that they pursue come from engineers.
>
> (Chakanyuka and Musiiwa, 2016: online)

The tertiary institutions are thus enablers of smart cities initiatives, as illustrated in the preceding paragraphs. However, as stated by Professor Jonathan Moyo, the enabling environment matters the most for the ideas to be nurtured and flourish. Although the government sets to fund STEM programmes in tertiary institutions, there have been allegations of misuse of these funds or instances where they benefit students who can afford to pay their fees.

8.3.7 Budgetary Complacency and Institutional Constraints

In 2018, the national budget of Zimbabwe was US$5.75 billion, which is ironic that the equivalent of three years' national annual budget would be required to develop a single smart city. In attempting to overcome these financial hurdles, local governments use different platforms to raise funds and the first is usually central government bonds. Still, these often result in the central government trying to dictate the pace and nature of the developments. An official from the City of Harare pointed out that most of the urban development projects in Harare are stifled by the lack of and mismanagement of funds and lack of clear roles between different institutions. The official commented on the funds from the central government for road development and maintenance in the city, which has been mandated to the Zimbabwe National Road Administration (ZINARA) and not the local authority.

This depicts centralisation of urban management, contrary to smart governance, where urban governance has to be decentralised and participatory. Smart city initiatives for Harare include the rehabilitation of roads and introducing the Bus Rapid Transit (BRT) system to cater to public transport services (City of Harare, 2012). However, the funding arrangement between ZINARA and the City of Harare makes this challenging as ZINARA has been defaulting the disbursement of funds to the city of Harare, albeit having had collected the funds from the road users (Zhangazha, 2017). The poor state of the roads in Harare has been attributed to ZINARA's failure to disburse funds to rehabilitate the road network in the capital city (Madzimure, 2020).

Therefore, local governments may seek to engage in private–public partnerships or smart bond financing mechanisms. The City of Harare has been involved in some PPPs as a policy option to finance housing development in the city (City of Harare, 2013). Through the PPP programmes, the Harare City Council, Old Mutual Zimbabwe and CABS entered into a PPP agreement in 2012 that involved constructing low-cost housing on land provided by the council with funding being provided by

CABS. This initiative in housing development is crucial in securing smart housing development as tenure security is guaranteed to the beneficiaries while also making great strides towards housing the low-income earners, thus satisfying the spatial justice aspect of smart city initiatives. Ironically, the reality on the ground is that the intended beneficiary of this project has not been able to access the 'low-cost' housing, which has been beyond the reach of many low-income earners. Subsequently, the housing shortages among the poor continue to increase, escalating the urban inequalities and making it difficult to close the housing shortage gap. This situation thus brings to attention the willingness of various stakeholders to uplift the lives of the poor without robbing them because, as pointed out by Watson (2015), a smart city is one in which justice is upheld.

8.4 CONCLUSION

The chapter described and evaluated the smart city concept and development using Harare as an exemplar of an African city. Key questions emerging evolved around the capacity of these cities to implement successful smart city developments that would spur sustainable development, considering the complex socio-economic and political challenges that characterise the region. Emerging from the analysis is that, albeit the challenges plaguing Harare, there are both opportunities and constraints to the development of smart cities.

The opportunities include a youthful population that is technologically literate to adapt to the smart life. The university student massification is another enabling factor that increases the technological literacy among the citizens while also providing living labs for innovation and nurturing of different skills. Some government policies that have been introduced, such as indigenisation and youth empowerment programmes that provide start-up capital to the youth, enable specific smart city programmes to be implemented. The city twinning initiatives have also been instrumental as they provide platforms for collaboration and co-production in smarting the city. Moreover, the presence of a wide array of social media platforms that citizens use enables disseminating information and services while also serving as a platform for citizen participation. Although platforms such as WhatsApp, Twitter and Facebook are currently being utilised, there is still room for more advanced uses.

Yet, some constraints frustrate the smart city developments. The political significance of Harare somehow hinders some development initiatives, such as in instances when the central government interferes with and disrupts the local government activities. The poor governance characterised by gross corruption, mismanagement of public funds and authoritarian rule is a far outcry from the ideals of a smart city that places good governance at the centre of all the undertaken developments. The archaic and rigid statutes limit the implementation of some smart initiatives. At the same time, lack of human capacity, infrastructure, finances and political will frustrate the efforts to adopt smart initiatives in settlement planning and development. Further research relating to this chapter may examine the smart city approaches and strategies that enhance the social aspects of city development beyond the provision of technology that produces enclaves for the elites, thus perpetuating socio-economic and class segregation.

REFERENCES

Ajibade, I. (2017). Can a future city enhance urban resilience and sustainability? Apolitical ecology analysis of Eko Atlantic city, Nigeria. *International Journal of Disaster Risk Reduction*, 26: 85–92.

Albino, V., Berardi, U. and Dangelico, R. M. (2015). Smart cities: Definitions, dimensions, performance, and initiatives. *Journal of Urban Technology*, 22: 1–19

Baffoe, G., Ahmad, S. and Bhandar, R. (2020). The road to sustainable Kigali: A contextualised analysis of the challenges. *Cities*, 105. doi: 10.1016/j.cities.2020.102838

Caragliu, A., Bo, C. C. and Nijkamp, P. (2011). Smart cities in Europe. *Journal of Urban Technology*, 18(2): 65–82.

Chakanyuka and Musiiwa (2016). STEM funding for varsity students. *The Sunday Mail*, 2 October. Available from: www.sundaynews.co.zw/stem-funding-for-varsity-students/ (Accessed 26 December 2020)

Chifera, I. (2015). Harare seeking links with China's Guangzhou City. *Voice of Africa Zimbabwe*, 21 September. Available from: www.voazimbabwe.com/a/zimbabwe-harare-twin-city/2972809.html (Accessed 23 December 2020)

Chigudu, A. and Chavunduka, C. (2020). The tale of two cities: The effects of urbanization and spatial planning heritage in Zimbabwe and Zambia. *Urban Forum*. doi: 10.1007/S12132-020-09410-8

Chigudu, S. (2019). The politics of cholera, crisis and citizenship in urban Zimbabwe: 'People were dying like flies'. *African Affairs*, 118(472): 413–434

City of Harare (1992). *The Harare Combination Master Plan 1992*. Harare: City of Harare.

City of Harare (2012). *City of Harare Strategic Plan 2012–2025*. Harare: City of Harare.

City of Harare (2013). *Housing Procedure Manual*. Harare: City of Harare.

Datta, A. and Odendaal, N. (2019). Smart cities and the banality of power. *Environment and Planning D: Society and Space*, 37(3): 387–392

Deloitte & Touche (2014). *Smart Cities Technology: Africa is Ready to Leapfrog the Competition*. Johannesburg: Deloitte & Touche.

Grobbelaar, S., Tijssen, R. and Dijksterhuis, M. (2017). University-driven inclusive innovations in the Western Cape of South Africa: Towards a research framework of innovation regimes. *African Journal of Science, Technology, Innovation and Development*, 9(1): 7–19

Huet, J. (2016). *Smart Cities: The Key to Africa's Third Revolution*. Paris: BearingPoint Institute.

Inkinen, T., Yigitcanlar, T. and Wilson, M. (2019). Smart cities and innovative urban technologies. *Journal of Urban Technology*, 26(2): 1–2

Kaika, M. (2017). 'Don't call me resilient again!': The New Urban Agenda as immunology... or ... what happens when communities refuse to be vaccinated with 'smart cities' and indicators. *Environment and Urbanization*, 29(1): 89–102.

Kamete, A. Y. (2004). Home industries and the formal city in Harare, Zimbabwe. In: Hasen, K. T. and Vaa, M. (eds), *Reconsidering Informality: Perspectives from Urban Africa*. Uppsala, Sweden: Nordic Africa Institute, pp. 120–138.

Kummitha, R. and Crutzen, N. (2017). How do we understand smart cities? An evolutionary perspective. *Cities*. doi: 10.1016/j.cities.2017.04.010

Madzimure, J. (2020). City gets $52m for road rehab. *The Herald*, 15 February. Available from: www.herald.co.zw/city-gets-52m-for-road-rehab/ (Accessed 25 December 2020)

Manirakiza, V., Mugabe, L., Nsabimana, A. and Nzayirambaho, M. (2019). City Profile: Kigali, Rwanda. *Environment and Urbanization ASIA*, 10(2): 290–307

Matamanda, A. R. (2020). Living in an emerging settlement: The story of Hopley Farm Settlement, Harare Zimbabwe. *Urban Forum*, 31: 473–487

Matamanda, A. R. (2021). Mugabe's urban legacy: A postcolonial perspective on urban development in Harare, Zimbabwe. *Journal of Asian and African Studies*, 56(4): 804–817

Moime, D. (2016). Kenya, Africa's Silicon Valley, Epicentre of Innovation. Available from: https://vc4a.com/blog/2016/04/25/kenya-africas-silicon-valley-epicentre-of-innovation/ (Accessed 22 December 2020)

Moura, F. and Silva, J. A. (2019). Smart cities: Definitions, evolution of the concept and examples of initiatives. In: Leal Filho, W. et al. (eds), *Industry, innovation and infrastructure. Encyclopedia of the UN Sustainable Development*. Cham: Springer.

Muchadenyika, D. (2020). *Seeking Urban Transformation: Alternative Urban Features in Zimbabwe*. Harare: Weaver Press

Ndawana, E. (2020). The military and democratization in post-Mugabe Zimbabwe. *South African Journal of International Affairs*, 27(2): 193–217

Ndoro, N. (2020). Expensive data tariffs a challenge for info-hungry Zimbabweans. *Nehanda Radio*, 31 March. Available from: https://nehandaradio.com/2020/03/31/expensive-data-tarrifs-a-challenge-for-info-hungry-zimbabweans-misa/ (Accessed 23 December 2020)

Piesse, M. (2019). *Zimbabwe's Economic Mismanagement Leaves Harare with Limited Water Supplies*. Nedlands: Future Directions International Pty Ltd.

Rich, R., Westerberg, P. and Torner, J. (2019). *Smart City Rwanda Master Plan*. Nairobi: UN-Habitat

Senkhane, M. (2016). African cities vie for 'tech hub of Africa' throne. *African Cities* #5, UCLG-A, Rabbat. 56–61

Shamsuddin, S. and Srinivasan, S. (2020). Just smart or just and smart cities? Assessing the literature on housing and information and communication technology. *Housing Policy Debate*, 27: 1–24

Siba, E. and Sow, M. (2017). Smart city initiatives in Africa. *Africa in Focus*, 1 November. Available from: www.brookings.edu/blog/africa-in-focus/2017/11/01/smart-city-initiatives-in-africa/ (Accessed 22 December 2020)

Smart Africa. (2017). *Smart Sustainable Cities: A Blueprint for Africa*. Kigali: Smart Africa, Republic of Rwanda

Vázquez, A. N. and Vicente, M. R. (2018). Exploring the determinants of e-participation in smart cities. In: *E-participation in Smart Cities: Technologies and Models of Governance for Citizen Engagement*. Public Administration and Information Technology, vol. 34. Cham: Springer, pp. 157–178. doi: 10.1007/978-3-319-89474-4_8

Watson, V. (2014). African urban fantasies: Dreams or nightmares? *Environment and Urbanization*, 26(1): 213–229

Watson, V. (2015). The allure of 'smart city' rhetoric: India and Africa. *Dialogues in Human Geography*, 5(1): 36–39

Weber, V. (2019). Smart cities must pay more attention to the people who live in them. *World Economic Forum*, 16 April. Available from: www.weforum.org/agenda/2019/04/why-smart-cities-should-listen-to-residents/ (Accessed 22 December 2020)

World Health Organisation (2001). *Guidelines for City Twinning*. Geneva: World Health Organisation

Yigitcanlar, T., Han, H. and Kamruzzaman, M. (2019). Approaches, advances, and applications in the sustainable development of smart cities: A commentary from the guest editors. *Energies*, 12: 4554. doi: 10.3390/en12234554

Zhangazha, W. (2017). Zinara, Harare city fathers haggle as motorists cry foul. *The Independent*, 23 January. Available from: www.theindependent.co.zw/2017/01/23/zinara-harare-city-fathers-haggle-motorists-cry-foul/ (Accessed 25 December 2020)

9 Smartness in Developing Liveable Informal Settlements
The Case of Hopley in Harare

Morgen Zivhave and Collins Dzvairo

CONTENTS

- 9.1 Introduction 121
- 9.2 Theoretical Framework... 123
  - 9.2.1 Urban Informality, Smart Cities and Liveability.............................. 123
  - 9.2.2 Liveability Indicators.. 124
  - 9.2.3 Informality and Liveability 125
- 9.3 Methods 125
- 9.4 Results... 127
  - 9.4.1 Demographic Data of Respondents 127
  - 9.4.2 Liveability Dimensions and Their Attributes 128
- 9.5 Discussion... 133
  - 9.5.1 Smartness in Developing Liveable Informal Settlements 133
  - 9.5.2 Nuances in the Liveability and Smartness of Informal Settlements.. 135
- 9.6 Conclusion 135
- References.. 136

9.1 INTRODUCTION

The world is urbanised with 54% (4 billion) of the population living in towns and cities by 2015 (UN-Habitat, 2016) and it continues to urbanise at unprecedented rates, especially in cities in the global South (Mahabir et al., 2016). Unfortunately, the massive urban growth has coincided with inadequate central and local government capacity to plan, manage and guide urban development (ibid), leading to the rise of informal settlements. Globally, over one billion people live in informal settlements and this is projected to rise to two billion by 2030 (ibid). This means that a large world urban population lives in precarious conditions without proper infrastructure, affecting the quality of their lives. At the same time, there is 'growing awareness of the deterioration of liveability particularly in urban environments because of the pressure of rapid development and growing population' (Leby and Hashim, 2010:70). Liveability mirrors the wellbeing of residents, which is critical for cities where urbanisation often presents limitations (ibid). This study focuses on the liveability of informal settlements; a unique gap where dominant publications on Zimbabwe have concentrated on informal settlement challenges, characteristics, land tenure and upgrading (UN-Habitat, 2013; Chirisa and Muhomba, 2013; Sandoval et al., 2019; Matamanda et al., 2020). Globally, studies on liveability from Nigeria, the United States and Malaysia have focused primarily on the city and planned neighbourhoods (Omuta, 1988; Leby and Hashim, 2010).

Globally, urbanisation has been associated with economic development and improving people's lives as shown in Asia, and especially, China through the country's economic transformation and industrialisation (UN-Habitat, 2016). However, sub-Saharan Africa, with the exception of South Africa, has witnessed worsening living conditions in cities, and especially, the informal settlements. Though informal settlements account for 30% of the global urban population (Mahabir et al., 2016), sub-Saharan Africa is expected to house 55.9% of this population by 2050 (UN-Habitat, 2016; Table 9.1). This chapter interrogates the liveability of informal settlements and how this affects the smartness of these neighbourhoods. The study questions whether informal settlements within largely incapacitated cities of the global South can be liveable at all.

The rate of urbanisation in Zimbabwe is generally low by global and African standards. About 32% of the Zimbabwean population live in urban areas with an urbanisation rate of 4% per annum [Zimbabwe National Statistics (ZIMSTAT), 2017] (see Table 9.1), which is lower than the 37.9% for sub-Saharan Africa and 49% for the developing countries. Similarly, urban informality in Zimbabwe was low during the first few decades of independence (UN-Habitat, 2003) because of stringent rules on development control and lack of public land in peri-urban areas (Tibaijuka, 2005). However, informality is not new to the country as shown by the cases of Epworth and Chirambahuyo before independence (Patel, 1988). The low informality in Zimbabwe was uncharacteristic of the global South that experiences high rates of urbanisation but this changed after the year 2000. The country's Fast Track Land Resettlement

TABLE 9.1
Urbanisation and Informality in Zimbabwe

Region	Urban population	Urban population living in informal settlements (2014)
Developing regions	49% (2015)	29.7%
North Africa	51.2% (2010)	11.9%
Sub-Saharan Africa	37.9% (2015)	55.9%
Zimbabwe	32.4% (2015)	25.1%

Source: Dodman et al. (2013), UN-Habitat (2016), Slum Almanac (2016), ZIMSTAT (2017), and Sandoval et al. (2019)

Programme (2000–2003) changed these dynamics by providing cheap land in peri-urban areas that spurred massive informal urbanisation.

However, even with the growing level of informality, the concept of improved quality of life in cities is highly regarded by both residents and officials in Zimbabwe and hence the significance of this study. The ill-conceived and disastrous Operation Murambatsvina of 2005 was a government programme that demolished illegal structures countrywide to restore order and aesthetics in cities. During this operation, 700,000 people lost their accommodation and livelihoods and 2.4 million were affected (Tibaijuka, 2005). Ironically, this was a strategy by a government preconceived with 'restoring' liveability in cities.

9.2 THEORETICAL FRAMEWORK

The smart city and liveability concepts measure the quality of living in cities and neighbourhoods. Studies on these concepts sought to understand the relationships between the socio-economic, functional and environmental attributes of the built environment, often with a bias towards attractive settlements (Leby and Hashim, 2010). However, such frameworks have not captured the contexts of urban informality that have limited infrastructure and services. Contextualising liveability for informal settlements would require measuring instruments that largely depend on resident perceptions than purely expert analyses based on pre-defined criteria. This section presents a theoretical framework on informal settlements, smart cities and liveability.

9.2.1 Urban Informality, Smart Cities and Liveability

Informal settlements are neglected parts of the city where infrastructure is poor (Cities Alliance, 2006) that are associated with sub-standard living conditions (Mahabir et al., 2016). Matamanda et al. (2020) add that informal settlements are often located in environmentally hazardous areas where residents have no legal claim to their land and their housing does not comply with statutory plans. The terms slum and informal settlements are often used interchangeably (UN-Habitat, 2003); however, this study adopts the term informal settlements and treats slums and informal settlements as synonymous. Informal settlements are characterised by the lack of basic infrastructure and services which translate into burdensome transaction costs for residents (UN-Habitat, 2013, Mahabir et al., 2016) that negatively impact on the environment and human life (Mahabir et al., 2016). Thus, the quality of life of informal settlement dwellers is contestable when viewed through the lenses of the smart and liveable city.

Hancke et al. (2013:393) define a smart city as a 'city which functions in a sustainable and intelligent way by integrating all its infrastructure and services into a cohesive whole and using intelligent devices for monitoring and control'. This definition has a bias towards the use of technology towards improving service delivery and reducing cost. However, Watson (2013) adds that the smart city debates links the opportunities of maximising the benefits of technology and environmental sustainability. Arguably, this expands the smart city concept to include the relationship between people and their environment. Moreover, Harrison and Rubin's (2018) balanced definition argues that smart cities prioritise people and human capital rather than assuming that technology will transform and improve cities. The foregoing argument balances technology, growth and sustainability. The aforementioned assertions acknowledge that smart city arguments incorporate the human and social dimensions of the urban environment, which takes us to the next concept of liveability.

Liveability upholds the principles of quality and character of a place, health of communities and sustainability and quality of life (Myers, 1987). Liveability also embraces the human needs of food, shelter, security and a sense of belonging to a community (National Research Council, 2002). The Economist Intelligence Unit (2019) uses the concept of liveability to assess cities with the best and worst living conditions globally. The liveability variables are critical to residents' quality of life and are central to settlement planning. However, since liveability is broadly divided into themes of people and place, what is liveable is relative. Thus, liveability measures a resident's perception of their quality of life (Leby and Hashim, 2010) and is a subjective evaluation of their environment. A community's cultural dynamics and living standards shape its impression on the quality of infrastructure, transportation, service provision and even urban design (National Research Council, 2002). As a result, the concept is complex and there are contestations on the mix of variables of what constitutes liveability (Heylen, 2006). These arguments do not discount a common understanding of liveability as The Economic Intelligence Unit, for instance, produces annual liveability assessments for cities globally. Liveability encompasses elements of homes, neighbourhoods, cities and metropolitan areas that are used to measure safety, economic opportunities, health, convenience, mobility and recreation (Leby and Hashim, 2010). The next section expands on the variables of liveability.

9.2.2 Liveability Indicators

The indicators of liveability can be traced back to Lynch's (1998) criteria of a good settlement and an earlier publication by Jacobs (1961) on the principles of active retail spaces and mixed residential space as aspects of urban liveability. In Europe, researchers in England measured liveability using variables such as environmental quality, physical location, functional place quality and safe place (Leby and Hashim, 2010). Within sub-Saharan Africa, Omuta (1988) measured the liveability of neighbourhoods in Benin City in Nigeria using five indicators that are employment, housing, amenities, nuisance and socio-economic factors. Globally, the Economist Intelligence Unit (2019) uses 30 qualitative and quantitative factors divided into five broad categories that are stability, healthcare, culture and environment, education and infrastructure.

The common threads from the aforementioned studies are that variables for liveable spaces include safety, clean, affordable and efficiently administered spaces, with functional infrastructure, effective public transport and employment opportunities. Literature appraisal from Omuta (1988); Heylen (2006); Leby and Hashim (2010); and The Economist Intelligence Unit (2019); liveability has four main dimensions as follows: (i) the social dimension such as community life and sense of place that measures social relations, diversity and tolerance; (ii) the physical dimension such as environmental quality, open spaces and maintenance of the built environment that measures the residential environment where people live; (iii) the functional dimension such as availability and proximity of amenities (such as schools, clinics, public transport system and shopping centres) and employment opportunities, which give vitality to a place and (iv) the safety dimension such as crime and accidents that gives a settlement feeling and sense of safety.

This study measured the liveability of an informal settlement using the four dimensions of social, physical, functional and safety. That said, Valcarcel-Aquiar et al. (2018) call for the balancing of liveability indicators. They argue that high standards of living can be achieved at the expense of deteriorating natural environment and agglomeration may negatively affect green spaces as the environment provides raw materials for built-up areas (ibid). This study balances the four dimensions in measuring settlement liveability.

9.2.3 Informality and Liveability

As discussed earlier, liveability is based on the concepts of quality of life and place that reflect on the environment. In this regard, liveable communities provide affordable housing, infrastructure, clean air and attractive livelihoods. However, these qualities are often compromised in informal settlements. In Nairobi, residents of informal settlements have limited assets and engage in low-income jobs such as casual labourers, security guards, entertainers and shop assistants (UN-Habitat, 2008). Another concern for informal settlements is their impact on the environment. Chirisa and Muhomba (2013) found that Epworth in Harare has uncontrolled sand abstraction for brick moulding and construction purposes causing land degradation. This is also evident in Chirundu in the border between Zimbabwe and Zambia where informal settlements are severely affected by land degradation (Munyoro and Nyaushamba, 2016). Again, informal settlements are associated with poor health attributed to overcrowding, poor housing and unsanitary conditions (UN-Habitat, 2003; Chirisa and Muhomba, 2013; Munyoro and Nyaushamba, 2016).

Though the stereotyped characterisation of informal settlements is contested by the National Research Council (2002); the average informal settlements lack some or all of the liveability attributes and hence the essence of this study. In summary, the liveability of informal settlements depends on the availability of infrastructure, environmental conditions and livelihood opportunities.

9.3 METHODS

Liveability can be measured through two main ways, that are; (i) an 'objective' evaluation where experts analyse the physical parameters that assess the quality of the environment 'independently' or (ii) a subjective evaluation where resident perception assesses the environmental quality of a neighbourhood through questionnaires and interviews (Leby and Hashim, 2010). This study is qualitative and adopted the subjective approach for measuring neighbourhood liveability. This study also adopted the case study research design and in particular, Hopley in Harare. The settlement was purposely chosen as one of the new neighbourhoods that are still growing and constitute an information-rich case as characterised by Yin (2009). We focused on the neighbourhood as the unit of analysis, borrowing from Leby and Hashim's

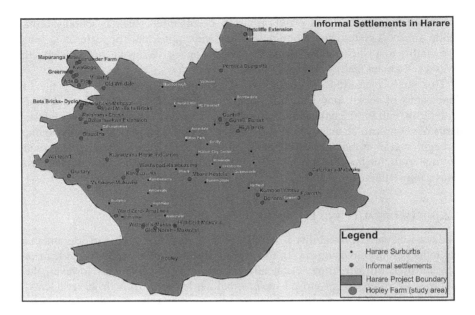

FIGURE 9.1 Hopley Settlement.

Source: Dialogue on Shelter, Zimbabwe Homeless People's Federation and City of Harare, 2014

(2010) argument that it constitutes the basis when residents make decisions on where to live.

Hopley is a settlement 12 km South of Harare's central business district, along the new Chitungwiza Road (Figure 9.1). The settlement emerged after the Operation Murambatsvina of 2005 as a temporary shelter for residents evicted from Hatcliffe Extension, Porta Farm and Mbare (Dialogue on Shelter, Zimbabwe Homeless People's Federation and City of Harare, 2014). Hopley has over 6,000 households with an average size of five people where more than 70% of the residents are poor (ZIMSTAT, 2012). The study gathered data between September and October 2019. Within Hopley, the study sampled 60 households. With these households, the researcher reached the point of data saturation, where interviewing more participants produced the same results. In addition, key informants from the Tariro Youth Centre and Kushinga Primary School within Hopley, the City of Harare, the Department of Spatial Planning and Development, Ministry of Local Government and Public Works and Dialogue on Shelter were interviewed. These institutions provide critical social services, infrastructure and guide urban development towards promoting settlement liveability. The study also complemented interviews with participant observations and reviewed progress reports from key stakeholders in the area. In line with ethical considerations, the study interviewed school children in the presence of their guardians.

9.4 RESULTS

This section presents findings on the assessment of the liveability of Hopley. The parameters used for assessing settlement liveability were social, physical, functional and safety. The researcher also reviewed the demographic data such as the residents' occupation, tenure security and type of housing and how this affects liveability.

9.4.1 Demographic Data of Respondents

The study conducted 60 household interviews selected from the six zones of Hopley. A total of 19 respondents were male and 41 were females between the ages of 11 and 51 years (Table 9.2). The sample comprised of school pupils, construction workers, household workers, transport workers, small business operators, civil servants and farmers from the neighbourhood. The respondents' minimum education was primary (grade seven) and maximum was tertiary (university degree) (see Table 9.2).

TABLE 9.2
Summary of the Demographic Data of the Respondents

Variable	Modalities	Total	Percentage
Sex	Male	19	32
	Female	41	68
Years	<35	33	55
	35–60	27	45
	>60	0	0
Educational background	Primary	9	15
	Secondary	37	62
	Tertiary	14	23
Occupations	Skilled construction worker	21	35
	Household work	9	15
	Transport worker	7	12
	Small business or trading	14	23
	Government service	6	10
	Urban agricultural	3	5
Tenure security	Lease agreement with government	6	10
	Rented	26	43
	No land access agreements	28	47
Type of housing	Pole and dagga	3	5
	Two roomed concrete structure	43	72
	>2-roomed concrete structure	14	23

Source: Dzvairo (2019)

Hopley largely comprises of skilled and semi-skilled labour force. The largest percentage of the sample of Hopley residents (35%) were skilled and semi-skilled artisans, followed by small businesses owners (23%). Household workers such as maids, gardeners and guards constituted 15% and transport workers such as drivers, touts of commuter omnibuses and cart pullers made up the remaining 12% of the interviewees. The residents of Hopley have different land tenure arrangements. These include lease agreements with the Ministry of Local Government and Public Works or the City of Harare within the approved land use plan. However, residents from the unplanned zones of Magada have undocumented land occupation.

9.4.2 Liveability Dimensions and Their Attributes

As mentioned earlier, the study measured the liveability of Hopley based on the perception of residents. The four liveability variables used for data collection were ranked in Table 9.3.

9.4.2.1 Social

Hopley residents ranked the sense of place and sociability as follows: five households identified sense of community as important while two preferred close relationships

TABLE 9.3
The Resident Perceptions on Liveability Dimensions

Liveability dimensions	Attributes	Frequency	Rank
Social	Sense of community	5	
	Sociability of people or neighbours	2	
Sub-Total		**7**	**3**
Physical	Maintenance of streets	4	
	Refuse collection service	7	
	Availability and maintenance of open spaces	3	
Sub-Total		**14**	**2**
Functional	Availability and proximity to health facilities (including water and sanitation)	13	
	Availability and proximity to schools	7	
	Availability and proximity to shops	3	
	Availability of employment	10	
Sub-Total		**33**	**1**
Safety	Personal safety from crime	4	
	Availability of police protection	2	
Sub-Total		**6**	**4**
Total for all dimensions		**60**	

Source: Dzvairo (2019)

among residents (Table 9.3). The two attributes of social liveability provide the social capital that empowers communities to negotiate with the city and non-governmental organisations towards improving the neighbourhood. However, these social dimensions among residents are affected by widespread poverty, limited source of employment and lack of educational opportunities. As a result, some youths from the area are involved in bad behaviour such as teenager prostitution, especially at 'Kwa Anthony'. Interviews show that residents stigmatise this place and stereotype youth behaviour based on these practices. In addition, surrounding communities have prejudices that Hopley residents contribute significantly towards the spread of sexually transmitted diseases, a label that affects the community self-worth.

9.4.2.2 Physical

Both interviews and observations demonstrated settlement limitations on the general quality of the built environment. Hopley residents ranked the three attributes of physical liveability as follows: seven households prioritised refuse collection because of the negative impacts of garbage, three households chose clean streets and three identified the maintenance of open spaces as crucial towards physical liveability. The physical environment in Hopley is severely affected by illegal dumpsites due to the lack of refuse collection and economic activities such as sand mining that have defaced the neighbourhoods. Existing open spaces and vacant land are used for these illegal activities, which are breeding grounds for flies and mosquitos (Figure 9.2). Besides illegal dumpsites and sand mining activities, open spaces on wetlands are used for urban agriculture, especially in Zones 3 and 5. The illegal activities have turned the settlement into 'detestable scenes', practice by residents out of desperation.

The physical environment of a neighbourhood is often defined by its road network. Observations show that many secondary roads in Hopley are unconnected and characterised by dead-ends. As a result, many individual properties are not accessible by road (Figure 9.3). The disjointed road network makes movement within Hopley a nightmare. Consequently, residents and visitors can hardly manoeuvre within the neighbourhood and park their vehicles outside the settlement.

Moreover, the road network is predominantly gravel, characterised by depressions and potholes. Consequently, many roads are slippery and impassable during the rainy season (Figure 9.4). The bad road infrastructure has resulted in the settlement being shunned by most 'public' transport operators. Most formal and informal operators (the *combis and mushikashikas*) avoid the settlement and the few that ply the neighbourhood charge twice or two and half times the rates charged if one is dropped off along a nearby highway.

9.4.2.3 Functional

Neighbourhoods render service to residents and participants highly ranked the functional attributes of Hopley such as availability and proximity to health facilities, schools, shops and availability of jobs. Thirteen households recognised health facilities as important in a liveable neighbourhood. Seven households prioritised the quality and distance to schools and three preferred shopping facilities located within the neighbourhood. Employment opportunities were selected by ten households who valued jobs as means to access food and shelter.

FIGURE 9.2 Open spaces used as dumpsites and sand mining.

Source: Dzvairo (2019)

The Case of Hopley in Harare

FIGURE 9.3 Disjointed road network in Hopley.
Source: Dzvairo (2019)

FIGURE 9.4 Road conditions in Hopley during the rainy season.
Source: Dzvairo (2019)

Hopley does not have reticulated water and sewerage systems. Three boreholes installed by a previous commercial farmer dried up as the settlement grew. About 35% of interviewed households access clean water from a borehole and water tanks established by Medicine San Frontier, a non-profit organisation in the area. The remaining 65% are located too far from the borehole and water tanks and rely on shallow wells. However, the shallow wells are exposed to contamination from pit latrines and Blair toilets that are common methods of sewage disposal (Figure 9.5). The toilets are built onsite with poor-quality material such as wood and farm bricks. Residents with no ablution facilities resort to open defecation. A study by Dialogue on Shelter and the Zimbabwe Homeless People's Federation (2019) found that 7% of the 101 households interviewed in Hopley that had no toilets resorted to open defecation.

Educational facilities are key functions in neighbourhood planning. The City of Harare established the Tariro Primary and Secondary Schools in the settlement. Besides these public schools, children attend private 'colleges' such as the Kushinga and Tamuka Primary and Secondary Schools. Early childhood development centres in Hopley utilise different premises that include individual homes. Only a few households can afford to send their children to better schools in surrounding planned neighbourhoods. In addition, the city runs two skills training centres in the settlement. The City of Harare and Silveira House established the Hopley Livelihoods Centre and Taririo Youth Community Centre that offer skills development in various trades such as performing arts and entrepreneurship. In addition, these centres also offer a conducive environment for socialisation of residents.

FIGURE 9.5 Toilets in Hopley.

Source: Dzvairo (2019)

As an unplanned settlement, most economic activities in Hopley are informal. The highest paying livelihood activity in the settlement is brick moulding, responding to the high demand within and the surrounding settlements of Budiriro, Retreat, Glenview and Fidelity Housing Scheme. Consequently, the physical environment is marred by scars from brick moulding and sand mining. However, some households depend on wage employment and small business operations. The informal businesses include vegetable and food vending, grocery and hairdressing. A local carpenter observed that residents are too poor to purchase household furniture from conventional shops; as a result, he provides cheaper and tailor-made furniture for them.

Planned neighbourhoods normally provide shopping centres with low-order goods for residents' daily needs. The informal traders provide these daily groceries in Hopley. The settlement has mixed-use development with a mix of shops, food kitchens, *shebeens* (an informal private house or shop selling alcohol) and snooker shades operating along the main road. The low rentals on commercial plots in Hopley allow informal start-ups to operate profitably with low working capital.

9.4.2.4 Safety

The safety dimension measured the general feeling of security in the neighbourhood and assessed the resident's perceptions of crime. Four households ranked personal safety as important and two identified the presence of police as offering protection. Some residents noted that they lacked valuable assets that would attract thieves but still valued personal safety.

However, there were mixed responses to the general safety of the neighbourhood. Some households felt that Hopley's high-density residential development promoted natural surveillance. For instance, only two out of 60 interviewed households had their family members attacked at night. In addition, school children from the interviewed households felt safe when travelling to school because of the large number of people on the streets. However, some households argued that the settlement was not safe and attributed illegal activities such as *shebeens* to promoting cultural dislocation, crime, child abuse and domestic violence in the neighbourhood. The next section discusses the findings from Hopley within the larger discourse of liveability.

9.5 DISCUSSION

9.5.1 SMARTNESS IN DEVELOPING LIVEABLE INFORMAL SETTLEMENTS

This study presents informal settlement resident perceptions on the different attributes of neighbourhood liveability. Of the four dimensions, namely social, physical, functional and safety, the Hopley residents identified the functional dimension to be the most critical attributes towards settlement liveability. This is unsurprising since Hopley has poor health and educational facilities, shopping centres and limited jobs. These variables are critical components for a good settlement. The poor functionality of Hopley mirrors the performance of other informal settlements in Harare because of rapid urbanisation and the corresponding poor service delivery. Makunde et al. (2018) argue that settlements in Harare and across the country perform badly because of the current national economic crisis that incapacitated the local government service delivery. Suffice to say, informal settlements present grimmer neighbourhood conditions than planned ones.

In line with limited health facilities, water-borne diseases related to lack of clean water and proper sanitation are common in Hopley, especially during the rainy season. For instance, diarrhoea, singled out as the most prevalent disease among school-going children, had the highest incidences in the settlement. An earlier Harare study identified poor waste disposal as a principal cause for underground water pollution (Ndoziya, 2015). Settlements such as Budiriro in Harare and Seke in Chitungwiza have experienced underground water pollution from poor waste and sewage disposal. The challenges of water and sewerage service provision in cities and informal settlements are expected to rise. The level of service provision in Harare is lower than the national figures. About 61.2% of the population in Harare were connected to piped water and 68.3% were connected to sewerage in 2010 (UN-Habitat, 2016). These service levels are lower than the national urban population where 71% had piped water and 74.4% were connected to sewerage (ibid). A study by the Economist Intelligence Unit (2019) ranked Harare as one of the ten least liveable cities globally, confirming these poor performance indicators. Thus, the poor liveability indicators for Hopley correlate with broader Harare experiences.

Educational institutions are strongly valued in Zimbabwe where human capital development is perceived as a path towards employability. In informal settlements, private education service providers have filled the gap left by the central and local governments. However, many of the schools have inadequate classrooms and sport fields required for holistic childhood development. However, Hopley residents resort to private schools as they charge low tuition and are conveniently located in the neighbourhood.

The livelihood and income opportunities of Hopley residents are limited. Besides informal trading, Hopley residents provide cheap household labour as builders, painters and carpenters to their middle-class counterparts within or in nearby settlements. A small section of Hopley residents are low-ranked civil servants employed as cleaners, drivers and security guards. This pattern mirrors informal settlements as a source of cheap labour for the formal sector. The study thus confirms that informal settlement dwellers subsidise middle-class and formal business by providing cheap labour because of their social standing and low education (Mahabir et al., 2016).

As a result of limited livelihood opportunities, many Hopley households have inadequate income to provide basics for their families. In addition, the households have job insecurity and lack social security benefits because of the informal nature of their work. The wages are irregular and many households are in constant debt. Broadly, these findings mirror urban poverty in Zimbabwe that stands at 46.5% (Sandoval et al., 2019).

The other significant dimension of liveability in Hopley is the physical environment. This is a critical indicator of neighbourhood performance and local authorities are often criticised for the poor maintenance of streets and failure to collect refuse. The non-existence of paved streets in Hopley affects mobility and making journeys relatively longer and expensive. The poor connectivity and maintenance of streets also affect emergency services such as ambulances and fire engines. However, an unexpected finding was that the poor road network and connectivity in Hopley reduced the number of through traffic and over-speeding motorists, thereby reducing accidents. With limited access to recreation facilities and open spaces, children use roads as playgrounds for football and cricket and even adults relax along these roads.

Open and public spaces are essential breathing spaces for cities for both active and passive recreation. Nonetheless, usable open spaces often do not exist in many informal settlements. Open spaces in Hopley are fragmented, degraded and unusable patches of land that are left out by property development. The challenge of open spaces mirrors a study by Leby and Hashim (2010) in Malaysia, where residents did not value them. In the Malaysian study, open spaces were conflicting land uses for unintended uses such as garbage dumping (ibid). The failure of refuse collection in Hopley mirrors the challenges of Zimbabwean cities where open spaces are used for sand mining, brick moulding and illegal waste dumping.

A unique factor that affects the functional and partly the physical dimensions of liveability in informal settlements are jurisdictional issues. The City of Harare and the Ministry of Local Government and Public Works officials confirmed their joint and conflicting planning responsibility within Hopley. This perceived lack of a clear responsibility challenges the management of the settlement and the provision of services.

9.5.2 Nuances in the Liveability and Smartness of Informal Settlements

The study of Hopley shows that informal settlements generally perform badly in all four liveability variables. In response to this scenario, the Government of Zimbabwe and the City of Harare have prioritised informal settlement upgrading as a strategy towards improving liveability. The City of Harare and the Ministry of Local Government Public Works have plans towards settlement upgrading in the area and have produced a draft Harare South Consolidated Land Use Plan to guide the development of Hopley.

However, one could question whether informal settlement upgrading improves neighbourhood liveability. The dominant argument would support such efforts as they improve the physical environment. Informal settlement upgrading also improves functionality by providing health, education and shopping centres. However, informal settlements have different social structures and economies and planning decisions should be cognisant of these underlying dynamics. Findings from Hopley show that besides the poor liveability dimensions, the residents have a mixed feeling towards settlement upgrading. Residents support regularisation as a way to improve tenure security. However, tenants fear gentrification and eventual displacement. Moreover, some residents run illegal day-care centres, schools and *shebeens* and are anxious about the potential displacement by the regularisation. These findings thus present an area for future research on how informal settlement upgrading and improved liveability would reduce the negative impacts of displacement.

9.6 CONCLUSION

Studies on liveability have been conducted differently between neighbourhoods, cities and across countries because of the contestations in the attributes of the concept. Notwithstanding these variations, this study provides useful information on assessing the smartness of informal settlements. The resident perceptions of liveability are the critical indicators that guide the policy and programmes of neighbourhoods by the local and central governments. Policies and programmes towards upgrading informal settlements such as Hopley to smart neighbourhoods should prioritise functional dimensions such as health, educational, shopping centres and employment opportunities and physical dimensions such as maintenance of streets, refuse collection and management of open spaces. This study also shows that informal settlement upgrading efforts are often caught between conflicting interests of protecting the environment and the livelihood activities that directly depend on the environment such as brick moulding. Upgrading of the informal settlements improve some of the parameters of smart settlement; however, this should be done in a way that balances improving living conditions and guarding against gentrification and displacement of existing residents and their livelihoods.

The central and local governments of Zimbabwe have traditionally promoted liveable and smart cities based on the functional and physical dimensions. However, the current rapid urbanisation and urban poverty have compromised the ability of neighbourhoods to uphold these functions. Contemporary planning approaches should balance the attainment of liveability settlement as well as protect the interest of residents. While the liveability conditions of informal settlements can be measured with similar variables as formal neighbourhoods, the improvement of the conditions of informal settlements may not necessarily bring them to the level of planned settlements, since some informal settlement residents can be displaced while some can lose their livelihoods.

REFERENCES

Chirisa, I. and Muhomba, K. (2013) Constraints to managing urban and housing land in the context of poverty: A case of Epworth settlement in Zimbabwe. *The International Journal of Justice and Sustainability*, 18 (8), 950–964.

Cities Alliance (2006) *Cities alliance for cities without slums: Action plan for moving slum upgrading to scale*. Cities Alliance, Washington DC.

Dialogue on Shelter and Zimbabwe Homeless People's Federation (2019) *Hopley settlement profile and survey final report*, Harare.

Dialogue on Shelter, Zimbabwe Homeless People's Federation and Harare City Council (2014) *Harare slum profiles report*, Edition 2, Harare

Dodman, D., Brown, D., Francis, K., Hardboy, J., Johnson, C. and Satterthwaite, D. (2013) *Understanding the nature and scale of urban risk in low and middle-income countries and its implication for humanitarian preparedness, planning and responses. International Institute for Environment and Development*. Human Settlement Discussion Paper Series, Climate Change and Cities 4, London.

Dzvairo, C. (2019) The liveability and impact of informal settlement on surrounding environments. The case of Hopley Farm, Harare. BSc dissertation. Department of Architecture and Real Estate, University of Zimbabwe. Harare.

Hancke, G., Silva, B. and Hancke, G. (Jnr.) (2013) The role of advanced sensing in smart cities. *Sensors*, 13 (1), 393–425.

Harrison, P. and Rubin, M. (2018) Urban innovation: Theory and practice. In P. Harrison and M. Rubin (eds) *Urban innovations: Researching and documenting innovative responses to urban pressures*, pp. 2–35. Department of Planning, Monitoring and Evaluation, Republic of South Africa, Pretoria.

Heylen, K. (2006) *Liveability in social housing. Three case studies in Flanders*. Paper presented at the ENHR Conference 'Housing in an expanding Europe: Theory, policy, participation and implementation'. 2–5 July 2006, Ljubljana, Slovenia.

Jacobs, J. (1961) *The death and life of great American cities*. Random House, New York

Leby, J.L. and Hashim, A.H. (2010) Liveability dimensions and attributes: Their relative importance in the eyes of neighbourhood residents. *Journal of Construction in Developing Countries*, 15 (1), 67–91.

Lynch, K. (1998) *Good city form*. Cambridge, MA: MIT Press.

Mahabir, R., Crooks, A., Croituru, A. and Aqouris, P. (2016) The study of slums as social and physical constructs: The challenges and emerging research opportunities. *Regional Studies, Regional Science*, 3 (1), 399–419.

Makunde, G., Chirisa, I., Mazorodze, C., Matamanda, A. and Pfukwa, C. (2018) Local governance system and the urban service delivery in Zimbabwe. *Issues, Practices and Scope*, 3, 1–13.

Matamanda, A., Mafuku, S.H. and Mangara, F. (2020) Informal settlement upgrading strategies: The Zimbabwe experiences. In F.W., Leal, A. Azul, L. Brandii, P. Ozuyar, and T. Wall (eds) *Sustainable cities and communities*, pp. 316–327. Springer, Zurich.

Munyoro, G. and Nyaushamba, G. (2016) The impact of illegal settlement on economic development: A case study of Chirundu border town, mash west, Zimbabwe. *Researchjournali's Journal of Economics*, 4 (4), 1–11.

Myers, D. (1987) Community-relevant measurement of quality of life: A focus on local trends. *Urban Affairs Quarterly*, 23 (1), 108–125.

National Research Council (2002) *Community and quality of life: Data needs for informed decision making*. The National Academies Press, Washington, DC.

Ndoziya, A.T. (2015) Assessment of the impact of pit latrines on groundwater contamination in Hopley Settlement, Harare, Zimbabwe, Master's Thesis, University of Zimbabwe, Harare.

Omuta, G.E.D. (1988) The quality of urban life and the perception of liveability. A case study of neighbourhoods in Benin City, Nigeria. *Social Indicators Research*, 20 (4), 417–440.

Patel, D. (1988) Some issues of urbanization and development in Zimbabwe. *Journal of Social Development in Africa*, 3 (2), 17–31.

Sandoval, V., Hoberman, G. and Jerath, M. (2019) *Urban informality: Global and regional trends*. DRR Faculty Publications. 16. Florida International University, Miami, Florida.

Slum Almanac (2016) *Tracking improvement in the lives of slum dwellers*. UN-Habitat, Nairobi.

The Economist Intelligence Unit (2019) *The global liveability index 2019. A free overview*. The Economist, London.

Tibaijuka, A.K. (2005) *Report of the fact-finding mission to Zimbabwe to assess the scope and impact of operation Murambatsvina by the UN special envoy on human settlements issues in Zimbabwe*. UN-Habitat, Nairobi.

UN-Habitat (2003) *The challenge of slums—global report on human settlement*. United Nations Human Settlement Programme, London.

UN-Habitat (2008) *The state of the world cities reports 2008 09*. UN-Habitat, Nairobi.

UN-Habitat (2013) *State of the world's cities 2012/2013. Prosperity of cities*. UN-Habitat, Nairobi.

UN-Habitat (2016) *Urbanization and development: Emerging futures. World cities report, 2016*. UN-Habitat, Nairobi.

Valcarcel-Aquiar, B., Murias, P. and Rodriguez-Gonzalez, D. (2018) Sustainable Urban Liveability: A practical proposal based on a composite Indicator. *Sustainability*, 11 (86), 1–18.

Watson, V. (2013) African urban fantasies: Dream or nightmares. *Environment and Urbanisation*, 26 (1), 215–231.

Yin, R.K. (2009) *Case study research: Design and methods.* 4th Edition, Applied Social research Methods Series, Vol. 5. Sage Publications, London.

Zimbabwe National Statistics (2012) *2012 census report.* Government Printers, Harare.

Zimbabwe National Statistics (2017) *Inter-censual demographic survey 2017.* Zimbabwe National Statistics Agency, Government of Zimbabwe, Harare, Zimbabwe.

Section IV

Goals and Practices of Sustainable and Smart Spatial Planning

10 The Urban Laboratory
A Case of Data Mining and Management for the Successful Hosting of Smart Settlements in Zimbabwe

Innocent Chirisa, Valeria Muvavarirwa and Fungai N. Mukora

CONTENTS

10.1 Introduction 141
10.2 Literature Review 142
    10.2.1 Urban Laboratory Concept 142
    10.2.2 Urbanisation Data and Satellite Imagery 143
    10.2.3 Infocracy 143
    10.2.4 Data mining 144
    10.2.5 UrbLab Cases 144
10.3 Results and Lessons for Zimbabwe 146
10.4 Discussion 148
10.5 Conclusion and Practical Implications 149
References 150

10.1 INTRODUCTION

This chapter aims at highlighting the essentiality of the Urban Laboratory (UrbLab's) architecture and its functionality in dealing with urban development in Zimbabwe. This can only be achieved through programmatic data mining and management, which is made difficult because of the contextual phenomenon in energy and power challenges in the country to date. The chapter is grounded on the idea that rural and urban settlements in the country are rich but lack adequate data and information in making practical policies that positively impact development. Therefore, to make well-informed decisions, there is a need for an UrbLab that focuses on gathering and dissemination of data. It will be the centre at which information is gathered, verified and shared with people in various places, especially in this era of globalisation, making it easier to access relevant and adequate information that is essential for problem definition, policymaking and for setting SMART goals in dealing with the most essential challenges plaguing the country, such as infrastructure provision, maintenance and disposal and a centre for making sure services such as water, sewer and housing, are delivered to the people efficiently.

The UrbLab is best suited to be accommodated by the University of Zimbabwe because of its competitive advantage against other possible hosts. The literature in this chapter shows examples of where such an initiative has been attempted and the subsequent results, thereby providing a framework as to the parameters required in the project. The UrbLab provides a platform to deal with risks and disasters before they occur in various aspects. The success of the project hinges upon the social and political will of the people and the technology available. It presumes the presence of professionals in the built environment that are agile and responsive to information as critical in defining the events and processes in the management of human settlements.

In constructing this paper, document analysis was mainly used to come up with the research on UrbLabs. This comprised of various publications on urban laboratories and websites on the 22@Barcelona and the University College London. These gave information on the plans and outcomes of projects carried out. The websites also provide some sort of advertisement for encouraging other institutions to participate and join in sustainable settlement development. This provides lessons to other institutions as to the procedures to be taken and what to expect.

10.2 LITERATURE REVIEW

10.2.1 URBAN LABORATORY CONCEPT

An UrbLab has various definitions and aspects to it but the main theme is that it focuses on the development of a city or an urban area and tries to help in solving problems. An UrbLab makes it easier to better manage urban challenges that constantly undermine the efforts made by governments and policy makers (Dorstewitz, 2014). Urban labs are multi-disciplinary components that foster participation. They tend to facilitate the integration of data from various small components and combining it into large useable information that different departments can access. Therefore, UrbLabs are vital to the proper growth and development of cities.

UrbLabs avail information that is relevant to a specific project. It refers to the use of information to govern development, into fitting into the world vision expressed through the sustainable development goals (SDGs) (BuRalli et al., 2018). This creates a platform for gathering the data freely. Informed decisions require up to date and relevant data thereby making them reliable. They help in the development of sustainable settlements, economically, socially and environmentally. The UrbLab recognises that development in the form of construction and all other aspects does not occur in a vacuum but there are a variety of stakeholders and the effects of certain developments have a ripple effect on other areas (Singh and Parmar, 2019). For instance, a prohibition order can stall development. This has an effect on the financial side of the project, and thereby the owner of the project will be disadvantaged and leaving the area abandoned can turn it into a hub of all sorts of criminal activities, thus impacting negatively on the people living nearby.

10.2.2 Urbanisation Data and Satellite Imagery

Urbanisation has been a critical topic in issues to do with urban development. Urbanisation refers to the growth in the proportion of a country's population living in urban centres of a particular size (Brenner, 2013). The rapid growth of urban populations has made it difficult to gather adequate information relevant for development because data required for this task is inadequate, inconsistent and unreliable. The data inconsistencies have made it difficult to understand the full impact of urbanisation. Most human settlement data is compiled through censuses. Most governments use demographic data from censuses to plan for the future population, the males and females requiring jobs, the number of children who will require schooling and houses for the people, that are vital for national planning (Goswami et al., 2019). There have been technical advancements that have made it easier to come up with human settlement data. This has been through the use of German radar satellites that gathered over 180,000 images in a period of three years from 2010 to 2013 (Leinss, Round and Hajnsek, 2017). These images have been used to come up with the Global Urban Footprint from the digital terrain models. Two radar satellites cover the earth over 500 km and can record data during the night with no challenges even when it is cloudy. The satellites are also equipped to detect the smallest of objects from a greater altitude, such as chimneys and road signs.

Information gathered by satellites has provided valuable information to the World Bank, universities, researchers, governments and think tanks. This has helped in decision-making and policy analysis. It allows both quantitative and qualitative comparisons to be undertaken on settlement patterns at various levels from the local to the global scale (Esch et al., 2017). This is all in a bid to determine how urban development can impact on economic growth relating to employment, poverty reduction through various projects and even carbon emissions. This goes beyond urban development as adequate data on the growth of settlements helps in understanding and getting a picture on how to map the settlements and track their growth (Melchiorri et al., 2019).

South Africa's human settlement data is obtained by the Department of Human Settlements. The information regarding human settlement is mostly utilised by the housing department. The data is used to come up with housing policies as housing is one of the most challenging issues in most countries. The South African Housing Policy of 2015 puts emphasis on seven strategic thrusts (Lemanski, 2017). These include the need for stable housing environments and mini-maxing, that is in reference to minimising resources but obtaining maximum gain. There is also a drive to provide sustainable credit facilities to help people in obtaining accommodation (Taruvinga and Mooya, 2018). The need for coordination is also highlighted in the policy including the facilitation of prompt housing and service delivery.

10.2.3 Infocracy

Infocracy refers to the actions and procedures that are controlled by information. Relying on information enables people to act wisely, with an informed judgement (Rocha et al., 2018). Judgements of an informed nature offer reliability and that offers the contractor and owner an opportunity to take responsibility in the planning and outline of procedures. Information-guided development has a greater chance of success and acceptance as it gives adequate information for or against the proposed steps (Krellenberg et al., 2019). The infocracy also helps in fostering confidence as it has data from confirmed experts (Berlo, Wagner and Heenen, 2017).

10.2.4 Data Mining

Data mining is an interdisciplinary sub-area of computer science and statistics. Its main aim is to extract relevant data from a data set and transform it into information of a comprehensive structure for future use (Parra-Royon et al., 2018). The goal of data mining is to extract patterns and knowledge from vast data amounts. It is not exactly the extraction that has the main focus but also data processing that includes collection, extraction, storing, analysing and structuring (Eldén, 2019). Data mining is a phrase that has been used since the 1960s by statisticians and economists, referring to the bad part of analysing data without a preconceived hypothesis (Shmueli et al., 2017).

Since knowledge discovery is a step in data mining, there are a number of processes that are followed (Trivedi et al., 2017). These include the selection of knowledge, pre-processing of the data, transformation of the data, data mining and the evaluation or interpretation of the data (KS and Kamath, 2017). These stages guide the knowledge discovery process.

10.2.5 UrbLab Cases

The University College London (UCL) UrbLab (UCL UrbLab) is characterised as a cross-disciplinary centre that carries out research in London and internationally. It is a centre where investigations are carried out and propositions on viable routes are tested out and shared, in a bid to solve critical challenges in the country and globally (Ali, 2017; Harrison and Hoyler, 2018). The UrbLab is non-discriminatory as it looks at a range of issues including physical, social and technological and issues to do with the built environment. Its multi-disciplinary approach ensures the participation of various groups of people, such as business people and non-academics in urban research and dealing with urbanism (Acuto, Parnell and Seto, 2018). The UCL UrbLab was established in 2005 and has been a part of the Erasmus Mundus Urban Laboratory International network that comprises eight universities across the globe (Couling, 2015; Faggella et al., 2012). It seeks to strengthen and spread competence and influence in the dominions of the built environment.

The Barcelona UrbLab (22@Barcelona) was established to fulfil the 22@Barcelona goal, which entails the creation of an innovative city that encourages innovation in the business fabric (Pairolero, 2013). The whole city was to be used as an UrbLab and this made available the city as a ground for companies and businesses to come up with projects and test the infrastructure and services for future capacities in a live environment (Abellán, Sequera and Janoschka, 2012). In initiating the project of turning the city into an UrbLab, there have been a variety of initiatives that have been undertaken. Since the conception of the project in 2008, there has been the

The Urban Laboratory

establishment of public outdoor street lighting units. These have been equipped with eco-digital facilities like LED technology. This has improved the visibility of the streets at night, thereby enhancing security in those areas (Gascó, 2016).

There has been the creation of metres to read gas, electricity and water telematically. This has been initiated in about 150 housing units in three separate buildings. The metres test the utility used up by each household and also collectively as a building (Gavaldà and Ribera-Fumaz, 2012). This has helped to reduce the trips that are taken in going to read the metres at different points and has also helped in reducing the manpower needed in such excursions. The information helps in coming up with adequate data referring to the amount of power used by each household and hence coming up with an estimate as to how much gas or electricity or water will be needed to use resources efficiently. It also helps in clearing up the areas where leakage in terms of electricity and water use are concerned, making it a vital tool in service provision (Droege, 2011).

Due to the technological advancements, Barcelona has adopted the use of electrically charged vehicles (Bakici, Almirall and Wareham, 2013). This has resulted in the establishment of charging points for cars and also the production of solar-powered motorbikes. This is some form of clean energy as the vehicles do not use fuel but electricity, thus reducing the carbon footprint and conserving oil. There has also been the adoption of solar-powered vehicles for the police (Soret, Guevara and Baldasano, 2014). These technological advancements have also resulted in the adoption of traffic lights fitted for the blind. These enable them to press a button that encourages that they cross intersections safely. They are retrofitted with volume adjustments that reduce the constant and persistent noise of traffic lights. The lights are also equipped with cameras that help in traffic control and monitor traffic flow and any traffic misdemeanours. This has also extended to the installing of fibre optics in homes to increase access to the internet and encourage globalisation. There has also been a project for the creation of bicycle lanes. These have helped in the reduction of the need for private vehicles, which have been causing a lot of congestion, thereby, improving traffic circulation.

By 2030, Africa and countries from other continents seek to have achieved sustainable human settlements (du Plessis, 2018). These settlements cater for human needs in a variety of ways, making them sustainable to the people. The vision encourages walkability as a means to attain sustainability by creating residential units that are closer to work areas, making it easy for people to walk to and from work. It also supports the creation of affordable housing to enable the lower and non-existent income groups an opportunity to afford housing. This creates a market that is able to accommodate all income groups as a means of attaining a goal to provide adequate housing to all groups of people (Melchiorri et al., 2018). Adequate housing according to the United Nations (UN) refers to housing that has a secure tenure because a secure tenure enables the people to feel safe in an area and safe to develop the area and homes further. The housing is also serviced in terms of providing electricity, water and sewer, roads and bridges. This is because services enable an area to be liveable as infrastructure is available and is the backbone of every settlement. Adequate housing also requires that there be other services, such as schools, clinics and cultural centres in the area that allow people to be in the same vicinity and are not required to move out of the residential area in search of schools and health facilities (Barenstein, 2016).

Adequacy of housing also looks at the habitability of a structure. It looks at whether the structure can protect the residents from different weather elements, be it the heat, harsh cold or even wind and storms. Some housing may be cheap but unliveable, for instance, in Epworth, an informal settlement, some of the structures do not protect the residents and their resources from rain. It also looks at making housing accessible to all people, including disadvantaged groups, such as women, youths and people with disabilities. These groups of people face poverty without access to decent or adequate housing. Adequate housing also looks at the location of housing (Lauster, 2019) so that people can access functional services, such as educational and health services in their neighbourhoods. Cultural adequacy is another aspect that is taken into consideration in describing adequate housing (Gan et al., 2019). This looks at the respect of cultural differences. Singapore's high-rise buildings have no residential occupancy on the ground floor, which is then reserved for various activities including; community halls and cultural centres, where people from different races and cultures can interact. This, therefore, promotes unity and participation.

10.3 RESULTS AND LESSONS FOR ZIMBABWE

The establishment of UrbLabs is vital in development. They encourage properly informed decision making with regard to development. The lab provides the people with the information that they need to make better decisions. The lab is not just useful to individual people who want to build their own structures but is also used by the city and even the whole country. This reduces the chances of double allocations and prohibition orders being issued because the proper and allotted use will be clearly known. This reduces chances of cost and time wasting. UrbLab development creates an environment ripe for development and investors that are willing to participate because of the information available to the people. Investors feel more comfortable investing in projects when there is adequate information on different interests. They do not enter into business blindly because they expect returns. The information is then crosschecked against projections to see if they are favourable for investment and development.

The development of UrbLabs is an answer to many problems plaguing many cities in the world. The 22@Plan provided an answer to various challenges. The plan constituted the rezoning of Poblenou, Sant Marti district in Barcelona, Spain and saw the development of economic activities on about 4 million m^2 and approximately 40,000 housing units. The housing facilities were also situated in and around the housing area as a means to facilitate sustainability through creation of a compact city. This encourages walking to work and home and thus limit the carbon footprint on Earth as an answer to the climate change and global warming questions. This was adopted in Barcelona by making cycling attractive compared to driving to different destinations. UrbLabs necessitate the flow of information to where it is needed. They provide insight as where to develop and which authority to seek for carrying out development. This reduces the chances of double allocation of the same piece of land for different uses, causing confusion and it also provides information as to who owns the land and reduces disputes. The lab is vital to development as the epicentre of developmental information. The lab is established to carry and deliver information, not just local information but regional and global information as well. This gives people information on prospective projects and lessons from differently executed ideas. This also gives educational institutions materials to use for learning and also an option to come up with how they can go about employing better methods that are contextually fitting to a country like Zimbabwe.

The concept of the UrbLab is hinged upon a certain paradigm towards sustainable informed development. Therefore, there is a need to educate people to ascertain a change of mind. People are of different philosophies in life and deem different things necessary and important. In order for the UrbLab to be a success, there is need for a collective understanding and pooling of resources, particularly information sharing. Changing people's mindsets and perceptions is not an easy feat and, in this case, there is need to persuade the whole city of the concept and extend to the whole country. Since UrbLabs exist for the betterment of people's quality of life, collective participation is required. Mass media in the form of national radio and television can be used to inform the people on the benefits of providing information for sustainable development.

Development does not take place in a vacuum; therefore, there is a clear need for multi-pronged approaches. Focusing on one sector may diminish the impact of proper development because the other sectors may remain stagnant. The UrbLab operates at the centre as a lynch pin of all the other stakeholders for development. The lab connects learning institutions such as universities, industries, manufacturers, the government and its citizens. This combines the efforts of all participants and creates linkages and networks that facilitate development. This increases the chances of success, with every participant aiming for the best possible outcome. The government may task students to come up with designs suitable for sustainable development and may then task different manufacturers to develop samples to test. The city council becomes the first customer of all the trial products, inspiring confidence in the citizens and eliminates the chances of potential losses to private developers and consumers if the product is not well received. The city acts as a confidence booster and allows the citizens to use the evidence from the city as a demonstration.

Improvement in infrastructure and other facilities is not an overnight phenomenon and it requires strategising and planning to attain success. Barcelona's innovation began during the preparation of the 2004 Olympics. This saw the need to revitalise and create networks to accommodate the temporary population swell. The projects involved the transformation of brown fields, such as in the Forum area, with transformations taking place through the improvement of technology without the reallocation of other services, such as water and sewer treatment plants. This shows that transformations do not necessarily require a complete destruction and the construction of services and infrastructure but brown fields can be transformed in situ without any displacement. This helps to reduce costs and saves land for other uses. However, Zimbabwe faces difficulties due to the poor technical expertise available to the manufacturers and contractors. Most projects in the country are left untouched due to fear of undertaking costly endeavours that will cripple the country. There is, however, a need to learn from other countries like Spain with cities, such as Barcelona, that show that development does not mean improvement of green fields but can also be improvement of already existing environments.

10.4 DISCUSSION

UrbLabs are functional facilities that cities need to decide on the course of action that governments and authorities do. They help in reducing time spent in the gathering of various information needed in decision-making (Voytenko et al., 2016). The data will be readily available under one roof, making it convenient for stakeholders to access and use it. A few crosschecking instances may be necessary to verify the information, but it will require less time and energy than starting an investigation afresh. The value of UrbLabs is of significant proportions as they facilitate the growth and development of urban areas and in extension, the country. The lab provides intel to people in need of it for various purposes. The lab also helps in solving issues, such as crime. This has been studied in Chicago as they use an evidence-based approach to problem solving.

The whole city of Barcelona has been turned into an UrbLab. This has been in the hope of rejuvenating the city. City revitalisation gives the country an opportunity to grow and further develop. This occurred through planning techniques, such as rezoning, subdivisions and consolidations. The Poblenou industrial land in Sant Marti, Barcelona was dilapidated and no longer producing revenue. So, it was converted into an area encouraging growth and expansion (Lee and Hwang, 2018). The district growth includes construction of housing and centres for commerce. A collaboration with small to medium enterprises initiates the trust of local community business people who participate in the city development. The conversion of the whole city into an UrbLab increases the chances of transparency and accessibility of information since municipal documents are public records that anyone has the right to access.

The establishment of an UrbLab at the University of Zimbabwe has more advantages because of the university's access to people from various disciplines, who have the opportunity to add more knowledge on the subject and have great impacts on the UrbLab. The university has the capacity to influence a lot more people and encourage participation, thereby encouraging inter-dependence amongst the people, creating a web of ideas directing the development of the cities. The university's strategic plan ensures that the students also participate in the collection and analysis of data. An UrbLab is a crucial asset to have as it gives the decision makers a baseline to problem identification and problem solving. It helps in identifying what has been done to combat a particular problem and helps to assess whether the initiative has helped in the long run. The challenges that are faced in Zimbabwe, such as water shortages, burst sewer pipes, litter and electrical cuts all stem from poor management that is mostly derived from poor prioritisation of problems. The challenge of water shortages is mostly answered by increasing water tariffs and even denying some households access to water. This is usually the wrong answer. The same is putting rent controls in a time when there is a high demand for housing.

These problems can only be sufficiently solved if and when the authorities realise the depth of the problem and its roots. For instance, water shortages maybe because Harare's water supply extends to Norton and other new areas sprouting up and so there is a need for new water treatment plants and more sources of water. The same can be attributed to the housing challenges of high rents. This may be due to an increase in the cost of buying or constructing a house, which in turn leads people to find alternative sources of housing that also hike rentals. This makes the living situation in the country unbearable. Therefore, an UrbLab is essential in Zimbabwe to get adequate access to information before implementation of programmes and projects to increase the chances of a positive result.

10.5 CONCLUSION AND PRACTICAL IMPLICATIONS

The chapter sought to justify the need for data mining and management using UrbLabs in different countries, including Zimbabwe. The UrbLabs are responsible for the collection, compilation, analysis and distribution of data. It is vital to make informed decisions that enable different developers and other stakeholders to make decisions necessary for development. The UrbLab encourages the development of sustainable settlements through data mining and management that provide information to the people easily (Chronéer, Ståhlbröst and Habibipour, 2018). However, this is hindered in this day and age by the diminishing value of money. Zimbabwe is facing huge financial challenges that hinder the capacity of financial institutions and the people from lending and borrowing money. The banks do not have the capacity to lend money because of bad debts. The residents find it hard to borrow substantial loans as they do not have the capacity to repay. Some people end up borrowing money from loan sharks who tend to demand exorbitant interests and destroy the people and their livelihoods. Without money for planning processes, the people cannot afford to make any long-term plans that require huge capital inputs, such as home building.

Various issues have arisen from the chapter. It shows that sometimes it is difficult to establish UrbLabs. This makes it convenient to use the lab for information and reduces many unnecessary trips to various local authorities with several bureaucratic hoops to go through, that at times stifle development because they are too rigid. This saves time and money that are vital resources (Ciaffi and Saporito, 2017). The centralisation of issues makes it difficult to get relevant information, especially taking into account the cultural dynamics of organisations. Developing UrbLabs takes a lot of effort and financial input, which is hard for developing countries to do. This chapter also shows the importance of establishing UrbLabs because of their vitality in development.

The lab helps the country and its various organs to make informed decisions. Informed decisions give a higher chance of success to projects. This reduces the chances of some projects being halted in the middle of implementation. The lab also assesses the reception of the project from past experiences Therefore, Zimbabwe needs to come up with a framework that allows the University of Zimbabwe to communicate with various sectors through multi-disciplinary participation in order to create a functional UrbLab that provides data to anyone.

It can be concluded that data mining and management is vital for development because it provides information regarding to future plans and any other issues that may affect development. The development of UrbLabs is important but they can be effectual with social and political will, accepting the initiative. It is also vital to note that in order for data collection, coding and distribution to be possible in a single space, there is need for technological expertise to be present when adopting it, because the collection and dissemination of data requires better equipped platforms at retrieval. The establishment of an UrbLab is a step towards embracing globalisation as a way forward in policy-making, planning and adoption of ideas. Accepting that not one entity can know everything helps to accept help and insight from various sources, all in the interest of the common good. The UrbLab stores information at various levels, from a small community to a global extent, thereby giving people different perspectives and with them new ways to tackle issues with experience from those who have already tried it (Scholl et al., 2018). An UrbLab helps to reduce stress and strain on the policy makers, investors and various stakeholders.

REFERENCES

Abellán, J., Sequera, J. and Janoschka, M., 2012. Occupying the Hotel Madrid: A Laboratory for Urban Resistance. *Social Movement Studies*, *11*(3–4), Pp. 320–326.

Acuto, M., Parnell, S. and Seto, K.C., 2018. Building a Global Urban Science. *Nature Sustainability*, *1*(1), P. 2.

Ali, A., 2017. Identifying Urban Laboratory as a New Method for Tackling Urban Development. In *1st International Conference on Towards a Better Quality of Life*. https://ssrn.com/abstract=3163432 or doi: 10.2139/ssrn.3163432

Bakici, T., Almirall, E. and Wareham, J., 2013. A Smart City Initiative: The Case of Barcelona. *Journal of The Knowledge Economy*, *4*(2), Pp. 135–148.

Barenstein, J.E., 2016. The Right to Adequate Housing in Post-Disaster Situations: The Case of Relocated Communities in Tamil Nadu, India. Chapter 9 in Daly, P. and Freener, M.R. (eds) *Rebuilding Asia Following Natural Disasters: Approaches to Reconstruction in the Asia-Pacific Region*, Pp. 236–260. Cambridge University Press, Cambridge.

Berlo, K., Wagner, O. and Heenen, M., 2017. The Incumbents' Conservation Strategies in The German Energy Regime as An Impediment to Re-Municipalization—An Analysis Guided by The Multi-Level Perspective. *Sustainability*, *9*(1), P. 53.

Brenner, N., 2013. Theses on Urbanisation. *Public Culture*, *25*(1 (69)), Pp. 85–114.

Buralli, R.J., Canelas, T., Carvalho, L., Duim, E., Itagyba, R.F., Fonseca, M., Oliver, S.L. and Clemente, N.S., 2018. Moving Towards the Sustainable Development Goals: The Unleash Innovation Lab Experience. *Ambiente & Sociedade*, *21*. doi: 10.1590/1809-4422asoc17Ex0001vu18L1TD

Chronéer, D., Ståhlbröst, A. and Habibipour, A., 2018. Towards A Unified Definition of Urban Living Labs. In *The ISPIM Innovation Conference—Innovation, The Name of The Game*, Stockholm, Sweden on 17–20 June 2018.

Ciaffi, D. and Saporito, E., 2017. Shared Administration for Smart Cities. In Eleonora, Riva Sanseverino, Riva Sanseverino, Raffaella, and Vaccaro, Valentina (eds) *Smart Cities Atlas Western and Eastern Intelligent Communities*, Pp. 243–248. Springer, International Publishing, Cham, Switzerland.

Couling, N., 2015. Constructing Cultural Territory. In *Erasmus Mundus Symposium* (No. CONF). University College, London, September 16–19, 2015. UCL Urban Laboratory, London. Laboratories: LABA.

Dorstewitz, P., 2014. Planning and Experimental Knowledge Production: Z Eche Z Ollverein as an Urban Laboratory. *International Journal of Urban and Regional Research*, *38*(2), Pp. 431–449.

Droege, P., 2011. *Urban Energy Transition: From Fossil Fuels to Renewable Power*. Elsevier, Amsterdam, The Netherlands.

Du Plessis, A., 2018. The Judiciary's Role in Shaping Urban Space in South Africa As Per the Sustainable Development Goals. *South African Journal of Environmental Law and Policy*, 24(1), Pp. 5–43.

Eldén, L., 2019. *Matrix Methods in Data Mining and Pattern Recognition* (Vol. 15). Society for Industrial and Applied Mathematics. Elsevier, London.

Esch, T., Heldens, W., Hirner, A., Keil, M., Marconcini, M., Roth, A., Zeidler, J., Dech, S. and Strano, E., 2017. Breaking New Ground in Mapping Human Settlements from Space—The Global Urban Footprint. *ISPRS Journal of Photogrammetry and Remote Sensing*, 134, Pp. 30–42.

Faggella, M., Monti, G., Braga, F., Gigliotti, R., Capelli, M., Spacone, E., Laterza, M., Triantafillou, T., Varum, H., Safi, M.D. and Subedi, J., 2012, September. EU-NICE, Eurasian University Network for International Cooperation in Earthquakes. In *15WCEE 15th World Conference of Earthquake Engineering, Lisboa, Portugal*, 24–28 September 2012.

Gan, X., Zuo, J., Wen, T. and She, Y., 2019. Exploring the Adequacy of Massive Constructed Public Housing in China. *Sustainability*, 11(7), P. 1949.

Gascó, M., 2016, January. What Makes a City Smart? Lessons from Barcelona. In *2016 49th Hawaii International Conference on System Sciences (HICSS)*, Pp. 2983–2989. IEEE, Washington DC.

Gavaldà, J. and Ribera-Fumaz, R., 2012. Barcelona 5.0: From Knowledge to Smartness? *IN3 Working Paper Series*.

Goswami, D., Tripathi, S.B., Jain, S., Pathak, S. and Seth, A., 2019. Towards Building A District Development Model for India Using Census Data. *Proceedings of the 2nd ACM SIGCAS Conference on Computing and Sustainable Societies*, Pp. 259–271. doi: 10.1145/3314344.3332491.

Harrison, J. and Hoyler, M. Eds., 2018. *Doing Global Urban Research*. Sage, Loughborough.

Krellenberg, K., Bergsträßer, H., Bykova, D., Kress, N. and Tyndall, K., 2019. Urban Sustainability Strategies Guided by The SDGs—A Tale of Four Cities. *Sustainability*, 11(4), P. 1116.

KS, D. and Kamath, A., 2017. *Survey on Techniques of Data Mining and Its Applications*. International Journal of Engineering Research & Management Technology, Muradpur Urf Sholda, India

Lauster, N., 2019. What's Livable? Comparing Concepts and Metrics for Housing and Livability. In *A Research Agenda for Housing*. Edward Elgar Publishing, Cheltenham.

Lee, I. and Hwang, S., 2018. Urban Entertainment Center (UEC) As A Redevelopment Strategy for Large-Scale Post-Industrial Sites in Seoul: Between Public Policy and Privatization of Planning. *Sustainability*, 10(10), P. 3535.

Leinss, S., Round, V. and Hajnsek, I., 2017, July. Single Pass InSAR Missions for Monitoring Hazardous Surging Glaciers. In *2017 IEEE International Geoscience and Remote Sensing Symposium (IGARSS)*, Pp. 934–937. IEEE, Fort Worth, Texas.

Lemanski, C.L., 2017. Citizens in the Middle Class: The Interstitial Policy Spaces of South Africa's Housing Gap. *Geoforum*, 79, Pp. 101–110. Elsevier.

Melchiorri, M., Florczyk, A., Freire, S., Schiavina, M., Pesaresi, M. and Kemper, T., 2018. Unveiling 25 Years of Planetary Urbanisation with Remote Sensing: Perspectives from the Global Human Settlement Layer. *Remote Sensing*, 10(5), P. 768

Melchiorri, M., Pesaresi, M., Florczyk, A.J., Corbane, C. and Kemper, T., 2019. Principles and Applications of the Global Human Settlement Layer as Baseline for the Land Use Efficiency Indicator—SDG 11.3.1. *ISPRS International Journal of Geo-Information*, 8(2), P. 96.

Pairolero, J., 2013. Barcelona, The New Districts: 1992 Olympics, 2004 Forum of Cultures, & 22@ BCN.

Parra-Royon, M., Atemezing, G. and Benítez, J.M., 2018. *Data Mining Definition Services in Cloud Computing with Linked Data*. ArXiv Preprint ArXiv, New York.

Rocha, V.T., Brandli, L.L., Kalil, R.M. and Tiepo, C., 2018. The Urban Planning Guided by Indicators and Best Practices: Three Case Studies in the South of Brazil. In Azeiteiro, U. M., Akerman, M., Leal Filho, W., Setti, A.F.F and Brandli, L.L. (eds) *Lifelong Learning and Education in Healthy and Sustainable Cities,* Pp. 87–101. Springer, Cham, Switzerland.

Scholl, C., De Kraker, J., Hoeflehner, T., Wlasak, P., Drage, T. and Eriksen, M.A., 2018. Transitioning Urban Experiments: Reflections on Doing Action Research with Urban Labs. *GAIA-Ecological Perspectives for Science and Society,* 27(1), Pp. 78–84.

Shmueli, G., Bruce, P.C., Yahav, I., Patel, N.R. and Lichtendahl Jr, K.C., 2017. *Data Mining for Business Analytics: Concepts, Techniques and Applications in R.* John Wiley & Sons, Hoboken, NJ.

Singh, B. and Parmar, M., 2019. *Smart City in India: Urban Laboratory, Paradigm or Trajectory?* Routledge India, New Delhi.

Soret, A., Guevara, M. and Baldasano, J.M., 2014. The Potential Impacts of Electric Vehicles on Air Quality in The Urban Areas of Barcelona and Madrid (Spain). *Atmospheric Environment,* 99, Pp. 51–63.

Taruvinga, B.G. and Mooya, M.M., 2018. Neo-Liberalism in Low-Income Housing Policy—Problem or Panacea? *Development Southern Africa,* 35(1), Pp. 126–140.

Trivedi, S.K., Dey, S., Kumar, A. and Panda, T.K. Eds., 2017. *Handbook of Research on Advanced Data Mining Techniques and Applications for Business Intelligence.* IGI Global, Pennsylvania.

Voytenko, Y., McCormick, K., Evans, J. and Schliwa, G., 2016. Urban Living Labs for Sustainability and Low Carbon Cities in Europe: Towards A Research Agenda. *Journal of Cleaner Production,* 123, Pp. 45–54.

11 Opportunities and Constraints of Solar Harvesting as a Sustainable and Resilience Strategy in Zimbabwe
The Case of Nyangani Renewable Energy, Mutoko District

Enock G. Mukwekwe, Leonard Chitongo and Godwin K. Zingi

CONTENTS

11.1 Introduction	154
11.2 Methodology	155
11.2.1 The Study Area	155
11.3 Conceptual Framework	156
11.3.1 The Smart City Concept	156
11.4 Literature Review	156
11.4.1 Evolution of the Smart City Concept	156
11.4.2 Smart Energy Concept	157
11.4.3 Solar Energy Efficiency in Comparison with Other Methods	157
11.4.4 Global, Regional and Local Overviews of Solar Harvesting	158
11.5 Results and Discussion	158
11.5.1 The Utility of Solar Electricity as a Resilience Strategy in the Mutoko District	158
11.5.2 The Efficiency of Solar-Generated Energy Relative to Conventional Methods	161

11.5.3 Opportunities Faced When Providing Electricity in the Mutoko District.. 162
11.5.4 Constraints Faced When Providing Electricity in the Mutoko District.. 163
11.6 Conclusion and Recommendations.. 165
References... 166

11.1 INTRODUCTION

Climate change among other issues including sustainability challenges have advocated for the revisit to the marginalised solar technology around the globe, which is more pronounced in the developing world (Zhang and He, 2013). It is from the 2015 Paris Climate Agreement that there is an invigoration of the need for renewable energy. According to UNDP (2018), having access to energy is one of the most important issues affecting developing countries. Availability and access to energy in its different forms is important in sustaining diverse economies and enabling change in the welfare of people (Ganda and Ngwakwe, 2014). However, in the event that there is a shortage of energy or power supply, there is stagnation of the economy and destruction of people's livelihoods (Dinçer and Meral, 2010). It is through the need to have a reduction in the amount of carbon released in the atmosphere that there is a global call for the reduction of the impact of climate change and this can be achieved through diversifying the sources of energy and its value chain (Sheu et al., 2012). Therefore, it is imperative to increase access to clean energy in the remote and rural areas. It is in this context that there is a need to adopt solar energy in the context of developing countries.

The Agenda for Sustainable Development, especially with the focus on SDG7, calls for universal access to energy, which is regarded as clean and efficient for the benefit of the members of society (FAO and WWC, 2015). It is through the SDG7 that there is potential to foster economic growth, reduce poverty levels, empower women and improve the conditions of living. In particular, target 7.1 promises that there is need for energy which is affordable, reliable and modern by 2030, target 7.2 indicates the need for increase in the share of renewable energy by 2030 and target 7.3 indicates that there should be doubling of the global rate of the level of efficiency in terms of energy by 2030 based on the previously stated targets (FAO and WWC, 2015).

Since the turn of the new millennium, Zimbabwe has had a myriad of economic challenges (Chitongo et al., 2020). Consequently, both domestic and foreign debt grew, swallowing up urgently needed resources for key social amenities (Dziva, 2014). This affected essential service provisions including electricity. Zimbabwe is one of the countries with high access to radiation from the sun yet the country contributes less than 1% to the global solar energy (Mutua and Kimuyu, 2015). Thus, there is need for energy that is renewable and cost effective and this becomes one of the solutions, which can be used to address the energy challenges in Zimbabwe. There are high costs that are associated with electricity and they make it inaccessible and as such electricity is mainly available in the urban areas implying that the rural areas and hard-to-reach areas are marginalised (Karekezi and Kithyoma, 2003).

This chapter examines the efficacy of adopting renewable energy as an urban resilience strategy. The study adopts a case study approach in which Mutoko district was selected. Specifically, this chapter assesses the opportunities and challenges that are associated with solar harvesting. Objectively, the study identifies the utility of solar-powered electricity as a resilience strategy in Mutoko district. Furthermore, it examines the effectiveness of solar-powered energy relative to conventional methods. The researcher concludes by assessing the opportunities and constraints faced when providing electricity to the Mutoko district. The sustainability and smart growth of Mutoko requires environmentally reliable and affordable sources of energy. Therefore, this research is vital for local economic development since energy efficient systems are a prerequisite for socio-economic transformation.

11.2 METHODOLOGY

A mixed-method approach was utilised to gather information where both quantitative and qualitative data were used. The study is a case study of solar harvesting in Mutoko district. Data were obtained using key informant interviews, questionnaires, field observations and documentary review. The sampling unit was the household. Out of a population of 1332 stakeholders, a sample of 100 households was randomly selected. The sample was determined by the rule of thumb that states that a sample should be approximately 10% of the total population (Creswell, 2014). Key informant respondents were purposively selected which involved officials from the Mutoko Rural District Council and Nyangani Renewable Energy (NRE). Quantitative data were presented in the form of figures and tables whilst a thematic content analysis was used to present and analyse qualitative data.

11.2.1 THE STUDY AREA

Mutoko growth point was chosen as a case study for this research. Mutoko growth point is in Mutoko District, Mashonaland East Province and it is located about 143 km to the northeast of Harare (ZimStat, 2012). The growth point consists of 29 wards each with an elected councillor. According to Bhatasara (2015), the larger component of Mutoko lies in agro-ecological region two receiving between 450–600 mm annual rain and can frequently be affected by droughts. Most of the rainfall is received between November–March and the other months constitute the dry season. In terms of temperature, summer daily temperature ranges from 19 to 32°C, winter season temperatures are slightly lower than 14°C and is susceptible to regular seasonal droughts. Precipitation occurs principally between November and March followed by a seven-month dry season (Bhatasara, 2015). The research covers Mutoko growth point's Chinzanga medium and low-density residential areas. Chinzanga is in ward 20 and about 15 km to the northwest of the Mutoko central business district (Figure 11.1).

The vegetation in Mutoko is mainly sparse tree woodland comprising of *Brachystegia boehmii* (mupfuti) and *Julbernardia globiflora* (munhondo). *B. boehmii* becomes dominant on the less well-drained soils and *J. globiflora* appears on the deeper well-drained soil profiles (Bhatasara, 2015). Dominant grasses are *Hyparrhenia* species with *Heteropogon* and *Loudetia* species.

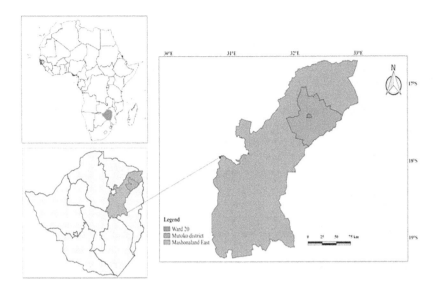

FIGURE 11.1 Location of the study area in Mutoko district, Zimbabwe.

Source: Mafumbabete et al. (2019)

11.3 CONCEPTUAL FRAMEWORK

11.3.1 THE SMART CITY CONCEPT

Lombardi et al. (2012) note that a smart city is a status given to a city that fuses information and communication technologies (ICTs) to improve the quality and execution of urban administrations; for example, energy, transportation and utilities so as to attain sustainability. The proponents of the smart city concept date back as early as 1994. A smart city interfaces human capital, social capital and ICT framework so as to address open issues, to accomplish a reasonable turn of events and increase the quality of life of its residents. A smart city is a worldwide pattern of urban methodologies planned for recuperating the nature of occupants living in urban territories and utilising development and high innovations to take care of issues created by high-population expansion (OECD, 2018). It assists with the fathoming issues of urbanisation, particularly contamination of natural resources, land utilisation, endless suburbia, transport blockage, energy needs and troubles in getting to open administrations and contains a differentiated arrangement of open activities: structures fabricating better transportation frameworks to support imaginative advancement and information for planning energy-sparing strategies (OECD, 2018).

11.4 LITERATURE REVIEW

11.4.1 EVOLUTION OF THE SMART CITY CONCEPT

Analysis of international literature concerned with the smart city suggests that the present concept is the result of three trends of urban research, that of the digital city, the green city and the knowledge city (Neirotti et al., 2014). ICT, knowledge and environment are seen as inextricably linked with the implementation of more innovative cities.

From a historical perspective, the first smart city is arguably Los Angeles, which in the 1970s was a frontrunner in the use of big data. A policy report found by Vallianatos (2015) in 1974 shows that the development of Los Angeles was being shaped by computer data. Back then, an administrative division called Community Analysis Bureau was employing the state-of-the-art computer technologies to process and organise huge amounts of data on different themes such as housing, traffic, crime and poverty. The overarching aim of this activity was to inform policy-making and urban planning. It is because of such initiatives that Los Angeles is often considered the first example of a computer city (Vallianatos, 2015).

Not long after the experimental urbanism of Los Angeles, in the late 1980s, is the case of Singapore which, at that time, was being advertised by the local government as an 'Intelligent Island' (Batty, 2012). In practice, the city was being rewired by means of then hyper-modern fibre optic cables to create a data network. In less than a decade, the Singaporean urban-digital network was already producing and circulating large quantities of data following the same dynamics and rationale through which contemporary smart-city projects operate. The ICT urban infrastructure of Singapore was being used, for instance, to decentralise business activities by allowing some categories of workers to accomplish their tasks from home, to increase communication between citizens and the government via online portals and to implement systems of automatic payment through smart cards and scanners (Arun and Teng Yap, 2000). By the time Singapore's plans of techno-urban renewal had been put into practice, the term smart city was beginning to be used in a growing number of urban agendas to signify the modernisation of the infrastructure of the city through the integration of ICT (Vanolo, 2014).

11.4.2 SMART ENERGY CONCEPT

In the face of massive developments in most cities, there is a now an orientation towards sustainable energy provision in a bid to attain sustainability. The present carbon-based energy system is experiencing a significant change driven by the increased worries over the life span and security of energy supply, just as energy-related emissions of carbon dioxide and air contaminations (World Bank & International Energy Agency, 2015). The developmental pattern of this progress is towards a smart energy system of things to come that is portrayed by a broad organisation of clean energy advances and savvy energy management innovations.

11.4.3 SOLAR ENERGY EFFICIENCY IN COMPARISON WITH OTHER METHODS

Most solar panels on the market have an efficiency ranging 14 to 16%, although 22.5% is the best range of efficiency one can find on the market commercially. Current innovations have increased the range of solar panel efficiency up to 44.5%, which researchers have achieved with the aid of stacking a couple of substances that take in diverse components of the sun spectrum (Aljazeera, 2015). Efficiency is an imperative issue for non-renewable energy assets since they include fuel as an input cost. On the other hand, renewable resources do not have such inputs cost thus making them more efficient in this respect. Efficiency is the extent of how much of a resource's energy potential is converted into electric energy. Solar panels have an estimated 80% efficiency. In comparison, car engines convert 20% of the energy in gasoline into motion, with the rest being waste heat and greenhouse gases. Coal attains between 33% to 40% efficiency, with the rest being simply wasted heat and greenhouse (IEA, 2008).

11.4.4 GLOBAL, REGIONAL AND LOCAL OVERVIEWS OF SOLAR HARVESTING

Some nations like Germany, Italy, Japan, Spain and the USA were among the forerunners in incentivising support for the development of their solar energy market (Shackleton et al., 2007). Due to the level of incentives that were being offered, there was an increase in worldwide growth for the solar market which rose from 4 GW in 2004 to 70 GW in 2011 which saw about 58% for the period running 2006 to 2011 (Zhang and He, 2013). As such, solar energy was accountable for the high generation of electricity in the whole of the European Union as compared to other energy generation technologies. All over the world, solar energy is amongst the highly untapped climate smart energies (Pavlović et al., 2006), although there was a 29.6% increase in solar energy generation in the year 2016. However, Africa has one of the high contributions to the global population yet its contribution to global energy is low and this can be attributed to low investment opportunities.

11.5 RESULTS AND DISCUSSION

11.5.1 THE UTILITY OF SOLAR ELECTRICITY AS A RESILIENCE STRATEGY IN THE MUTOKO DISTRICT

There are several benefits derived from the adoption of solar energy (Figure 11.2). A majority of the respondents indicated that the solar project generated employment opportunities for many people.

The following excerpts were indicated by the key informants:

MUTOKO RURAL DISTRICT COUNCIL (RDC) OFFICIAL: Through the launch of the solar system project we have managed to create local employment for the youth during the installation phase and now during the process of data collection. When the project was under the foreign currency regime, the youths were paid in forex and there have been changes in the livelihoods of the people of Mutoko. The NRE Company employed 300 local youths in the installation phase for 6 months. They had an average salary of 600 United States dollars per month in addition to this salary they were equipped with valuable skills in solar panel installation. The first year has seen the provision of the solar streetlight through the understanding of the partnership. Oliver Newton road, which is the secondary distributor from the Harare-Nyamapanda highway, has been provided with 60 solar powered streetlights.

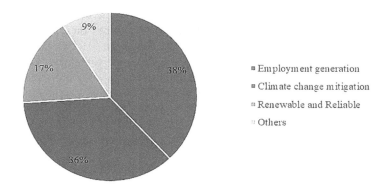

FIGURE 11.2 The utility of solar electricity as a resilience strategy in Mutoko district.

Source: Authors

ENVIRONMENTAL MANAGEMENT AGENCY (EMA) OFFICIAL: The coming of solar harvesting to Mutoko is a welcome development in the context of climate change mitigation as there are less emissions which are produced in the process of solar energy generation. These are the green technologies advocated by the Paris Agreement.

ZIMBABWE ELECTRICITY SUPPLY AUTHORITY (ZESA) OFFICIAL: This is a welcome development for the people of Mutoko as they can have access to renewable and reliable energy. They can have electricity for most part of the day as compared to the past. Here in Mutoko these days I can work 8 hours plus everyday depending on the amount of work at hand. I have a few electricity cuts challenges. This community development aims to overcome poverty and disadvantages by investing in the physical infrastructure of neighbourhoods, building family income and wealth, improving access to quality education, and promoting social equity.

NYANGANI RENEWABLE ENERGY (NRE) OFFICIAL: We have brought about a new dimension in the sense that we have blended solar energy production with agriculture as we have sheep, which can graze underneath the solar panels. This incorporates elements of climate smart agriculture. We have also managed to make technology transfer for the local people. Some of the boys are already making use of the skills by installing residential solar panels and geysers. Others still with NRE have been trained on how to operate and maintain the plant, which is now functioning.

The results of the study concur with Montag et al. (2016) who noted that the use of solar power provides reliable green energy. The provision of clean, efficient and reliable energy and is in line with the targets of SDG7 for universal access to energy, which is clean and efficient for the benefit of society (FAO and WWC, 2015). The European Court of Auditors (2020) report argues that the ultimate aim of a people-centred smart city should be hinged on sustainability, economic growth and enhancement of the quality of life. In this study, economic development has been in the form of socio-economic development, which has seen the provision of solar streetlights through partnership between the NRE and the MRDC. Oliver Newton road, which is the secondary distributor from the Harare–Nyamapanda highway, has been provided with 60 solar-powered streetlights. These are the indirect benefits of the solar energy project, which are being channelled towards community development. Similar results of employment generation were reported from India where local professional and semi-skilled employees benefited from jobs in a solar farm. Over 2,500 skilled and unskilled jobs were created throughout the development phase (Batchelor et al., 2019).

According to the EMA official, the use of solar energy reduces emissions. This is in line with Mutua and Kimuyu (2015) who indicated that solar energy in the context of climate change is prioritised as it has low carbon emission levels as compared to carbon-based fuels such as coal. Thus, there is reduction in the amount of carbon released into the atmosphere and this meets the global call for the reduction of the impact of climate change and this can be achieved through diversifying the sources of energy and its value chain (Sheu et al., 2012).

An innovative approach was developed involving the breeding of sheep on the solar farm (Figure 11.3). NRE has a flock of sheep at their site that graze the grass and keep it short so that the panels are not hindered from capturing sun light. The strategy of using sheep to maintain grass height rather than using energy demanding machines that emit greenhouse gases is a smart agriculture approach that reduces the costs and enhance climate proofing at the local level.

The findings of the study are in line with the fuel staking theory, which indicates that the use of energy in a household is dependent on factors other than income and the responsible factors can be in the form of social, economic, cultural or even

FIGURE 11.3 Sheep used to graze under the solar panels.

Source: Photo by Mukwekwe

personal preferences. In the end, it has been proven through research that a fuel stacking approach is applicable even to higher income households (Masera et al., 2000; Ogwumike et al., 2014; Bisu et al., 2016; Chitiyo et al., 2016).

11.5.2 THE EFFICIENCY OF SOLAR-GENERATED ENERGY RELATIVE TO CONVENTIONAL METHODS

Majority of the respondents argue that solar-powered energy is more efficient than conventional methods. Al-Saidi and Lahham (2019) observed that in developing countries with high scarcity of water and arable land, farmers are encouraged to use solar energy and are increasingly provided with the valuable ability of grid connection and selling excess produced energy at a subsidised price.

Key informants from ZESA and NRE concur that solar energy has a lower efficiency in terms of energy conversion compared to conventional methods such as electricity, natural gas, coal and fuel wood. Solar panels have an efficiency ranging between 14% and 16%, although 22.5% is the best range of efficiency one can find on the market commercially (Sheu et al., 2012). Current innovations have increased the range of solar panel efficiency. There is a lot of potential for harnessing solar energy in Zimbabwe. The findings are supported by the findings of the UNDP (2018) which indicated that Zimbabwe is one of the countries in Africa with high access to radiation from the sun yet the country contributes less than 1% to the global solar energy (Mutua and Kimuyu, 2015).

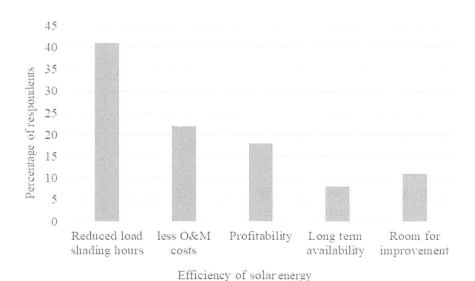

FIGURE 11.4 The efficiency of solar-powered energy relative to conventional methods.
Source: Original source by Author

Key informants highlighted that:

NRE OFFICIAL: For the plant to be able to feed electricity in the national grid the grid must be live, meaning there must be electricity running. As a result, the Mutoko line enjoys special preference since there will be need to receive electricity from the solar plant. Mutoko as of today enjoys less than 10 hours of load shedding which is a great improvement from the average of 18 hours it used to endure before the integration of the solar plant into the electricity network. Solar panels have been able to convert sunlight into electricity with an efficiency of between 15 and 22%. Coal has an efficiency of 33%–40%, but unlike solar energy, it pollutes the environment.

A ZETDC OFFICIAL: Fortunately, or unfortunately the surplus electricity from the solar power plant is not only for Mutoko as it shares with other districts. Maintenance and operational costs are kept low because sheep maintain the grass height when they graze thus reducing the maintenance cost of the plant. Unlike fossil fuels, however, solar energy will be a viable energy option so long as there is access to direct sunlight. Whereas fossil fuels will disappear for millions of years, the sun should be around for just as long.

MUTOKO RDC OFFICIAL: The solar plant in Mutoko has 14 employees running it at any given time unlike the hundreds at Kariba dam and Hwange power station. This huge difference in the number of employees results in the solar plant having lesser operation and maintenance costs. Profound remote data analysis has reduced the need for on-site personnel and measurement systems. The daily routine for solar power maintenance typically involves visual inspections and torque tests. By using remote data analysis system issues can be better predicted and mitigated.

A ZESA official indicated that the surplus solar energy from Mutoko is fed into the national grid. Apart from the NRE solar power project, the Nyabira power plant is producing 2.5 MW of electricity that is fed into the national power grid and this is only 20% of the anticipated output of 12.5 MW. In India, solar photovoltaic (PV) and wind energy represent 90% of capacity growth in India due to decreased costs (Al-Saidi and Lahham, 2019). Despite recent progress, solar power accounts for about 1% of the world's energy mix.

11.5.3 Opportunities Faced When Providing Electricity in the Mutoko District

There are various opportunities for providing extra energy in solar form (see Figure 11.5).

The results show that there are shared expenses and expertise and as such, the local council and ZEDTC are gaining experience in solar energy provision and installation from NRE partners. Some of the opportunities include special treatment by ZESA, ability to get experience in training and ability to produce energy for the increasing population.

Solar Harvesting in Zimbabwe 163

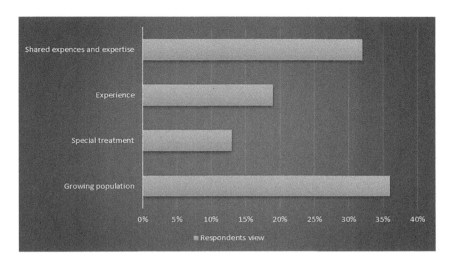

FIGURE 11.5 Opportunities for providing solar energy.
Source: Figure by Mukwekwe (author)

The following are excerpts from the interviews:

NRE OFFICIAL: We assisted ZETDC with the purchase of transformers and installation needed at our site. There is also a special treatment that is enjoyed and it would make a lot of sense if we had a special line with limited power cuts to make sure that NRE transmit all the power it can in the national power grid.

MUTOKO RDC OFFICIAL: The partnership between NRE and Mutoko Rural District Council is a learning curve for the local authority, which can be made use of in future years if the local authority chooses to venture into electricity production. We are learning a lot of important things from our partners NRE Company and if all goes well, as a local authority, we want to establish our own power plant.

ZETDC OFFICIAL: We are gaining experience in solar electricity production with is something new, because we are used to thermal and hydroelectricity. ZETDC is learning modern green energy production from NRE, which is the way forward into a sustainable resilient future. It would make a lot of sense if we had a special line with limited power cuts to make sure that NRE transmit all the power it can in the national power grid.

11.5.4 CONSTRAINTS FACED WHEN PROVIDING ELECTRICITY IN THE MUTOKO DISTRICT

The policy environment, financing and monopoly affected the provision of solar energy in Mutoko. Twenty-eight per cent (28%) of the respondents mentioned that

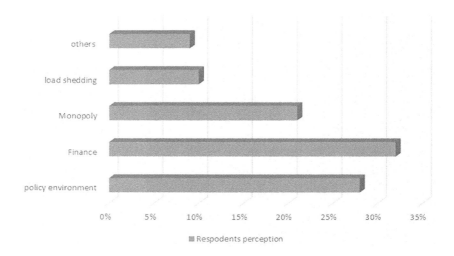

FIGURE 11.6 Constraints faced by electricity providers in the Mutoko district.

Source: Author

the policy environment has the potential of scaring investors (Figure 11.6). There seems to be the absence of sound legislation to protect the green energy sector and the various acts in place do not address the actual environmental issues attached to solar energy production.

Some of the informants were of the opinion that there is not enough foreign currency being injected into solar production. Sentiments were also that the government had low credibility in terms of honouring loans and this caused other partners to restrain from assisting on such projects, thus affecting state-run entities as much as ZETDC.

Some of the excerpts from the interviews indicated the following:

> Mutoko RDC official: *The government has removed payment of duty from solar system equipment imports through Statutory Instrument 147, 2010, and the Rural Electrification Fund Act (2002) has provision for renewable energy. However, there had been no practical implementation of a supportive policy and, in respect of the Rural Electrification Fund, press reports suggest this has not only failed to meet rural electrification targets but was the centre of a significant corruption scandal in 2014. Before installing the solar panels, the employees had to undergo a 1 week training course on how to handle the panels, the correct hole size for the poles, the correct angle for the panels, etc. This took up some valuable production time.*
>
> NRE official: *The past and present policies such as the Indigenisation legislation and land reform policy have negative impacts on investor confidence since these solar power plants require a lot of capital, usually in millions of dollars. Investors therefore require a stable environment, which is far from*

what is present in Zimbabwe, which recently witnessed political unrest. NRE had to purchase transformers that were supposed to be approved by ZETDC and were delayed for over 2 months. ZETDC was the one, which pegged the price of the electricity produced, which always has many challenges for NRE. To worsen the situation, the ZETDC delays payment for the electricity resulting in huge balance and cash flow problems for NRE.

ZETDC official: *A monopoly has merits and demerits for ZETDC. The demerits include the huge demand of electricity that has to satisfy without the aid of competing firms. Also, all the infrastructure cost, encompassing installation cost and research and development costs fall on the shoulders of one already failing company. But this would not have been the case if we had many firms providing electricity in the district, which would share the expenses, ideas and work force. This is because the NRE solar power plant cannot pump in electricity into the national power grid when it is not live with electricity.*

The results of the study showed that there are a number of technicalities, which are faced in the provision of solar energy to the Mutoko RDC. However, these technical barriers seem to affect the first solar power plant in the district which is NRE.

11.6 CONCLUSION AND RECOMMENDATIONS

The global aspiration to achieve sustainable development is attainable with the adoption of renewable energy. The use of solar energy will help to build sustainable and inclusive cities. There is need to invest more in solar infrastructure to achieve safe, cheap and liveable environments. The link between sustainable development and adoption of renewable energy use is in line with the New Urban Agenda. Several challenges in renewable energy provision were noted ranging from institutional, technical, environmental and financial. The study noted the need for public–private partnerships to ensure sustainable and efficient solar energy production. Since the NRE does not have adequate funds to provide infrastructure services, there is a need to adopt public–private partnerships with ZETDC and the Mutoko Rural District Council. Furthermore, partnerships with the government, private sector and local community can lead to sustainable and resilient planning. The utility of the solar electricity project as a resilience strategy in Mutoko is derived from the employment generated for the local people, climate change mitigation through reduced release of carbon dioxide and creation of green jobs, renewability, reliability and ability to accommodate other ventures such as sheep breeding. All the aforementioned utilities show that solar harvesting is used as a resilience strategy for the Mutoko RDC. The solar power plant is said to require relatively lower operational and maintenance costs, thus requiring little input in return for huge output. The main challenge was the growing population in the Mutoko RDC, which meant increased demand for solar power against its capacity to supply. The study recommends an integrated holistic approach among all the stakeholders to make the project efficient and effective. This calls for the local authority to provide more land, ZETDC to perform the regulatory role and NRE to provide finance and expertise.

REFERENCES

Aljazeera. (2015) Australia's rising solar power 'revolution' https://www.aljazeera.com/features/2015/1/13/australias-rising-solar-power-revolution accessed 27 February 2019.

Al-Saidi, M., & Lahham, N. (2019). Solar energy farming as a development innovation for vulnerable water basins. *Development in Practice*, 29(5), 619–634. Doi: 10.1080/09614524.2019.1600659

Arun, M., & Teng Yap, M. (2000). Singapore: the development of an intelligent island and social dividends of information technology. Urban Studies, 37(10), 1749-1756.

Batchelor, S., Brown, E. D., Scott, N., Jon Leary, J., et al. (2019). Two birds, one stone—reframing cooking energy policies in Africa and Asia. *Energies*, 12(1591), 1–18. Doi: 10.3390/en12091591.

Batty, M. (2012). Smart cities, big data. Environment and Planning B: Planning and Design, Volume 39, 191-193.

Bhatasara, S. (2015). *Understanding Climate Variability and Livelihoods Adaptation in Rural Zimbabwe: A Case of Charewa, Mutoko*. Grahamstown, South Africa: Rhodes University.

Bisu, D. Y., Kuhe, A., & Iortyer, H. A. (2016). Urban household cooking energy choice : An example of Bauchi metropolis, Nigeria. Energy, Sustainability and Society, 6(15), 1–12. Doi: 10.1186/s13705-016-0080-1

Chitiyo, K., Vines, A., & Vandome, C. (2016). *The Domestic and External Implications of Zimbabwe's Economic Reform and Re-engagement Agenda*. London: Chatham House.

Chitongo, L., Chikunya, P., & Marango, T. (2020). Do Economic Blueprints Work? Evaluating the Prospects and Challenges of Zimbabwe's Transitional Stabilization Programme. *African Journal of Governance & Development*, 9(1), 7–20. Retrieved from https://journals.ukzn.ac.za/index.php/jgd/article/view/1622

Creswell, J. W. (2014). *Research design: Qualitative, Quantitative and Mixed Methods Approaches*. 4th edn. Los Angeles: Sage Publication Inc.

Dinçer, F., & Meral, M. E. (2010). Critical factors that affecting efficiency of solar cells. *Smart Grid and Renewable Energy*, 1, 47–50. Doi: 10.4236/sgre.2010.11007.

Dziva, C. (2014). Achievability of MDGs Under the Inclusive Government in Zimbabwe: Successes, challenges and prospects. *International Journal of Politics and Good Governance*, 5(5.2), 1–23.

European Court of Auditors. (2020). Biodiversity on farmland: CAP contribution has not halted the decline. 1–58. www.eca.europa.eu/Lists/ECADocuments/SR20_13/SR_Biodiversity_on_farmland_EN.pdf

FAO & WWC (2015) *Towards a Water and Food Secure Future. Critical Perspectives for Policy-makers*. Rome: Food and Agriculture Organisation of the United Nations.

Ganda, F., & Ngwakwe, C. C. (2014). Problems of sustainable energy in sub-Saharan Africa and solutions. *Mediterranean Journal of Social Sciences*, 5(6), 453–463. Doi: 10.5901/mjss.2014.v5n6p453.

IEA, (2008). Worldwide Trends in Energy Use and Efficiency. Key Insights from IEA Indicator Analysis. Paris, France: International Energy Agency.

Karekezi, S., & Kithyoma, W. (2003). *Renewable Energy Development: Prospects and Limits*. Dakar: NEPAD.

Lombardi, P., Giordano, S., Farouh, H., & Yousef, W. (2012). Modelling the smart city performance. *Innovation: The European Journal of Social Science Research*, 25(2), 137–149.

Mafumbabete, C., Chivhenge, E., Museva, T., Zingi, G. K., & Ndongwe, M. R. (2019). Mapping the spatial variations in crime in rural Zimbabwe using geographic information systems. *Cogent Social Sciences*, 5(1), 1661606.

Masera, O., Saatkanp, B., & Kammen, D. (2000). From linear fuel switching to multiple cooking strategies: A critique and alternative to the energy ladder model. *World Development*, 28(12), 2083–2103.

Montag, H., Parker, D. G., & Clarkson, T. (2016). *The Effects of Solar Farms on Local Biodiversity : A Comparative Study*. Blackford: Clarkson and Woods and Wychwood Biodiversity. ISBN: 978-1-5262-0223-9. https://solargrazing.org/wp-content/uploads/2021/02/Effects-of-Solar-Farms-on-Local-Biodiversity.pdf. Assessed: 22 December 2021.

Mutua, J., & Kimuyu, P. (2015). *Exploring the Odds for Actual and Desired Adoption of Solar Energy in Kenya*. Gothenburg, Sweden: SIDA.

Neirotti, P., De Marco, A., Cagliano, A. C., Mangano, G., & Scorrano, F. (2014). Current trends in Smart City initiatives: Some stylised facts. *Cities*, 38, 25–36.

OECD. (2018). Housing Dynamics in Korea: Building Inclusive and Smart Cities, OECD Publishing, Paris, France, http://dx.doi.org/10.1787/9789264298880-en. Assessed: 21 December 2021.

Ogwumike, F. O., Ozughalu, U. M., & Abiona, G. A. (2014). Household energy use and determinants : Evidence from Nigeria. *International Journal of Energy Economics and Policy*, 4(2), 248–262.

Pavlović, T. M., Radosavljević, J. M., Pavlović, Z. T., & Kostić, L. T. (2006). Solar energy and sustainable development. *Physics, Chemistry and Technology*, 4(1), 113–119. Doi: 10.2298/FUPCT0601113P

Shackleton, C. M., Gambiza, J., & Jones, R. (2007). Household fuelwood use in small electrified towns of the Makana District, Eastern Cape, South Africa. *Journal of Energy in Southern Africa*, 18(3), 4–10.

Sheu, E. J., Mitsos, A., Eter, A. A., Mokheimer, E. M. A., Habib, M. A., & Al-Qutub, A. (2012). A review of hybrid solar—fossil fuel power generation systems and performance metrics. *Journal of Solar Energy Engineering*, 134, 1–17. Doi: 10.1115/1.4006973

Vallianatos, M. (2015). Uncovering the Early History of "Big Data" and the "Smart City" in Los Angeles. Boom California: https://boomcalifornia.com/2015/06/16/uncovering-the-earlyhistory-of-big-data-and-the-smart-city-in-la/

Vanolo, A. (2014). Smartmentality: The Smart City as Disciplinary Strategy. Urban Studies, 51 (5), 883-898.

UNDP (2018) *Barrier Analysis of Small Grains Value Chain in Zimbabwe*. Harare: Zimbabwe Resilience Building Fund (ZRBF) Management Unit.

World Bank & International Energy Agency (2015) Sustainable Energy for All 2015 : Progress Toward Sustainable Energy, https://openknowledge.worldbank.org/handle/10986/17138 assessed 22 December 2021.

Zhang, S., & He, Y. (2013). Analysis on the development and policy of solar PV power in China. *Renewable and Sustainable Energy Reviews*, 21, 393–401. Doi: 10.1016/j.rser.2013.01.002

ZimStat. (2012). *Zimbabwe Population Census 2012*. Harare, Zimbabwe: Zimbabwe National Statistics Agency.

12 Infrastructure Projects Design Versus Use in Local Authorities
A Case Study of Banket Small and Medium Enterprises Mall in Mashonaland West Province of Zimbabwe

Moses Chundu

CONTENTS

12.1 Introduction 169
12.2 Theoretical Framework... 171
12.3 Literature Review 173
12.4 Methodology... 175
12.5 Results and Discussion 176
    12.5.1 Sample Characteristics 176
    12.5.2 Reasons for Zero Uptake of Market Stalls 177
12.6 Conclusion 181
References... 181

12.1 INTRODUCTION

African governments, including the Government of Zimbabwe (GoZ), are seeking to implement smart city initiatives that address challenges arising from rapid urbanisation and simultaneously promote investment and national economic development. A key constraint that features in small and medium enterprise (SME) growth studies is the lack of infrastructure (Maunganidze, 2013; Karedza et al., 2014; Bomani, Fields and Derera, 2015). Poor infrastructure is reported to be one of the major factors influencing investment decisions (Mourougane, 2012). One of the policy interventions answering to smart infrastructure for smart urban economy has been the construction of formal SME infrastructure (vending malls, factory shells, bus termini, etc.) in various local authorities[1]. There has, however, been a very low uptake of such infrastructure, raising questions of what could have gone wrong with the interventions. The new school of thought in infrastructure development (user-design perspective) could have the answer to this seemingly confusing phenomenon.

Zimbabwe has experienced rapid urbanisation fuelled largely by dying rural economies leading to unsustainable rural–urban migration. The deindustrialisation that has taken place in the urban centres has meant that most of those migrating to urban areas have had to join the informal SME sector for survival. This has posed an infrastructure challenge for local authorities faced with declining revenues and hence the many attempts to formalise the operations of SMEs by housing them in formal structures. It is in this context that the Banket vending mall under investigation was constructed.

The vending mall was built in 2013, sponsored by the then local Member of Parliament (MP) as a community-development initiative. The MP belonged to the then ruling ZANU PF party. Information gathered points towards the funding of the project having been bankrolled by one of the diamond-mining companies in the Manicaland Province probably as part of its corporate social responsibility. It was constructed in the Banket Town under the Zvimba Rural District Council jurisdiction. The mall built to accommodate 198 vendors is located along the Harare–Chirundu highway in the commercial district of Banket. It was designed to have a vast open space filled with small tables one metre wide to accommodate the vendors as well as basic ablution facilities at the end the building. Figure 12.1 shows the mall and the intended beneficiaries trading next to the mall seen behind the mall in thatched rudimentary structures. There are no other buildings immediately in the vicinity with the nearest shopping mall and service stations located within a 500-metre radius.

The working hypothesis is that authorities are developing these projects without properly consulting end-users on location and user interface and secondly, where there is consensus on the location of the facility as well as its features, low uptake could be a result of suboptimal user fees. Sub-optimal fees imply a cross subsidy that tends to affect the other unrelated areas of service delivery and hence the general deterioration of service delivery in most local authorities.

This research focuses on vendors in the retail sector in the farming town of Banket where a state-of-the-art vending mall along the highway has become a white elephant for over six years. In some cases, like the Harare Colcom commuter omnibus rank, it appears as a question of poor siting but in the case of Banket it appears more than a siting issue as targeted users are occupying rudimentary infrastructure within a few metres from the white elephant. With growing calls to provide smart infrastructure for the predominantly urban SMEs, it is important that the reasons for low space uptake on new infrastructure projects be understood before building more white

---

[1] For instance, in Harare, the local authority has tried to relocate vegetable and clothing vendors to the famous Coca Cola junction, a move they have strongly resisted citing low business volumes as most of their customers are commuters travelling to and from work. A beautiful SME mall lies unoccupied along the Simon–Mazorodze Road with very low uptake, again citing the inappropriateness of the infrastructure. Another white elephant is the Colcom commuter omnibus rank also meant to decongest the city centre but also heavily resisted as being inappropriately sited.

FIGURE 12.1 Banket vending mall with informal structures appearing behind.
Source: Author

elephants. The theoretical framework follows this introductory section, followed by a brief review of the literature around the concept of user-centred system design (UCSD). The methodology and presentation and discussion of results are followed by the chapter ending with concluding remarks on how the concept of UCSD has contributed to the agenda of developing sustainable and smart cities in Zimbabwe.

12.2 THEORETICAL FRAMEWORK

The current study is anchored on the concept of UCSD popularised by Norman (1986) who emphasised the importance of having a good understanding of the users (but without necessarily involving them actively in the process). The definition has evolved over the years including one by Karat who viewed UCSD as 'an iterative process whose goal is the development of usable systems, achieved through involvement of potential users of a system in system design' (Karat et al., 1996, p. 161). Gulliksen et al. (2010, p. 398), however, argue that a key anchor principle in UCSD is actually 'active user participation throughout the project, in analysis, design, development and evaluation' emphasising the role of multidisciplinary design teams in the process. Whilst the origin of the concept is in information communication and technology (ICT), its applicability has included general engineering space including civil engineering and other disciplines where a project design now requires participatory methodologies to be followed. Figure 12.2 gives a schematic framework for the user-centred design (UCD) process.

The UCD approach basically involves four distinct phases as represented in the boxes in Figure 12.2. First, designers work in teams to understand the *context* in which users may use a system/project/product followed by identification and specification

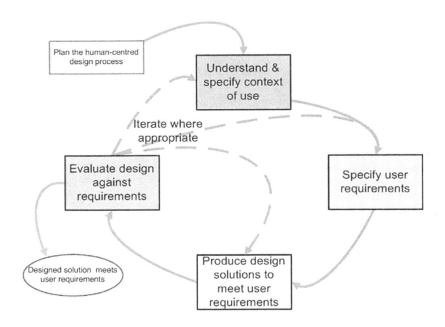

FIGURE 12.2 User-centred design framework.

Source: Interaction Design Foundation (2020)

of users' *requirements*. A *design* phase follows, in which the design team develops solutions, which should be *evaluated* before implementation phase. The evaluation at this stage is relevant to see how close the design is to a level that matches the users' specific context and satisfies all of their relevant needs. This process must continue until the evaluation results are satisfactory before going into the full implementation phase (construction phase in the case of infrastructure projects) (Interaction Design Foundation, 2020).

Usability is key because if users cannot achieve their goals efficiently, effectively and in a satisfactory manner, they are likely to seek other ways of achieving their goals. Usability is the outcome of a UCD process that examines *how* and *why* a user will adopt a product/service and seeks to evaluate that use. The process is carried out repeatedly through a series of cycles in the process improving the design towards meeting user expectations (Komninos, 2019). The term 'UCD' refers simply to a method of building products with an eye towards what users want and need. According to this line of thinking, we should be driving towards a world where government designs policies with an eye towards the individuals that stand to benefit from—or that could be hurt by—changes to public policy (Moilanen, 2019).

In evaluating usability, there are a number of project evaluation models that include the common Environmental Impact Assessment (EIA) for environmental sustainability and Cost Benefit Analysis (CBA) for economic sustainability. There is also a composite Triple Link Sustainability (TLS) model that involves evaluating

Infrastructure Projects Design

projects in terms of environmental, economic and social aspects of sustainability, where integration and optimal balance of all three dimensions and objectives are needed for overall sustainability (Sijanec and Zeren, 2010). The European Union (EU) carried out a program in Best Practice in Sustainable Urban Infrastructure developed under the Cooperation in Science and Technology program (COST C8), which resulted in a book summarising the questions, methods and tools for assessing the sustainability of the development of urban infrastructure and presenting 43 different good practice cases from 15 EU countries and Canada. Sijanec and Zeren (2010) advocated for a multi-criteria decision analysis (MCDA) adapted to complex projects with monetary and other objectives allowing for the disaggregation of projects into smaller components to facilitate a detailed appraisal before putting them together for consideration by policy makers.

12.3 LITERATURE REVIEW

The majority of literature on UCD has, to date, centred on ICT-related aspects as the concept was only recently adapted to other spheres of life including the built engineering space and infrastructure development in general. Baird (2009) advocated for the establishment of a set of user performance criteria beyond technical criteria to ensure the development of building stock that is sustainable through meeting the expectations of end users. The inclusion of user performance criteria in the sustainability assessment tool will cover the social dimension of building sustainability. Aspects for which the inputs of users are required include:

> Operational—space needs, furniture, cleaning, meeting room availability, storage arrangements, facilities, and image; Environmental—temperature and air quality in different climatic seasons, lighting, noise, and comfort overall; Personal Control—of heating, cooling, ventilation, lighting, and noise; and Satisfaction—design, needs, productivity, and health.
>
> (Baird, 2009, p. 1074)

A case study of a Hong Kong boundary crossing bridge by Ugwu et al. (2006) concluded that being able to do a sustainability appraisal early in the project cycle has the potential to generate significant savings. If not done at the design stage, it makes it difficult to take corrective measures on time, resulting in white elephants as the one under investigation in this study. Timely valuation leading to corrective measures at the design stage presupposes that at the initial design stage. There should be alternative design options that allow for quick adjustment towards sustainability should there be a need to adjust the designs. This is an important and critical part of design evaluation and sustainability-driven decision making. The appraisal requires that the public and end-users are made aware of the issues and are fully consulted as a critical stakeholder and the different attitudes and interests of the critical stakeholders are reconciled.

> Possible consequences such as impacts on current land use practices, which can be partially linked to income foregone, can increase the need for building awareness and

participatory approaches. While on the other hand, the need for such 'trust building' activities might be less important in state-owned and sparsely populated areas. A need for public awareness can also arise in projects aiming to change behaviours and attract public and/or private financial support.

(Naumann et al., 2011, p. 37)

Public infrastructure projects are said to be 'complex and time consuming due to diverse communication capabilities, diverse communication platform-preferences and non-recognition of all stakeholders' required for their successful implementation (Ahuja and Priyadarshini, 2015, p. 2). Ahuja and Priyadarshini (2015) emphasised the importance of effective sustainable communication management system citing the need to develop appropriate focus for each stakeholder group to understand their interests, involvement, interdependencies, influence and potential impact on the project.

A multi-dimensional approach is suggested that is sensitive to the diversity and magnitude of influence and impact groups to study them under categories such as Direct/Indirect, Single/Multiple, interest in the project, level of aggregation at each project stage, influence and impact on other specific stakeholder groups or individuals, power to influence and impact the project at different stages, and impact by the project.

(Ahuja and Priyadarshini, 2015, p. 4)

Two case studies were carried out; a successful one being the Delhi Metro Rail system, which was initiated in the early 1990s with the formation of the Delhi Metro Rail Corporation (DMRC) and the second one being the 5.8-km 'Open BRT' System designed for buses operated by the Delhi Transport Corporation (DTC) as well as private operators on upgraded dedicated lanes. Despite the high level of technical design, the DTC project secured neither political support nor public sympathy (Ahuja and Priyadarshini, 2015).

Willetts et al. (2010) argued that the delivery of sustainability within infrastructure projects is not just about the cross-cutting industry themes of energy reduction, resource conservation, waste minimisation and climate change mitigation, etc., but also about a far wider and long-term commitment to create a better and healthier infrastructure to support society in the long term. They proposed the following as key changes necessary to ensure the sustainability of infrastructure projects:

- Engineers need to have far greater involvement in the early engagement of stakeholders.
- Engineers should use their technical skills to educate and influence decision makers.
- Engineers need to be allowed to look beyond project/site-specific problems and begin to look at the larger issues and systems.
- Planners and engineers should work more closely to develop indicators and benchmarks relating to the delivery of sustainable infrastructure.

The first bullet is very critical in that it saves not only time but also resources in the development of infrastructure as issues are picked at the early design phase of

Infrastructure Projects Design

infrastructure development. It should become possible at this stage to pick both user requirements and context issues critical to inform the sustainability indicators. Though Willetts et al. (2010) give the impression, from the first three bullets, that it is only engineers who are involved in stakeholder engagement, it can be argued that for projects such as the mall, the multiple stakeholders involved require the involvement of more than engineers to also include planners and the local authority among others who also have to engage a diversity of stakeholders.

12.4 METHODOLOGY

The study made use of mixed research methodologies[2] from administering questionnaires to vendors and local authority officials to conducting focus groups interviews with community leaders, including the councillor and the leader of the vendor association in the area. A questionnaire was administered to 109 SME traders representing 55% of the targeted beneficiaries of the project. The one-page questionnaire had questions that allowed the researcher to collect the views of the vendors with precision whilst the focus group interviews allowed for individuals' interpretation of events which were key to understanding the full scope of the project, especially from the perspective of the developer who was the then sitting MP for the constituency to get balanced views on the project. Since the majority of the intended users of the new building decided to locate themselves next to the unoccupied building, there was not much need for sampling but to take the entire population as the sample. Purposive (Total Population Sampling (TPS[3])) non-probability sampling was therefore employed, targeting every vendor who was on site on the date of survey which is a normal day of trading. A target questionnaire response rate of 80% was deemed sufficient to give enough data representative of the population under study to enable the credible analysis of the phenomenon. A one-page questionnaire was administered to capture information about the entrepreneur, the business and the infrastructure project.

To get the perspective of the developers on the project and its current status, one-to-one interviews were targeted at the local authority officers directly linked to the project and also focus group interviews with senior community leaders. It proved very difficult to meet with council officials showing signs of discomfort with the subject under investigation. The reasons for their discomfort were later confirmed by the findings of the study discussed in the results section. The survey data from interviews

---

[2] Mixed methods research involves collecting, analysing and integrating quantitative (e.g., experiments, surveys) and qualitative (e.g., focus groups, interviews) research. This approach was chosen as it allowed for the triangulation to examine the same phenomenon allowing for a better understanding of the research problem than either of each alone. Focus group interviews allowed the researcher to verify facts collected in the questionnaire and seek a deeper understanding of the problem given the nature of investigation involving perceptions and attitudes (Creswell, 2014).

[3] Total Population Sampling (TPS) is a technique where the entire population that meet the criteria (e.g., specific skill set, experience) are included in the research being conducted. TPS is more commonly used where the number of cases being investigated is relatively small (Etikan, Abubakar Musa and Sunusi Alkassim, 2016).

was processed using the Statistical Package for the Social Sciences (SPSS) data processing software making use of largely descriptive statistics.

12.5 RESULTS AND DISCUSSION

12.5.1 Sample Characteristics

Out of a target of 109 respondents representing 55% of the targeted 198 beneficiaries of the project, a total of 99 questionnaires were completed by the vendors who were on site on a Friday, a typically busy shopping day where most of the vendors were expected to be on site. Ten vendors refused to be interviewed for various reasons ranging from fatigue with researches that lead nowhere to being busy and some fearing political persecutions.[4] Of those interviewed, 49.5% were male whilst 50.5 were female who dominated the clothing and food sectors. Males were the sole participants in hardware, electrical and saloon businesses and a representation in the vegetable sector of 6% and 5%, respectively. Table 12.1 gives the key descriptive statistics characterising the sample used in this study.

The age group of the respondents ranged from 15 to 73 years with an average age of 37.44 years. The businesses were earning anything between Z$8 and Z$2,000, with average daily takings of Z$239 reflecting the diversity of the industries represented in the business mix. Vegetable vendors, for instances, had very low value tradable and thin margins whilst those in electronics and hardware tend to enjoy low-volume, high-value merchandise and healthier margins. When read in conjunction with the gender dimensions of the ownership structure of the businesses, it brings in

TABLE 12.1
Descriptive Statistics

	N	Min	Max	Mean	Std. Deviation
How old are you (in years)?	97	15.00	73.00	37.44	10.38
What is your average daily takings (in Z$)?	88	8.00	2,000.00	239.09	389.66
For how long have you been operating from your premises (in years) ?	96	0.20	25.00	7.33	5.70
How much do you pay to the local authority per month (in Z$)?	72	2.00	110.00	19.04	19.03
For how long have you been in this business (in years)?	97	0.20	35.00	9.70	7.61

Source: Author

---

[4] Since the mall was owned by an MP belonging to the then ruling ZANU PF party, some vendors feared the information they volunteered could be used to persecute them. The background to the fear follows previous experiences where some members of the party could abuse the state security apparatus to torture and persecute anyone seen to be opposed to their ascendancy to power or to anything they do. For anyone who had experienced this kind of torture closure home, the trauma could not allow them to take any chances, no matter how much the researcher tried to assure them of anonymity and confidentiality.

the issues of equity in terms of how wealth is distributed in the community favouring the male species.

The amounts payable in vending licenses ranged from Z$2 to Z$110, again reflecting the variability in the value of goods traded in the various sectors. Some people had been in the vending business for as long as 35 years, with an average of 9.7 years whilst the oldest business on that site had been there for the past 25 years with newcomers having been there for 0.2 years. The average age of business of 9.7 years and the average period at the current site of 7.33 years suggest that most of these businesses were stable and have been there long enough to have participated in the design and construction of the project under study.

Legally, only 43% were registered, with 77% of these registered as vendors with the town board whilst 21% were registered with the committee that coordinates the affairs of the vendors. the balance of 2% indicated other forms of registration that were not specified. In terms of education levels, about 70% of the respondents had attained at least 'O' level of which 3% of them had 'A' levels, 2% diploma and 1% a master's degree.

A significant 97% of the respondents were actually owners of the businesses whilst the other 3% represented workers and those running the businesses on behalf of relatives. In terms of their previous engagement before running the current businesses, the survey results show a wide array of backgrounds dominated by former vendors (12%), former farm workers (10%) and new players previously unemployed or in school (15%). The other categories of note at 5% each include former construction workers, salespersons, administrators and bakers and cooks.

On where the vendors were previously operating from before settling at the current location, 28% indicated that they came from outside Banket town, demonstrating the high degree of rural to urban and urban to urban migration in search of survival. Out of the remainder of the vendors, 27% indicated that they were previously plying their trade from the local Kuwadzana bus terminus whilst 18% were operating on the Harare–Chirundu highway, with 15% citing streets in their neighbourhood as their previous vending sites. The balance 12% were operating from various other locations.

12.5.2 Reasons for Zero Uptake of Market Stalls

Regarding the appropriateness of the mall under investigation, 80% the respondents felt that the mall was not properly created for them, with 79% feeling it was not appropriate for their businesses. In fact, 71% of the vendors claimed they were not consulted when the project was developed with 96% of those who admit to having been consulted believing that their input and opinions were not taken into consideration. These figures are worrying for a project that was purportedly created to empower small businesses suggesting a complete breakdown in communication between the developers and the users. The MP had developed the infrastructure on council land without a proper partnership arrangement, hence the political identity associated with the project. If the MP had simply donated the infrastructure to the local authority as a community empowerment project, the governance would probably have changed the outcome towards better coordination of the development of

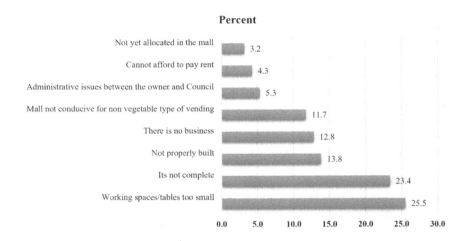

FIGURE 12.3   Reasons for not occupying the mall.

Source: Author

the project and hence its acceptance. The communication strategy emphasised by Ahuja and Priyadarshini (2015) seem to have been completely absent in this case.

Figure 12.3 gives the reasons cited for not occupying the mall. The bulk of the cited reasons had to do with design issues. The majority (26%) of respondents cited inappropriate working spaces with yet another 14% mentioning that it was not properly built, most likely referring to the same issue of small tables giving a combined 40%. The 23% who cited not complete as the reason for not taking up space in the mall also meant the same issues of small tables as they felt the developers needed to address the issue of small tables, raising the combined total number of respondents citing design issues to 63%.

The issue of what was meant by incomplete when everything looks complete was also magnified during the focus group interviews with community leaders. They claimed that a car park that was supposed to be part of the development was still missing and that the actual building was incomplete. The respondents also felt that the configuration of the mall was only suited for vegetable vending (constituting a mere 12% of the vendors[5]) and not the other aspects of business they are involved in like hardware and clothing.

Unfortunately, the local authority officials were not available for interviews. The views of community leaders, mainly committee members of the vendors' association who participated in the focus group meetings, point towards political factors being the main reason for the zero uptake of the facility. They claimed that the mall is owned by an individual who is a politician belonging to a faction of ZANU PF that was booted out of power in the November 2017 change of government. The local

---

[5] Notwithstanding that the mall configuration was more suited for vegetables vendors, these vendors had not taken up the space either probably because of the other factors cited by the respondents.

Infrastructure Projects Design 179

board, on the other hand, is made up of councillors mainly from the ruling ZANU PF and from opposition MDCA, the latter coming mainly from the urban part of Banket whilst the former largely from the surrounding commercial farming community. It is not clear how the local authority was involved but one member of the group dramatised the issue by asking a rhetorical question *'zvingaite here kuti iwe ungavaka chivakwa chako pamusoro pechoumwe munhu?'* (How can one build own structure on someone's property?) The statement by this community leader is quite telling in terms of the extent of poor corporate governance around the project. The land is clearly owned by the local authority whilst the building eventually became private property when it was initially developed as a community empowerment project. At the time of building the project, the MP was holding a powerful ministerial post, which he might have leveraged to proceed with the project without adherence to corporate governance procedures. At the end of the interview, it became clear why the town board officials were reluctant to grant the researcher an interview as the political issues looked contentious and officials could have feared the potential consequences of disclosing information.

The community leaders felt that the only way to break the impasse was for the town board to bring the beneficiaries and the owner of the building to the negotiating table and get to iron out the sticking issues in the interest of development. They, however, echoed the same technical and design issues raised by the vendors revolving mainly around the inappropriateness of the vending spaces. Whilst the makeshift stalls did not have any of the required specifications, they remained attractive, as the vendors did not have to pay much for them. Moving into the formal vending stall came with financial obligations, which the vendors felt was not worth it. Moreover, even in their rudimentary nature, the makeshift stalls were easily accessible as they were lined along the main pathway used by customers. The leaders felt the current stalemate is not healthy for the development of the town as it is sending a wrong signal to potential investors.

When further asked what they felt could have been done better from a design perspective, the respondents buttressed the same issues raised in Figure 12.3. Figure 12.4 shows that the majority vendors at 49% felt that the developers should have built one big hall without tables and allowed them to put in their mobile tables to specifications. The feeling was that the small brick and mortar vending tables were too restrictive whereas if allocated just the space they had a chance of using proper tables that have openings and therefore flexible. The other 30% were suggesting improvements in the same aspect of working spaces with 21% arguing that the tables should have been made bigger[6] whilst the other 9% felt there was need for more space in between tables. Thus, in total, 79% of the vendors pointed towards the need to address poor design aspects. This points towards the need for adaptability in building designs as a way of promoting sustainable infrastructure projects (Gosling et al., 2013). An open

---

[6] A number of those interviewed acknowledged that initially the tables were of the correct sizes but the developer decided to halve the size in a bid to double the number of users in the process making it unattractive. They argued that the move was motivated by the need to capture more votes by the then sitting MP since the project was politically motivated.

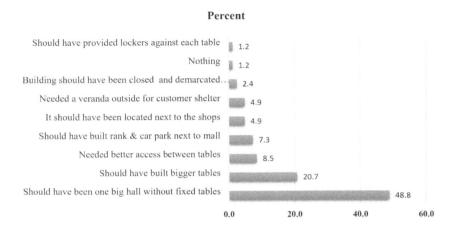

FIGURE 12.4 What in your opinion should have been done better?

Source: Author

plan would have made the current building more usable as opposed to fixing tables that turned out to be too small for end-user purposes.

The other related and significant suggestions on how best the mall could have been designed included the need to have located the mall closer to the bus terminus (cited by 7% of the respondents) and closer to the shopping mall[7] (5%) and the need to have a veranda on the vending mall (5%) to accommodate customers since there is not enough space in between tables to accommodate customers especially when it is raining. The majority of the vendors are clearly happy with the location of the facility going by the proportion that has continued to operate next to the mall. It could actually be that the developer must have seen revealed preference in terms of location and just proceeded to develop what he thought the vendors needed in the process missing on basic usability issues. This could have motivated the doubling of carrying capacity of the mall by halving the original working spaces to unstainable levels.

From the aforementioned results, it is clear that the unoccupied vending mall failed the beneficiary consultation test with the result that the infrastructure has become a white elephant. The issues being raised in Figure 12.4 could have been easily captured and informed the design and implementation of the project if appropriate sustainability measures were taken into consideration right from the onset. The complex political dynamics around this project could also explain the limited application of the UCD methodologies in the development of urban infrastructure in a typical developing country set-up characterised by weak governance structures.

---

[7] The logic here being that the bus terminus generates volumes of potential buyers, whilst proximity to bigger shops allows customers to buy especially vegetables on their way from the supermarket or look for items they would have failed to get in the supermarket. A few of the interviewed vendors had actually relocated to strategic sites closer to the nearest supermarket.

12.6 CONCLUSION

The study sought to understand the infrastructure puzzle of the Banket vending mall. The results have shown that end users were not consulted with the result that a facility exists that is not meeting the basic needs of the intended beneficiaries. The main bone of contention was around the size of the working spaces and political issues in terms of who is in charge of the project between the individual developer and the local authority. Beyond technical sustainability issues, there were also political dynamics where the politicisation of the civic project meant that the stalemate has taken long to resolve with little hope of a resolution any time soon.

As a recommendation, it is important that the local authorities embrace the new methodologies in user-designer perspectives to avoid creating more white elephants in urban infrastructure. Smart cities need to be built on the foundation of smart governance. For Zimbabwe, the need for consultation is even more relevant in light of the government's devolution agenda in the face of local authorities whose governance is either multi-party or exclusively run by the opposition whilst the responsible ministry is run by the ruling party. The country has seen far too many development projects suffer because of this dichotomy. The need to separate politics from community development cannot be overemphasised. The key stakeholder in the sustainable and smart infrastructure development agenda is the end user whose views must always be factored in project design and implementation for sustainability. The chapter has, thus, contributed to the discourse on the development of sustainable and smart spatial planning in Zimbabwe, focusing mainly on the domain of sustainable project management. This contribution is timely as it is coming at a time the government is mulling developing a new city west of the current capital of Harare.

REFERENCES

Ahuja, V. and Priyadarshini, S. (2015) 'Effective Communication Management for Urban Infrastructure Projects Project'. Available at: www.researchgate.net/publication/282 439575 (Accessed: 1 January 2020).

Baird, G. (2009) 'Incorporating User Performance Criteria into Building Sustainability Rating Tools (BSRTs) for Buildings in Operation', 1, pp. 1069–1086. doi: 10.3390/su1041069.

Bomani, M., Fields, Z. and Derera, E. (2015) 'Historical Overview of Small and Medium Enterprise Policies in Zimbabwe', *Journal of Social Sciences*. Available at: www.tandfonline.com/doi/abs/10.1080/09718923.2015.11893493 (Accessed: 25 March 2020).

Creswell, J. W. (2014) *Research Design: Qualitative, Quantitative, and Mixed Methods Approaches*. 4th edn. Thousand Oaks, CA: SAGE.

Etikan, I., Abubakar Musa, S. and Sunusi Alkassim, R. (2016) 'Comparison of Convenience Sampling and Purposive Sampling', *researchgate.net*. doi: 10.11648/j.ajtas.20160501.11.

Gosling, J. et al. (2013) 'Adaptable Buildings: A Systems Approach', *Sustainable Cities and Society*, 7, pp. 44–51. Available at: www.sciencedirect.com/science/article/pii/S2210670712000868 (Accessed: 15 May 2020).

Gulliksen, J. et al. (2010) 'Behaviour and Information Technology Key Principles for User-centred Systems Design'. doi: 10.1080/01449290310001624329.

Interaction Design Foundation (2020) 'What is User Centered Design?'. Available at: www.interaction-design.org/literature/topics/user-centered-design (Accessed: 25 March 2020).

Karat, J. et al. (1996) 'User Centered Design: Quality or Quackery?', in *dl.acm.org*, pp. 161–162. doi: 10.1145/257089.257232.

Karedza, G. et al. (2014) 'An Analysis of the Obstacles to the Success of SMEs in Chinhoyi', *European Journal of Business and Management*, 6(6), pp. 38–42. Available at: www.iiste.org.

Komninos, A. (2019) 'An Introduction to Usability | Interaction Design Foundation'. Available at: www.interaction-design.org/literature/article/an-introduction-to-usability (Accessed: 1 January 2020).

Maunganidze, F. (2013) 'The Role of Government in the Establishment and Development of SMEs in Zimbabwe: Virtues and Vices', *infinitypress.info*. Available at: www.infinitypress.info/index.php/jbae/article/view/193 (Accessed: 25 March 2020).

Moilanen, S. (2019) 'When to Use User-Centered Design for Public Policy'. Available at: https://ssir.org/articles/entry/when_to_use_user_centered_design_for_public_policy# (Accessed: 1 January 2020).

Mourougane, A. (2012) 'Promoting SME development in Indonesia', *OECD Economics Department Working Papers*, 995. doi: 10.1787/5k918xk464f7-en.

Naumann, S., Davis, M., Kaphengst, T., Pieterse, M. and Rayment, M. (2011) *Design, Implementation and Cost Elements of Green Infrastructure Projects*. Final report, European Commission, Brussels, 138.

Norman, D. A. (1986) 'Cognitive Engineering', in D. A. Norman and S. W. Draper (eds) *User Centered Systems Design*. Hillsdale, NJ: Lawrence Erlbaum Associates Inc., pp. 32–61.

Sijanec, Z. M. and Zeren, M. T. (2010) 'Sustainability of Urban Infrastructures', *Sustainability*, 2, pp. 2950–2964. doi: 10.3390/su2092950.

Ugwu, O. et al. (2006) 'Sustainability Appraisal in Infrastructure Projects (SUSAIP): Part 2: A Case Study in Bridge Design'. doi: 10.1016/j.autcon.2005.05.005.

Willetts, R. et al. (2010) 'Fostering Sustainability in Infrastructure Development Schemes', *Proceedings of the ICE: Engineering Sustainability*, 163(3), pp. 159–166. doi: 10.1680/ensu.2010.163.3.159.

13 Towards Responsive Human Smart Cities

Interrogating Street Users' Perspectives on Spatial Justice on Street Spaces in Small Rural Towns in South Africa

Wendy Wadzanayi Tsoriyo, Emaculate Ingwani, James Chakwizira and Peter Bikam

CONTENTS

- 13.1 Introduction 183
- 13.2 Theoretical Framework and Literature Review 184
  - 13.2.1 Theoretical Framework 185
  - 13.2.2 Literature Review: Public Space Concept and Attributes of Spatially Just Streets Spaces 186
- 13.3 Methodology 187
- 13.4 Findings 188
  - 13.4.1 Street Users' Perception of Safety 188
  - 13.4.2 Street Users' Perception of Accessibility 189
  - 13.4.3 Street Users' Perception of Legibility 189
  - 13.4.4 Street Users' Perceptions of Variety 190
  - 13.4.5 Street Users' Perceptions of Maintenance and Management 191
- 13.5 Discussion of Findings 192
- 13.6 Conclusions and Implications for Smart Street Planning 193
- References 194

13.1 INTRODUCTION

The concepts of smart cities and spatial justice are increasingly gaining prominence in global urban design discourses. Interrogating these concepts requires contextualisation because in most cases, the concepts remain as mere abstractions that seemingly run a parallel agenda unless human- or citizen-based interventions are included. According to Cathelat (2019:27), there is the dire need 'to promote cities that will provide their citizens with essential infrastructure and decent quality of life, in a context of sustainable development and by applying smart solutions. This development must be inclusive'. This shows there is an intrinsic relationship between the spatial justice tenets and smart planning, design and management.

Smart cities are defined as 'ecosystems that place emphasis on the use of digital technology, shared knowledge and cohesive processes to underpin citizen benefits in sectors such as mobility, public safety, health and productivity' (Cathelat, 2019:41). Spatial justice on street spaces is defined as the fair and equitable distribution of the valued resource of the street space, the fair consideration of activities and needs of the least-advantaged group of street space users' and the involvement of street space users in street space production processes (Tsoriyo, 2021:64). Often planning, design and management of street spaces prioritises the movement function of streets at the expense of the multiplicity of public space functions that is required by other non-vehicular users (Jacobs, 1961; Jalaladdini and Oktay, 2012; Hartman and Prytherch, 2015). As a result, these users' spatial needs or the right to the city's claims are infringed and this translates to spatial injustice (Lefebvre, 1996; Basset, 2013, Uwayezu and de Vries, 2018).

Clearly, interrogating users' perspective of spatial justice is an important imperative in advancing the smart cities agenda because cities can only be considered as smart if the users' needs are satisfactorily met (Oliveira and Campolargo, 2015). The purpose of this chapter is to highlight a localised and contextualised version of smart cities and spatial justice on street spaces reflecting on the perceptions of street space users in small rural towns (SRTs) of South Africa where technological advancement is limited. The specific objective of the study is to examine spatial justice from the context of street users' expectations versus experience on street spaces in the three SRTs of Thohoyandou, Musina and Louis Trichardt in the Vhembe District of South Africa. The research results enhance the decision-making processes of the local municipalities in the SRTs to prioritise inclusive, adaptive and human-oriented strategies that incorporate the total smart street concept in the planning and design and management of street spaces in SRTs.

The study is organised into five main sections. The first section is the theoretical framework and literature review section on streets as public spaces. The second section is a methodological overview of the study. This is followed by the third section that presents the key findings and the fourth section discusses these major findings. The fifth section concludes the study and provides implications for spatial justice on smart street planning, design and management in the three rural towns of Vhembe, which may also apply to other small towns in Africa.

13.2 THEORETICAL FRAMEWORK AND LITERATURE REVIEW

This section discusses the key theories underpinning this study, namely the Human Smart City Model (PLANUM, 2014), the Right to the City Theory (Lefebvre, 1996b) and the Space Production Theory (Lefebvre, 1991) and reviews the literature based on the concept of public space and the attributes of spatial justice on streets spaces.

13.2.1 THEORETICAL FRAMEWORK

13.2.1.1 Human Smart City Model

The human smart city manifesto is a model that seeks to circumvent the danger of designing smart cities that are not for humans (Oliveira and Campolargo, 2015; Cathelat, 2019). In other words, human smart cities should be more human needs oriented than they are physical or technical (PLANUM, 2014). The concept of human smart cities emerged as a realisation that the traditional smart city concept overemphasises the use of technology as a panacea to the urban problem (Depiné et al., 2017). However, the largest part of the urban problem lies in understanding the expectations of city space users, which then informs the right technological innovations in the planning, design and management of space and thus the human smart city paradigm.

Human smart cities are those cities that are designed and managed by citizens in mind or co-created to enhance the work, play, live and other needs of citizens (Concilio and Rozzo, 2016). The human smart street model seeks to understand citizens' perceptions and aspirations on streets and align these aspirations to those of their local municipalities in influencing the appropriate information and communication technology innovations that address challenges and enhance opportunities that street users face on street spaces. Creating human smart street spaces that are spatially just does not necessarily entail re-planning or doing everything at once. In some cases, street acupuncture where the focus is on small strategic areas rather than on huge expensive projects can meaningfully improve the experiences of street users (Hoogduyn, 2014).

13.2.1.2 The Right to the City Theory

Lefebvre's Right to the City Theory of 1968 is founded on the moral claim of spatial justice and the right to public space (Lefebvre, 1996b; Marcuse, 2012). The right to the city concept is fluid, complex and open to different interpretations (Tsoriyo, 2021). Often the question of whose right to the city is inevitable in all right to the city discussions (Brown and Kristiansen, 2009; Merrifield, 2011). The right to the city of urban dwellers is affected by the various processes of social production of space which orders urban spaces such as streets in a way that eradicates urbanity and deprives street users (particularly non-vehicular street space users) of places of social encounter (Lefebvre, 1996b). In this study, non-vehicular users of street spaces are the 'urban dwellers' or the 'working class'. Street space production mechanisms often favour the rights of vehicular users at the expense of those of non-vehicular users (Lefebvre, 1996; Jones et al., 2007; Loukaitou-Sideris and Ehrenfeucht, 2009; Hartman and Prytherch, 2015).

Non-vehicular users' social encounter or interaction with street spaces is characterised by dissatisfaction with the distribution of spatial qualities and activities that take place on streets (Loukaitou-Sideris and Ehrenfeucht, 2009; Hartman and Prytherch, 2015; Stratford et al., 2020). For example, sidewalks are important spaces that people transverse every day, in most cases, however, they are inadequately provided on street spaces in SRTs. This infringes on the users' right to 'move in, through, across and between different places' (Middleton, 2018:302). As such, the use-value of the sidewalks as public spaces that provide users with the capacity to move freely depreciates (Lefebvre, 1996; Cresswell, 2009:110). This becomes an infringement on the users' right to the city.

One key assumption of this study is that human smart cities are those that reflect the disparate needs or right to the city claims of the street space users. Thus, the ability of users to make their various right to the city claims translates to spatial justice. Spatial justice is unpacked further by the DS4Si (nd:4) as the formation of solidarities across differences between the three factors of spatial claim, power and link. These factors describe the various right to the city claims of space users whereby 'spatial claim entails the ability to live, work, or experience space, spatial power entails opportunities offered to succeed in and contribute to space and spatial link entails access and connection to and with other spaces' (Basset, 2013:5).

13.2.1.3 Theory of Space Production

The third theoretical underpinning of this study is the theory of space production or the spatial triad by Lefebvre (1991). In articulating this space trialetics, Lefebvre (1991:33) posits that space is perceived by users (spatial practice), conceived by technocrats (representations of space) and lived or imagined (representational space, where the reality of conflicts and negotiations over space exist). The spatial triad discourse reveals the complexity of space production. It shows that space is subject to change, produced socially, used by multiple publics and not only physical but a mental construction (Massey, 2005; Soja, 2010).

Soja (1996:56) describes 'lived space' as the 'Thirdspace'. The thirdspace is a space of all simultaneity where everything comes together on space, that is 'the abstract and the concrete, the real and the imagined, the knowable and the unimaginable, the repetitive and the differential' (Soja, 1996:57). In the thirdspace, visions, ideals and perceptions of public space depend on the place, culture, societal norms and demographic characteristics of space users such as age, knowledge, gender, individual experiences and emotions among other factors (Konzen, 2013; Osóch and Czaplińska, 2019). Consequently, conflicts and negotiations over how space should be used or appropriated are inevitable (Loukaitou-Sideris and Ehrenfeucht, 2009; Erdiaw-Kwasie and Basson, 2018). As a result, users' have disparate right to the city claims.

13.2.2 LITERATURE REVIEW: PUBLIC SPACE CONCEPT AND ATTRIBUTES OF SPATIALLY JUST STREETS SPACES

This section discusses the debates around the concept of public space to determine the qualities or attributes that make spatially just street spaces. Streets are important public spaces that should be freely accessed and used by the public as public spaces (Loukaitou-Sideris and Ehrenfeucht, 2009; Shrestha, 2011). The concept of public space does not have a single consistent definition (Madanipour, 2013). The interpretations of public space vary by society and place and change over time (Miller, 2007). Whilst some define public space in terms of physical space, others define the same space as psychological or mentally conceived (Lynch, 1984; Osóch and Czaplińska, 2019). In the contemporary world, defining public space has become a complex endeavour. Various factors are prompting the evolution of the meaning of public space from Jacobs (1961)'s conceptualisation and the contemporary meaning of such spaces. Some of the factors driving changes in the meaning, nature, manifestations and use of public space include globalisation of society, public space privatisation prompted by market liberalisation, innovations in communication and digital technologies in the electronic (Ng et al., 2010; Crawford, 2016; Yu, 2017).

Some of the characteristics of publicness are ownership, control and physical configuration (Varna, 2014). Ownership of space considers the ability of different groups to identify with certain spaces and influence what happens in space. Control measures the extent to which citizens can enjoy the rights/liberties promised by the state whilst on the street taking into consideration regulations imposed to mediate between the conflicting expectations of different users on the same public space. Physical configuration considers the issues of connectivity, visual permeability, civility and animation (the capacity of space to accommodate a range of activities to satisfy different human needs (Varna, 2011). Connectivity encourages the formation of a compact urban system, by meeting the various mobility needs of users while visual access provides users with the ability to see through a place. Where there is obscured visual access, space becomes unsafe and exclusive (Jacobs, 1961; Bentley et al., 1985). Civility relates to the management and maintenance of space to achieve some desired standards of cleanliness so that space remains attractive and friendly.

Bentley et al. (1985)'s responsive design elements of permeability, legibility, visual appropriateness, personalisation, variety, richness and robustness are key socio-spatial qualities that provide space users with diverse choices to make on street spaces. To a greater extent, responsive design elements characterise spatially just public spaces (Tsoriyo, 2021). A key argument of this study is that spatial justice on streets is attainable when street users can enjoy the various choices available on street spaces.

The question of whether there is a universally public space or not shows there are multiple publics (Fainstein, 2009; Madanipour, 2013). Thus, the prospects of a particular space being universally public is utopian, despite having guidelines of the key facets of publicness (Akkar, 2005; Cruz et al., 2018). The study adopts attributes of publicness namely safety, accessibility, identity, variety, maintenance and management as key socio-spatial qualities of spatially just street spaces.

13.3 METHODOLOGY

To address the aim of this study, data were canvassed through multiple data collection strategies that include street intercept questionnaire surveys, structured interviews and observations. Questionnaires were administered to 500 various non-vehicular street space users from the three case study towns of Thohoyandou, Musina and Louis Trichardt. The sample size was determined by the sample size formula recommended by Kothari (2004) as no proper sampling frame for the non-vehicular street space users in the three SRTs could be established. This category of users are very fluid as individuals will never assume a permanent state or position. The urban population proportion for the towns was then used to determine the number of users to be randomly sampled in each town. For Thohoyandou Town, 255 street users were sampled, 155 users in Musina and 90 users in Louis Trichardt. Data were also obtained from semi-structured interviews with eight key experts knowledgeable in street space planning, design and management and also from direct observations of street space user behaviour. Secondary data were in the form of municipalities' legislative documents such as the integrated development plans (IDPs) and spatial development frameworks (SDFs) accessed online.

Qualitative analysis of keywords, phrases and themes was done using thematic analysis as recommended by Babbie and Mouton (2012). Quantitative techniques that were utilised for data analysis are descriptive statistics in terms of frequency percentages and Wilcoxon rank-sum tests. Wilcoxon rank-sum test is a nonparametric equivalent of the t-test (Gibbons, 1993). It compares the differences in distributions or proportions between two independent groups. In this case, the tests are performed to assess if there is statistically valid evidence at a 5% significance level indicating differences in proportions of responses agreeing with the literature meaning of a given attribute and being satisfied with how the same attribute is being animated on street spaces as users interact with these spaces. In the case where the level of the test significance value is greater than 0.05, one fails to reject the null hypothesis that there is no difference in the proportions. This means that the level of expectation matches with the level of satisfaction and thus one can say there is spatial justice. If there is a difference between the distributions, however, it then translates to spatial injustice and it has implications on the type of interventions that are required to realise human smart street spaces.

13.4 FINDINGS

This section presents the key findings on the gap analysis between the users' expectations and the lived experience on the spatial justice attributes of safety, accessibility, legibility, variety and street maintenance and management. The main assumption is that if there is a gap between the two, then there is a need to reduce the gap through interventions that are smart city-oriented. Analysis of street users' needs prioritisation framework shows areas that the local municipalities in the SRTs should consider putting efforts towards.

13.4.1 STREET USERS' PERCEPTION OF SAFETY

Street space users' perceptions of street safety and security of the street were measured using the following indicators, which were derived from various literature: (i) presence of police, (ii) presence of other users, (iii) street lighting, (iv) the protection offered by other users, (v) presence of cameras and (vi) presence of fences. The main assumption is that if users' expectations on the operational measures of safety match their satisfaction, then the street space is spatially just in terms of street safety and security. Table 13.1 shows the results from the Wilcoxon rank-sum test for equality of distributions that were performed to test the following hypothesis.

> Null hypothesis (H_0): The difference between the distribution of agreeing with the perceived meaning of street safety and the satisfaction experience of safety and security equals 0.

TABLE 13.1
Safety and Security Distributions Comparison

Distributions compared	Z-statistic	P-value
The perceived meaning of street safety		
Experience in street safety	5.0380	0.0000

Source: Authors

The alternative hypothesis (H_1): There is a difference between the distribution of agreeing with the perceived meaning of street safety and the satisfaction experience of safety and security.

At a 5% significant level, a difference was established. There is a difference between the distribution of agreeing with the perceived meaning of street safety and the satisfactory experience of safety and security of street users. Where such differences exist, it shows that there is a gap between street users' expectation and their lived experiences. This indicates spatial injustice or failure of streets as public spaces in meeting the right to safety and security of street space users.

13.4.2 STREET USERS' PERCEPTION OF ACCESSIBILITY

Street space users' perception of accessibility was assessed using the main indicators of (i) wide- sidewalks, (ii) non-interference of sidewalks with parking, (iii) availability of a cycling lane and (iv) barrier-free spaces. The key assumption is that these indicators are key literature measures of assessing accessibility and if users are satisfied with these measures then the 'right-to-the-city' claims, i.e. claims for more freedom and access, are met. The right to access or spatial link reveals spatial justice in street spaces.

Tests for equality of distributions in agreeing with the perceived meaning of street accessibility and the satisfactory experience of accessibility were performed using the Wilcoxon rank-sum tests. The findings are highlighted in Table 13.2.

The null hypothesis (H_0), being that there should be a difference between the distributions, is rejected. Instead, the alternative hypothesis (H_1) is accepted. This states that there is a difference between the experienced access as compared to the perceived street accessibility, with a 5% significance level. This shows that there is a gap between street users' expectation of accessibility and their experience in reality as accessibility as a spatial justice variable is animated on street spaces. In the case study involving towns, the lack of adequate access to street space is interference of the right of ease of access by users, which is also an injustice of spatial link.

13.4.3 STREET USERS' PERCEPTION OF LEGIBILITY

Legibility perceptions of street space users were assessed using the indicators relating to (i) one's culture, (ii) familiarity with what features to find from start to end, (iii) well informed about street activities, (iv) memorable landmarks, (v) clear street

TABLE 13.2
Comparison of Distributions of Accessibility

Distributions compared	Z-statistic	P-value
The perceived meaning of accessibility		
Experience of accessibility	5.7440	0.0000

Source: Authors

TABLE 13.3
Comparison of Distributions of Legibility

Distributions compared	Z-statistic	P-value
The perceived meaning of legibility		
Satisfaction with the experience of legibility	2.2320	0.0256

Source: Authors

directional signs and (vi) ability to use a map. We argue that more legible street space is more spatially just than an illegible street. When street spaces are easily identifiable, a street user is empowered with the knowledge of where they are and how to go where they want. In this process, users claim their identity and right to their presence in a particular space. The tests for equality of distributions in agreeing with the perceived meaning of street legibility and the satisfactory experience of legibility were also performed using the Wilcoxon rank-sum tests. The findings are highlighted in Table 13.3.

The null hypothesis (H_0) is rejected in favour of the alternative hypothesis (H_1) which states that there is a difference between the distribution of agreeing with the perceived meaning of street legibility and the satisfactory experience of legibility at the 5% significant level. A gap between street users' expectations of legibility and their experience, in reality, is identifiable from the findings. This shows that there is an interference in the right of street users to better understand and enjoy more of their street environment as legibility is being animated on street spaces in the case study of towns.

13.4.4 STREET USERS' PERCEPTIONS OF VARIETY

Interrogation of the street space users' perception of variety was assessed using eight indicators, namely: (i) opportunities for work, (ii) opportunities for living in the street, (iii) opportunities for playing, (iv) connects to shops (v) connects to the park (vi) connects to malls (vii) connects to the bus terminus and (viii) connects to the market. Street spaces are multifunctional places that should offer users social, physical and economic benefits or rights through a variety of uses. The assumption is that street spaces that offer variety are more spatial just than street spaces that do

TABLE 13.4
Comparisons of Distributions of Variety

Distributions compared	Z-statistic	P-value
The perceived meaning of variety		
Satisfaction with the experience of variety	2.5590	0.0105

Source: Authors

not. The tests for equality of distributions in agreeing with the perceived meaning of street variety and the satisfactory experience of variety were performed and the findings are highlighted in Table 13.4.

The null hypothesis (H_0) is rejected in favour of the alternative hypothesis (H_1) which states that there is a difference between the distribution of agreeing with the perceived meaning of street variety and the satisfactory experience of variety at a 5% significant level. This gap between perceptions of meaning and the level of satisfaction shows that street users lack spatial claims to street space thereby infringing their rights to live, work or experience street spaces as multifunctional spaces that offer variety of uses.

13.4.5 STREET USERS' PERCEPTIONS OF MAINTENANCE AND MANAGEMENT

Interrogation on street space users' perception of maintenance and management was assessed using the indicators of (i) general cleanliness, (ii) replacement of street lights, (iii) storm water drain maintenance, (iv) availability of public toilets, (v) presence of street benches and (vi) pothole maintenance. These indicators are the most common performance measures of the basic level of maintenance and management that were used to describe the literature meaning of maintenance and management in this study. These indicators are also common in South African literature such as the New South African Redbook on Neighbourhood Planning and Design Guide (DHS, 2019). The maintenance and management of street spaces is a core function of local municipalities as mandated by the Municipal Systems Act 32 of 2000. However, maintenance and management are not the sole mandates of local municipalities, although this is typical in most small towns in developing countries. As street space users interact with street spaces, they experience the outputs from the maintenance and management of street spaces in numerous ways, which determine the (un)justness of their interaction with the street spaces. Similarly, the key argument is that if the users are satisfied with the measures of maintenance and management as they experience them on street spaces, then there is a spatially just relationship between the street users and the street spaces. Table 13.5 shows the findings from the Wilcoxon rank-sum test for the variable of maintenance and management.

The null hypothesis (H_0) is rejected in favour of the alternative hypothesis (H_1) which states that there is a difference between the distribution of agreeing with the perceived meaning of street variety and the satisfactory experience of variety at a 5% significant level. This gap between perceptions of meaning or expectations of users on maintenance and management measures and the level of satisfaction shows that

TABLE 13.5
Comparison of the Distribution of Maintenance and Management

Distributions compared	Z-statistic	P-value
The perceived meaning of maintenance and management		
Satisfaction with the experience in maintenance and management	17.7370	0.0000

Source: Authors

street users in the SRTs do not enjoy their right to basic services, which is an injustice to the spatial claims, links and power.

Street space maintenance and management is an equal responsibility of citizens as much as it is a responsibility of local municipalities. Discussions with officials from the case study of local municipalities concur that there is a need for more coordination amongst various municipality departments that deal with the regulation and management of streets. The findings also reveal that the effectiveness of citizens in regulating the use of street spaces depends on the extent to which they participate in decision-making. It was therefore critical to consider the participation aspect in creating spatially just streets in the three towns. The participation variable was operationalised by measuring (i) street user membership to any association involved in the use of streets, (ii) street user knowledge of the local municipality plan of the street and (iii) street user participation in any forums or seminars related to street issues. The study established that less than 10% of respondents across all towns had participated in any of the three stated scenarios. These findings revealed that users in SRTs lack awareness and are sometimes indifferent to their democratic right to participate in street space production processes. This is reflected in the responses that some users provided in their textual form such as '*I am not interested. I am not a politician. I am busy*'. The lack of a mobilised constituency of street users who pressurise for change in street spaces in the SRTs incapacitates planners to be more proactive in creating spatially just spaces (Fainstein, 2009:11). Therefore, users' participation enables planners to pursue and actualise a justice vision.

13.5 DISCUSSION OF FINDINGS

The study established that users' need for safety, access, legibility, variety, civility or maintenance and participation in decision making reflect the disparate right to the city claims of users. This shows that users have diverse needs on street spaces and do have the right to similarities and differences (Lefebvre, 1996b). The study confirms that there are statistically significant differences between users' expectations of how a particular attribute ought to be and the actual lived street space experiences. It shows that the street space users in SRTs are lacking in terms of making spatial claims, links and power that are the various forms of spatial injustices on street spaces (Basset, 2013). Some descriptions provided by users show that users disagree with the meanings of some indicators such as the presence of police and fence barriers as measures of safety; cycling lanes as improving accessibility; and the ability to use maps as a strategy to improve legibility. This reveals further that street users have different perceptions of the meaning of a spatial justice attribute. Street spaces are sites where many different stories emerge at once (Lefebvre, 1991; Soja, 1996; Massey, 2005). As a result, contestations and negotiations on the use of streets are inevitable (Loukaitou-Sideris and Ehrenfeucht, 2009; Uwayezu and de Vries; 2018).

Users hold indifferent or neutral perceptions towards some indicators, such as cultural identity as a requisite for street legibility and prioritisation of streets as places of living and play and street trading as a strategy to improve variety on street spaces. The indifference is explained by limited knowledge of public space planning as well as their rights to the city. Some differences in perceptions between towns are due to the different public space management systems and strategies in different municipalities. After all, spatial justice is a contextual concept where perceptions may also vary from place to place (Erdiaw-Kwasie and Basson, 2018).

From the observations made on the attribute of accessibility, areas that need improvement are in the sidewalk width and cycling lane. Opportunities for living and connection to the park are areas that need improvement to enhance variety as cities are not only places of work but also places of living and playing (Jones et al., 2007; Concilio and Rozzo, 2016; Stratford et al., 2020). Conversations with local municipality officials revealed that street traders' claims of their right to appropriation and right work are other eminent challenges that municipalities in SRTs need to address in framing spatially just street spaces.

13.6 CONCLUSIONS AND IMPLICATIONS FOR SMART STREET PLANNING

This study established that there are gaps between street users' expectations from a spatial justice attribute and their lived experience on street spaces in the SRTs. Therefore, users' spatial justice claims, links and powers which are also translated as right to the city claims are infringed. However, these users' needs and right to the city claims are also very diverse and they sometimes clash or conflict. The focus of spatial justice is to find a common ground amongst the conflicting claims and build solidarity across differences (Lefebvre, 1996b). Spatial justice as a vision is always evolving; however, it helps municipalities as key space producers in SRTs to identify and prioritise issues as informed by users' needs and challenges to promote the co-creation of just human smart streets. This study recommends the prioritisation of the maintenance and management attributes of street spaces. To a greater extent, adequate maintenance and management of street spaces enhance the realisation of other socio-spatial qualities of spatially just streets for example safety and legibility

As a starting point, municipalities in SRTs require a Public Space Design and Management Department that coordinates all activities that are specifically related to public spaces as currently many departments do management of public spaces in the local municipalities of the SRTs. This proposed department should lead in vision sharing of creating spatially just street spaces and initiatives that promote human smart cities. The human smart initiatives that should be considered as they are lacking in the SRTs include smart public toilets, streetlights, seating furniture, water, bins and participatory planning and maintenance of street spaces. For example, strategic placement of smart waste bins with sensors encourages proper waste disposal by users. Through big data management, bin sensors will communicate to the municipality when a bin is full and needs to be emptied. This is a paradigm shift from the traditional waste collection routines. To fund these initiatives, local municipalities need to shift their focus from relying on government grants to create sustainable partnerships with the local business community and community members.

The general public can also be involved more in the maintenance and management of street spaces by encouraging locally branded initiatives that promote civic responsibility such as the maintenance of a certain portion of a street or the maintenance of parklets. Through these initiatives, users can earn or lose points and the points are linked to other services such as free access to pre-paid toilets or parking, smart electricity or water recharging. These initiatives will ensure the co-creation or co-keeping and stewardship of spaces through the involvement of local actors. The co-creation of spaces with users involves taking small bold steps and experimentation with various possibilities. This is critical in navigating the complex phenomenon of human smart cities that are spatially just.

REFERENCES

Akkar, Z. M. 2005. Questioning the 'Publicness' of Public Spaces in Post-industrial Cities. *Traditional Dwellings and Settlements Review,* 16(2): pp. 75–91.

Babbie, E., & Mouton, J. 2012. *The Practice of Social Research.* Cape Town: Oxford University Press.

Bassett, S. M. 2013. *The Role of Spatial Justice in the Regeneration of Urban Spaces.* Champaign, IL: The University of Illinois at Urbana-Champaign.

Bentley, I., Alcock, A., Murrain, P., McGlynn, S., & Smith, G. 1985. *Responsive Environments: A Manual for Designers.* Amsterdam: Architectural Press.

Brown, Alison Margaret Braithwaite, & Kristiansen, A. 2009. *Urban Policies and the Right to the City: Rights, Responsibilities and Citizenship* [Project Report]. Paris: UNESCO, UN-Habitat. Available at: https://unesdoc.unesco.org/ark:/48223/pf0000178090

Cathelat, B. 2019. *Smart Cities Shaping the Society of 2030.* France: United Nations Educational, Scientific and Cultural Organization (UNESCO).

Concilio, G., & Rozzo, F. eds. 2016. *Human Smart Cities: Rethinking the Interplay between Design and Planning.* Switzerland: Springer International.

Crawford, M. 2016. Public Space Updates: Report from the United States. *The Journal of Public Space,* 1(1): pp. 11–16.

Cresswell, T. 2009. The Prosthetic Citizen: New Geographies of Citizenship. In Go, J. (ed.) *Political Power and Social Theory.* Bingley, United Kingdom: Emerald Group Publishing Limited: pp. 259–273.

Cruz, S. S., Roskamm, N., & Charalambous, N. 2018. Inquiries into Public Space Practices, Meanings and Values. *Journal of Urban Design,* 23(6): pp. 797–802.

Department of Human Settlements. 2019. *The Neighbourhood Planning and Design Guide: Creating Sustainable Human Settlements.* South Africa: Department of Human Settlements (DHS).

Depiné, Á., de Azevedo, I. S. C., Santos, V. C., & Eleutheriou, C. S. T. 2017. Smart Cities and Design Thinking: Sustainable development from the citizen's perspective. Paper read at Proceedings of the February 2017 Conference: IV Regional Planning Conference, Aveiro, Portugal.

Erdiaw-Kwasie, M. O., & Basson, M. 2018. Re-Imaging Socio-Spatial Planning: Towards a Synthesis between Sense of Place and Social Sustainability Approaches. *Planning Theory,* 17(4): pp. 514–532.

Fainstein, S. 2009. Spatial Justice and Planning. *Spatial Justice,* 1: pp. 1–13.

Gibbons, J. D. 1993. *Non-Parametric Statistics: An Introduction.* Newbury Park: SAGE Publications.

Hartman, L. M., & Prytherch, D. 2015. Streets to Live in: Justice, Space and Sharing the Road. *Environmental Ethics,* 37(1): pp. 21–44.

Hoogduyn, R. 2014. *Urban Acupuncture: Revitalising Urban Areas by Small Scale Interventions.* MSc Thesis. Sweden: Blekinge Institute of Technology Faculty of Engineering, Department of Spatial Planning.

Jacobs, J. 1961. *The Death and Life of Great American Cities.* New York: Random House.

Jalaladdini, S., & Oktay, D. 2012. Urban Spatial Spaces and Vitality: A Socio- Spatial Analysis in the Streets of Cypriot Towns. *Procedia- Social and Environmental Sciences,* 35: pp. 664–674.

Jones, P., Boujenko, N., & Marshall, S. 2007. *Link and Place: A Guide to Street Planning and Design.* London: Local Transport Today Ltd.

Konzen, L. P. 2013. *Norms and Space: Understanding Public Space Regulation in the Tourist City.* PhD Thesis. Lund University.

Kothari, C. R. 2004. *Research Methodology: Methods & Techniques.* New Age International (P) Ltd. doi: 10.1017/CBO9781107415324.004

Lefebvre, H. 1991. *The Production of Space.* Oxford: Blackwell.

Lefebvre, H. 1996a. Space: Social Product and Use-value. In Brenner, N. & Elden, S. (eds) *State, Space, World: Selected Essays.* Minneapolis: University of Minnesota Press, pp. 167–184.

Lefebvre, H. 1996b. The 'Right to the City. In Kofman, E. & Lebas, E. (eds) *Writings on Cities.* Oxford: Blackwell Publishing, pp. 147–159.

Loukaitou-Sideris, A., & Ehrenfeucht, R. 2009. *Sidewalks: Conflict and Negotiation over Public Space.* Cambridge: MIT Press.

Lynch, K. 1984. *Good City Form.* Cambridge: MIT Press.

Madanipour, A. 2013. *Whose Public Space? International Case Studies in Urban Design and Development.* Oxon: Routledge.

Marcuse, P. 2012. Whose Right (s) to What City? In Brenner, N., Marcuse, P., & Mayer, M. (eds) *Cities for People, Not for Profit: Critical Urban Theory and the 'Right to the City'.* London: Routledge, pp. 24–41.

Massey, D. 2005. *For Spaces.* London: Sage.

Merrifield, A. 2011. The 'Right to the City' and Beyond: Notes on a Lefebvrian Re-Conceptualization. *City,* 15(3–4): pp. 473–481.

Middleton, J. 2018. 'The Socialities of Everyday Urban Walking and the 'Right to the City'. *Urban Studies,* 55(2): pp. 296–315.

Miller, K. 2007. Introduction: What Is Public Space? In *Designs on the Public: The Private Lives of New York's Public Spaces.* Minneapolis; London: University of Minnesota Press, pp. Ix–Xxii. doi: 10.5749/j.ctttv5pq.4

Ng, M. K., Tang, W. S., Lee, J., & Leung, D. 2010. Spatial Practice, Conceived Space and Lived Space: Hong Kong's 'Piers Saga 'through the Lefebvrian Lens. *Planning Perspectives,* 25(4): pp. 411–431.

Oliveira, A., & Campolargo, M. 2015. From Smart Cities to Human Smart Cities. *Kauai Conference Paper: 48th Hawaii International Conference on System Sciences (HICSS).* pp. 2336–2344. doi: 10.1109/HICSS.2015.281.

Osóch, B., & Czaplińska, A. 2019. City Image Based on mental Maps—the Case Study of Szczecin (Poland). *Miscellanea Geographica,* 23(2): pp. 111–119.

PLANUM. 2014. The Human Smart Cities Cookbook. *Journal of Urbanism,* 28(1): pp. 1–58.

Shrestha, B. K. 2011. Street Typology in Kathmandu and Street Transformation. *Urbani Izziv*, 22(2): pp. 107–121.

Soja, E. W. 1996. *Thirdspace: Journeys to Los Angeles & Other Real & Imagined Places*. Chichester: Wiley.

Soja, E. W. 2010. Seeking Spatial Justice. London: University of Minnesota Press (Globalisation and Community Series).

Stratford, E., Waitt, G., & Harada, T. 2020. Walking City Streets: Spatial Qualities, Spatial Justice and Democratizing Impulses. *Transactions of the Institute of British Geographers*, 2020(45): pp. 123–128.

Tsoriyo, W. W. 2021. *Spatial (In)Justice and Street Spaces of Selected Small Rural Towns In Vhembe District of Limpopo Province, South Africa*. (Unpublished doctoral thesis). Thohoyandou: University of Venda.

Uwayezu, E., & de Vries, W. T. 2018. Indicators for Measuring Spatial Justice and Land Tenure Security for Poor and Low Income Urban Dwellers. *Land*, 7(84). doi: 10.3390/land7030084

Varna, G. M. 2011. *Assessing the Publicness of Public Places: Towards a New Model*. PhD Thesis. University of Glasgow.

Varna, G. M. 2014. *Measuring Public Space: The Star Model*. London: Routledge.

Yu, H. 2017. The Publicness of an Urban Space for Cultural Consumption: The Case of Pingjiang Road in Suzhou: *Communication and the Public,* 2(1): pp. 84–101.

14 Synchronising the Spatial Planning Legislative and Administrative Frameworks of Mining and Other Human Settlements in Zimbabwe

Audrey Ndarova Kwangwama, Willoughby Zimunya and Wiseman Kadungure

CONTENTS

14.1 Introduction and Background 197
14.2 Theoretical Framework 198
14.3 Literature Review 201
14.4 Methodology 203
14.5 Results and Discussion 203
    14.5.1 Legislative Framework and Siting of Mining Works 203
    14.5.2 Limits of Spatial Planning in Mining Settlements 204
    14.5.3 Towards Sustainable and Smart Mining Settlements 206
14.6 Conclusion and Recommendations 207
References 209

14.1 INTRODUCTION AND BACKGROUND

Developing sustainable and smart cities and towns is now an imperative in the face of rapid urbanisation accompanied by multiple urban challenges. To ensure sustainable urbanisation and smart urban development, governments are encouraged to adopt holistic planning frameworks. A fragmented planning policy is selective in nature as it uses different planning standards for particular human settlements. Comprehensive planning policy is one of the pathways for achieving urban sustainability and smartness as it ensures the universal application of national planning standards in the development of all human settlements including mining settlements.

Despite the importance of comprehensive policies as a pathway for promoting sustainable urban development, Zimbabwe as other settler mining economies founded on mining in Southern Africa still has a fragmented policy framework for guiding the development of towns. The country has different policies governing the establishment and development of mining-related townships which are guided by the Mines and Minerals (MM) Act (Chapter 21:05). This mining legislation gives little consideration to aspects that promote the sustainable existence of these urban centres beyond the life of mining operations, resulting in the emergence of ghost settlements after mine closures. Mining settlements are thus not governed by the Regional Town and Country Planning (RTCP) Act (Chapter 29:12) which is the centre piece legislation of guiding the development of other human settlements.

Kamete (2012) observes that mining towns do not comply with the national minimum housing standards and infrastructure requirements. This has created unique challenges in decommissioning mining operations as the affected mining settlements cannot be sustained, given their substandard conditions. Consequently, the poor urban development of the mining settlements is negatively affecting socio-economic development in the country in general and the quality of life of the affected residents in particular. It is therefore important that the policies guiding urban settlements are harmonised to promote urban sustainability and smart settlements in the context of sustainable development goals (SDGs) particularly SDG11 that focuses on making cities and human settlements inclusive, safe, resilient and sustainable.

Failing to have proper policies to guide sustainable urban growth is detrimental to the achievement of SDGs as no country has attained the upper middle-income status with poorly developed cities. Zimbabwe has opportunities to learn from other countries and local experiences on how an integrated policy framework can promote sustainable urbanisation. Therefore, this chapter seeks to examine the possibilities of developing a holistic planning policy framework that integrates the development of mining townships and other existing urban settlements. The chapter is organised into six sections that focus on the introduction and background, theoretical framework, literature review, methodology, results, conclusion and recommendations.

14.2 THEORETICAL FRAMEWORK

The focus of this chapter is on the synchronisation of spatial planning and administrative frameworks of mining and other human settlements. It is informed by two theories: the first on spatial planning and enclaves. Spatial planning refers to the

> methods used largely by the public sector to influence the future distribution of activities in space... with the aim of creating a more rational territorial organisation of land-uses and linkages between them, to balance demands for development with the need to protect the environment and to achieve social and economic objectives.
>
> (The Compendium of European Spatial Planning Systems, 1997: 23)

In trying to attain better co-ordination, spatial planning straddles three dimensions, including the horizontal dimension cross-cutting sectors, the vertical dimension where it cuts across various levels of jurisdiction and the geographical dimension that cross-cuts administrative boundaries (Cullingworth and Nadin, 2006).

Compared to the traditional land-use planning/physical planning, spatial planning has a broader focus in contributing to the achievement of sustainable development that has been regarded as more positive and comprehensive. Thus, the major objective of spatial planning is to regulate development and diverse land-uses to protect public interest and to achieve sustainable development defined as 'development that meets the needs of the present without compromising the ability of future generations to meet their own needs' (The Brundtland Commission 1987: 43). The principle of sustainable development in the context of spatial planning challenges the entrenched power of short-term costs and benefits of a development in preference to long-term costs and benefits brought in by longer-term visions of spatial planning (Adams, 1994). As a field of study, spatial planning can be traced back to town planning which emerged in the nineteenth century to address the health and sanitation problems from industrialisation in Western Europe. Early town planning pioneers argued for the promotion of well-ordered, healthy and attractive human settlements comprehensively designed from the onset. This was achieved through physical action taken by the state and made legitimate through national legislation.

Attaining social and economic objectives is critical in town planning. From a social perspective, town planning makes people's lives happier and enjoyable since it results in the creation of a physical environment that is conducive to good health allowing convenient and safe passage of people from place to place (Keeble, 1969). Planning also creates a proper spatial relationship between communities in a region and the constituent parts of a town, compactness of development and an efficient arrangement of communication routes resulting in economic activities being undertaken more efficiently. Spatial planning plays a vital role in ensuring efficiency, equity in the usage of land and other resources and ensures protection and conservation of the natural, built and historic resources and environments. As an anchor to economic growth, the spatial planning activity ensures the availability of land for development and co-ordinates the provision of physical infrastructure, including highways and utilities. The development of inclusive communities and the opportunity to live in attractive, safe and healthy neighbourhoods would be impossible without spatial planning.

The RTCP Act defines development as undertaking building or mining operations, making material change of land uses or buildings, displaying advertisements, disposing refuse or waste materials on any land and using vehicles and similar objects for residential purposes. Miles, Berens, Eppli, and Weiss (2007) also define development as a sequentially linked process that physically transforms a vacant piece of land to a new use with new buildings or other human-made improvements using land, labour, capital, management, entrepreneurship and partnerships. Both definitions give emphasis on land and buildings. However, it is critical to highlight that mining operations are also regarded as development. Any activity defined as development requires a permit from a local authority prior to its implementation. The permit requirements are outlined in the RTCP Act (Chapter 29:12) of 1996 and the General Development Order.

All municipalities, town councils and some rural district councils are designated as local planning authorities and derive their planning powers from the RTCP Act which delegates them powers to plan, control and regulate development in their areas of jurisdiction. Development control refers to the regulation of building and land uses to ensure that they comply with the approved statutory plans and are arranged in an orderly manner. The planning process involves the preparation of statutory plans that are used to regulate development. Developmental control is conducted within a spatial planning framework defined as a wide range of statutes, regulations, directions, policy statements, circulars, guidance and other official documents used when regulating planning activities (Cullingworth and Nadin, 2006). The purpose of a planning framework is to avoid inconsistent decision-making and to resolve conflicts arising from individual development proposals.

In the context of Zimbabwe's planning framework, the RTCP Act is the centrepiece of the spatial planning system, including development control (Wekwete, 1989a). The Act provides for regional planning, bestows planning powers on certain local authorities and provides for the preparation and implementation of master and local plans. However, there are other pieces of legislation that impacting spatial planning that include the MM Act (Chapter 21:05), the Environmental Management Act (Chapter 20:27), the Land Acquisition Act (Chapter 20:10), the Housing and Building Act (Chapter 22:07), the Water Act (Chapter 20:22) and the Roads Act (Chapter 13: 18). These pieces of legislation relate to specific central government sector ministries. For example, the MM Act is administered by the Ministry of Mines and Mining Development. The government has not synchronised these pieces of legislation resulting in the MM Act overriding the RTCP Act in the planning of human settlements. The MM Act was deliberately given powers to override other pieces of legislation given the importance that was attached to mining in general and gold mining, in particular, during the formative years of the settler economy.

The term 'enclave' is found in most contemporary languages: German *'enklave'*, French, Spanish and Italian *'enclave'*, Russian *'anklav'* and Swedish *'enclav'*. The root of the word is the Latin word *inclavatus* meaning 'shut in, locked up' and *clavis* meaning a 'key' (Vinokurov, 2007: 9). Enclave refers to 'any small distinct area or group enclosed or isolated within a larger one' (https:www.dictionary.com/browse/enclave). This definition is more related to the sociological and social sciences perspectives that regard enclave as a compact settlement different from the neighbouring area, which may manifest in the form of ethnic or religious groupings ranging from Chinatowns to ghettos (Vinokurov, 2007).

Rubbers (2019) provides a similar geographical viewpoint to the sociological perspective and considers enclaves as enclosed spaces that companies in the extractive sector, such as mining, use for production. This definition depicts the spatial inscription of material space delineated on a physical map. Vinokurov (2007) also defines enclave from an economics perspective where dominant multinational companies operating in a national economy are denoted as enclaves. Rubbers (2019) concurs with this perspective and highlights that within the political economy context, the term enclave refers to the poor linkages that foreign multi-national companies develop with domestic firms in developing countries.

Mining industries fall within the extractive sector and mining operations have been viewed as enclaves in areas they operate. Isik, Opalo and Toledano (2015) argue that natural resource concessionaires, such as mining and plantations, have traditionally operated on the basis of enclaves in the development of infrastructure. They provide their own sources of power, transport infrastructure and services. The rationale for this enclave approach has been to guarantee uninterrupted and reliable infrastructure on their operations. It has however been observed that large-scale investments in physical infrastructure have not always been aligned with national infrastructure development plans. Consequently, there have been missed opportunities to promote the shared use of infrastructure and exploitation of potential synergies to strengthen linkages between the extractive resources and the broader economy (Isik, Opalo and Toledano, 2015). Thus, SDG11 focusing on attaining sustainable and smart settlements has been compromised. Furthermore, houses in the mining compounds do not comply with the national spatial planning, housing and infrastructure standards.

14.3 LITERATURE REVIEW

The development of sustainable and smart cities is a topical development issue considering that the world is rapidly urbanising. African cities should be sustainable and smart to cope with various challenges, such as climate change, urban poverty, pandemics like COVID-19 and unemployment, among others associated with their rapid growth. A sustainable city comprises a well-functioning economy, ecological balance, social development and sound governance system. While urban sustainability is difficult to achieve, governments are encouraged to adopt strategies that foster smart urban development. The smart city concept has emerged as an innovative way of addressing urban issues related to rapid growth, thereby transforming and improving productivity, governance, service delivery, environmental sustainability, mobility and liveability in urban areas (Allam, 2018). However, the smart city strategies are only able to promote sustainable development when they are among other issues guided by sound and holistic spatial planning frameworks.

Mining as an activity has been a key economic driver globally. The high levels of industrial and overall economic development attained in Western Europe, North Eastern USA and the Witwatersrand Region in South Africa were anchored on mining (Sango, Taru, Mudzingwa and Kuvarega, 2006). Mining is a major source of revenue for governments, playing a significant role in employment creation, skill development and as a key source of attraction of Foreign Direct Investment (FDI). Zimbabwe is no exception as mining has been a significant contributor to the country's foreign exchange earnings, gross domestic product (GDP), FDI and employment creation contributing between 9.5–11.4% for the period 2009–2014, (Chakanya, 2016).

As already highlighted, the mining sector in Zimbabwe is regulated by the Ministry of Mines and Mining Development, the principal agency mandated to administer the MM Act of 2001 (Chapter 21:05) which has been amended over the years to date (Kondo, 2011; Zamasiya and Dhlakama, 2016). The regulation of this sector includes approval of siting of mining works, installation of machinery and infrastructure, adjudication of disputes and monitoring the safety of mining operations and other requirements. The approval of siting works although largely a spatial planning activity does not subject the establishment of mining settlements to the provisions of the RTCP Act, the centrepiece legislation guiding spatial planning in Zimbabwe. This has resulted in the emergence of unsustainable settlements with substandard infrastructure, especially in residential compounds that were established for the low-income workers.

Mining towns developed as capitalist centres of accumulation on the basis of private capital of colonial corporations. The state provided the regulatory framework for establishing the mining settlements and marketing the minerals. To control their workers and achieve higher productivity, colonial corporations, including mining companies provided housing (Rubbers, 2019). Residential mining compounds were built for low-income black workers using low inappropriate standards. Some of the developed houses described in vernacular as '*misana yenzou*' meaning elephant-shaped houses were dome shaped and not compliant with the national housing standards and unfit for human shelter. As the mining settlements evolved, they also experienced the unique problem of municipal governance as evidenced by the siting, sanitation and overcrowding challenges. Njoh (2007) argues that this problem emerged as mining areas fell outside the administrative jurisdiction of existing municipal corporations and township ordinances and hence these settlements developed in an unsustainable manner.

In Zambia, the colonial state legislation required mining companies to provide housing for their employees. The provision of housing and complementary services demanded the establishment of an innovative and competent municipal governance framework (Njoh, 2007). The colonial state in Zambia reacted to this inevitable need by passing the Mine Township Ordinance of 1933. This legislation established a mining municipal governance framework similar to the governance framework in municipal townships. Njoh (2007) points out that mine townships in Zambia had boards that bestowed them with powers to maintain and implement colonial legislation involving hygiene, public health and socio-economic development. The Mine Township Boards were also required to preserve order and assist in the achievement of wider motives of colonialism by guaranteeing the welfare of mine employees.

In Zimbabwe, spatial planning during the early colonial period was attained through grid-iron layouts, bye-laws and public health ordinances and covered municipal areas only. Mining settlements fell under the jurisdiction of the MC and were established as provided in the MM Act (Chapter 21:05). Consequently, mining settlements developed as unique enclaves under the jurisdiction of the mining company and the MC outside the regulation of local planning authorities. The statutory spatial planning framework was thus not applied to control and manage the buildings within mining settlements. For example, mining companies constructed houses, hospitals, schools, canteens and social clubs for their work force within the mining complex for their employees. No layout designs and architectural plans for the mining buildings were submitted to the local planning authority. All development proposals were regulated by the MC under the siting of works. However, when mining operations are decommissioned, they are handed over to the local planning authority for administration. This lack of synchronisation of spatial planning and mining legislative frameworks has resulted in the development of sub-standard structures, which are not smart and sustainable.

14.4 METHODOLOGY

The research is informed by the social interpretivist paradigm, the qualitative research approach and the case study research design. The chapter focused on MGM as a case study. Data collection was based on desk review, key informant interviews and observations. The research instruments were designed to collect data related to spatial planning and mining enclaves, the focus of this chapter. An extensive literature review was undertaken on mining and spatial planning. Documentary reviews informed the study with information on spatial planning and development of mining settlements as enclaves. Key informant interviews were conducted with a senior official of the Department of Spatial Planning and Development in the Ministry of Local Government and Public Works, a senior legal officer with the Ministry of Mines and Mining Development, planning officials from the Mazowe Rural District Council (MRDC) and a former senior employee of MGM. The study also used observations. An observation checklist was used to collect information on the type of infrastructure and quality of houses developed at MGM.

14.5 RESULTS AND DISCUSSION

14.5.1 Legislative Framework and Siting of Mining Works

The MGM, also known as the Jumbo Mine, is located in the Mashonaland Central Province approximately 50 km to the north of Harare. In terms of geology, the mine is located on the west central portion of the Harare greenstone belt with high-grade ore (Metallon Corporation, 2014). It is one of the oldest mining operations in Zimbabwe dating as far back as 1890, the time of British colonisation and occupation of the country. The mine has 247 claims stretching over 2,939 hectares of land (Metallon, 2014). The mine is owned by Metallon Gold which operated it until September 2018 when it was placed under care and maintenance due to operational challenges. At that juncture, the administrative jurisdiction of MGM was handed over to MRDC (Business Times, 2018). However, it should be highlighted that the land under the mining lease was not handed over but the existing buildings and infrastructure facilities only.

The key informant interview with the planning officer at MRDC revealed that the establishment of MGM was processed through Part XIII of the MM Act (Chapter 21: 05) which provides for the Control of Siting of Works of Mining Locations. Section 234 of the Act provides that a miner should obtain an approved plan from the MC prior to the erection of certain works on a mining location. Such works include installation of machinery or plant used for the treatment of ores, concentrates, tailings, slimes or other residues and construction of mine dumps, dams for storage of waste water or slimes, residential compounds for employees, buildings of a permanent nature, sewerage disposal works, recreation grounds and roads.

Key informant interviews with the Senior Legal Officer at the Ministry of Mines and Mining Development and the planning officials at the Department of Spatial Planning and Development and MRDC indicated that the mining site work application procedure entails that the miner submits their application to the MC illustrating the location of the proposed mine, existing works and natural physical geographical features, such as rivers, valleys and hills. On receipt of the application, the MC consults affected landowners and each holder of a contiguous mining location. The MC also consults the provincial planning officer of the Department of Spatial Planning and Development, the regional mining engineer, the provincial water engineer, the regional land inspector of the Environmental Management Agency and the regional agricultural extension officer of the Department of Agricultural Technical and Extension Services.

It has been observed that most of the mining siting works relate to spatial planning namely siting and construction of dumps, dams and residential compounds for employees, sewerage disposal works, recreational grounds and roads. The MC considers and processes all views received regarding the proposed mining works prior to making a decision: approving the plan, approving the plan with amendments and conditions he/she may deem necessary or refusing the application. The planning official at MRDC highlighted that the MC acts as the local planning authority for the establishment of mining settlements. This observation concurs with Kamete (2012) who in his analysis of the administration of the closed Mhangura Mine argues that the existing mining legislation bestows spatial planning powers to the MC who does not have the technical capacity on spatial planning.

The Senior Legal Officer at the Ministry of Mines and Mining Development highlighted that the ministry initiated legal reforms to amend the MM Act in 2015. The several proposed changes in The Mines and Minerals Amendment Bill include provisions for strategic mineral resources, composition of the Mining Affairs Board, prioritisation of indigenous mining, environmental protection, mining titles, establishment of computerised mining cadastre system, dispute resolution, mitigation of farmer–miner conflict, preparation of work plans, application of the use it or lose it principle and the payment of fees and levies to rural district councils by mining companies. None of these proposed amendments seek to harmonise the national planning, housing and infrastructure standards enshrined in the RTCP Act with the MM Act.

14.5.2 Limits of Spatial Planning in Mining Settlements

The current legislative framework bestowing spatial planning powers of mining settlements to the MC has created enclave-mining settlements outside the jurisdiction of local planning authorities that do not comply with minimum national spatial planning, building and infrastructural standards. The RTCP, the centrepiece of spatial planning, outlines the procedures of preparing subdivision layouts. The key informant interviews with an MRDC planning official revealed that as a local planning authority, they were not involved in the processing of the MGM subdivision layout plan. The preparation of the subdivision layout was undertaken by the miner, Metallon Gold, and submitted to the MC as mining siting works under Section 234 of the MM Act (Chapter 21:05) for approval. Any spatial planning issues which should have been corrected at the layout design stage were not picked up as the MC does not have expertise in spatial planning. Under the existing mining legislation, mining settlements fall outside the jurisdiction of local planning authorities and are only handed over to them for administration after decommissioning of the mining works.

The existing legislative arrangement on the siting of works excludes the local authority from not only layout designing but also from regulating and managing the construction of buildings through development control as enshrined in the RTCP Act. The planning official at MRDC deplored the failure by MGM to meet the minimum spatial planning standards on the construction of buildings, roads and infrastructure. Model building by-laws require a minimum building height (floor to ceiling) of 2.7 m. However, there are some houses in the residential compound that have a height of 2.3 m which falls below the required minimum height of 2.7 m. Typical houses in the residential compound with a height of 2.3 m include the infamous '*misana yenzou*' unfit for human shelter. The model building by-laws also stipulate a minimum room ventilation of 10%. However, the worker houses at the residential compound at MGM provide room ventilation of below 10%. The purpose of building standards stated in the model building by-laws is to provide a healthy living space for occupants. Thus, the failure of the workers' houses at the mining compound to meet the stipulated minimum ventilation requirements exposes workers to diseases, such as tuberculosis, due to poor ventilation that compromises the principles of sustainability and smart settlements.

Spatial planning standards outline the minimum sizes of road reserves in accordance with the road hierarchy in a human settlement. All planned human settlements are serviced by a major road system which provides an accessibility framework through which the settlement is spatially arranged and individual properties are accessed. As already indicated earlier on in the chapter, economy and efficiency are very critical elements of spatial planning design (Government of Zimbabwe, 1993). Key informant interviews with the MRDC Planning Official revealed that the settlement has a poor road hierarchy. The road hierarchy standards for low-income residential areas equated to the mining residential compound stipulate a range of 2.5 m for foot paths, 8.5 m for access roads, 12 m for distributor roads and 15 m for district distributors and 20 m for primary distributors.

Most of the roads in the residential compound of the mine resemble footpaths. There is no road hierarchy with most access road widths measuring 3 m as opposed to the required 8.5 m. Consequently, vehicular traffic movement is severely restricted. Some of the roads have dead ends, which promote poor accessibility to some properties. Such roads with dead ends could have been designed as cul-de-sacs, which provide the most economical layout design. The head of the cul-de-sac with a minimum turning circle of a radius of 13.5 m as stipulated in the spatial planning standards would enable traffic to turn promoting a smooth traffic flow in the residential compound.

Planning infrastructure standards stipulate that a high-density residential area should be serviced with reticulated water supply and sewerage. The roads should be constructed with storm water drains properly laid out. However, this is not the case of the MGM residential compound. Water at the worker residential compound is provided through communal taps. Water reticulation for individual properties is provided for houses of the management personnel of the mine. The workers at the mine use blair toilets as their houses do not have a reticulated sewage system. It could be argued that the miner, Metallon Corporation, took cheaper options for infrastructure provision that are below the required standards of a low-income residential area most probably to save costs. Such infrastructure should be upgraded to the required standards to be safe for human settlements.

In designing residential layouts, the frontage of the stand should be much shorter than the depth so as to rationalise on the provision of services. However, the layout of the residential compound at the MGM was conducted in the opposite manner. Some of the stands are wrongly laid out with the frontage much longer than the depth. Consequently, much fewer stands have road access. If water and sewage reticulation were to be provided for the residential compound, the current layout design would result in high servicing costs. If the MRDC had been involved in the approval of siting of works, such design errors would have been picked up and corrected prior to the construction of the residential compound.

Any human settlement requires shopping facilities and social services. Spatial planning layout standards categorically state that such facilities should be centrally located in a settlement to ensure easy access and comply with the minimum radius of 800 m for commercial facilities. Contrary to this zoning requirement, commercial and institutional facilities, such as shops and schools at MGM, are located at the periphery of the residential compounds. This inconveniences residents who have to walk relatively long distances to access shopping facilities. It is impossible for local authorities to upgrade existing mining settlements to meet the planning standards of the local planning authorities due to the high levels of required financial resources (Kamete, 2012). Furthermore, if local authorities were involved in the initial establishment of mining settlements, they would ensure full compliance with the spatial planning standards.

Energy provision is very critical for any human settlement. The main source of energy at MGM is electricity as provided by the Zimbabwe Electricity and Distribution Company (ZETDC). The mine and the residential compound are both reticulated with electricity. However, the main criticism on the reticulation of the residential compound is that the electricity lines run over rooftops which poses a major risk to residents who can be electrocuted. This case at MGM residential compound is however different from the problem of a clinic in Marange that was built by the operating diamond miners without electricity. The clinic was also built with substandard facilities and with lack of electricity—it could not be handed over to the Ministry of Health and Child Welfare for administration. Both situations, however, illustrate the importance of involving spatial planning in designing mining settlements and ancillary facilities.

14.5.3 Towards Sustainable and Smart Mining Settlements

This section discusses the emerging models in Zvishavane and how they can be built upon to develop sustainable and smart mining settlements in Zimbabwe.

Although there has been a problem of mining enclaves in Zimbabwe, there are however some mines that have realised the importance of integrating their workers' residences within the spatial framework of existing towns under the jurisdiction of a local planning authority. This arrangement enables the local planning authority to control and regulate development and this ensures that national spatial planning standards are complied with. In the event that the mine decommissions its operations and closes down, there would be no need of handing the mining settlement over to the local planning authority. It would have been integrated in the municipal governance system from the onset. A case in point is the Mimosa mine in Zimbabwe.

The Mimosa mine is located in the Midlands Province 35 km to the south-west of Zvishavane town. The mine built approximately 2,000 houses in the Zvishavane residential areas targeted for its workforce exceeding 2,000 employees (Makore and Zano, 2012). The homeownership schemes catered for all its employees ranging from general workers, supervisors and managers. The Mimosa mine approached the Zvishavane Town Council requesting land for residential development. The council approved the request and prepared residential layouts, which were approved by the Department of Spatial Planning and Development. The servicing of the land, that is, construction of roads, water and sewage reticulation pipes and storm water drains was supervised by the Zvishavane Town Council's civil engineers.

The Mimosa mine prepared building plans for the houses and submitted them to the council for approval as stipulated in the model building by-laws. In terms of financing the residential project, the Mimosa mine entered into an agreement with the FBC Bank. The bank financed the construction of the houses. The Mimosa mine employees would own the houses after ten years of service to the mine. The Zvishavane Town Council's building inspectors inspected the construction of the houses until the certificates of occupation were issued. Thus, the residential schemes are within the residential zones of the Zvishavane Town Council, complying with the national spatial planning and infrastructure standards as any other residential neighbourhoods in Zvishavane.

The Mimosa mine has also been involved in corporate social responsibility projects. For example, the mine constructed a 12 km stretch of tarred road connecting the Zvishavane–Bulawayo Road and the road to the mine. The mine also assists the Zvishavane Town Council with the maintenance and repairs of its water treatment plant through seconding of their engineers and equipment to the council. Patching of potholes in some of the council's roads used by the mine's commuter omnibuses has also been a major focus area of the mine's corporate social responsibility.

14.6 CONCLUSION AND RECOMMENDATIONS

The study has revealed that while Zimbabwe has a long history of urban development in the modern style dating back to the late nineteenth century it still has separate planning policies for controlling the building of urban settlements. The country still has a dual planning policy framework in the form of the MM Act guiding the development of mining towns as well as the RTCPA for controlling the development of other human settlements. This fragmented policy framework results in the creation of poor mining-related towns as compared to other towns that are guided by the RTCPA. The case study of MGM provides empirical evidence about the inability of the existing policies to promote smart urban development. In light of this analysis, the study has concluded that the long-term implication of this disjointed policy framework is the undermining of sustainable urban development as the country continues to experience the creation of substandard mining towns side by side with better urban settlements.

Further, the study has revealed that there are opportunities for harmonising the planning policies for controlling the development of both mining towns and other urban settlements in Zimbabwe. It is recommended that legal reforms in the Mines and Minerals Amendment Bill should include synchronisation of the spatial planning legislative and administrative frameworks of mining and other human settlements by removing all spatial planning–related provisions including the siting of works from the MM Act and providing them in the RTCP Act and to transfer the jurisdiction of all mining settlements from the MC to local authorities. Such amendments would involve the engagement of the Ministry of Mines and Mining Development, the Ministry of Local Government and Public Works and the Ministry of National Housing and Social Amenities.

The long experience of Zambia which has been relying on a single policy to guide the development of all urban settlements is quite informative of what can be achieved in Zimbabwe. The literature review indicated that Zambia has a comprehensive planning policy (Njoh, 2007) and hence is not experiencing any urban challenges related to issues in Zimbabwe. In addition, the local experiences of Zvishavane which has initiated the process of integrating mining townships into the town is also a good indication of opportunities that exist in harmonising the planning policies for urban settlement developments in Zimbabwe. However, while efforts by Zvishavane are commendable in promoting sustainable urban development, they should be supported by the central government. To this end, based on both scholarly and empirical evidence, it can be concluded that it is feasible to integrate the policies guiding the development of urban areas in Zimbabwe.

Based on the analyses and conclusions of this study, it is recommended that the development of mining enclave approach for urban development should be abolished. Zimbabwe should formulate an integrated planning policy to control the development of all urban settlements, including those related to mining. The development of mining enclaves should be abolished and where mining-related settlements are really necessary, they should comply with the new integrated planning policy in terms of infrastructure, building and other relevant standards. In addition, the development of such settlements should be accompanied by a sustainability plan. The emerging new model in Zvishavane could be implemented at the national level to address the problem of enclave mining settlements and contribute to the development of sustainable and smart mining settlements. Conclusively, this chapter argues that it is important for Zimbabwe to formulate a comprehensive planning policy that fosters sustainable urbanisation and facilitate realisation of the SDGs, which promote high living standards for all.

REFERENCES

Adams, D. (1994). *Urban Planning and the Development Process*. London, UCL Press.
Allam, Z. (2018). Contextualising the smart city for sustainability and inclusivity. *New Design Ideas*, 2(2): pp. 124–127.
Brundtland, G. H. (1987). *Our Common Future*. Report of the World Commission on Environment and Development. Accessed online: www.un-documents.net/our-common-future.pdf. Accessed 13 January 2021
Business Times (Saturday 1 September 2018). Breaking: Metallon closes Mazowe mines. Accessed online: https://businesstimes.co.zw/breaking-metallon-closes-mazowe-mines/. Accessed 13 January 2021.
Chakanya, N. (2016). Extractivism & sustainable alternative models of economic development. In: Friedrich-Ebert-Stiftung (Ed), *Extractives and Sustainable Development ii: Alternatives to the Exploitation of Extractives*. Harare, Friedrich-Ebert-Stiftung Zimbabwe Office: pp. 1–26.
Cullingworth, B. and Nadin, V. (2006). *Town and Country Planning in the UK*. London and New York, Routledge.
Government of Zimbabwe (1993). *Design Manual*. Harare, Ministry of Local Government, Public Works and National Housing, Department of Physical Planning. Harare.
Government of Zimbabwe (1996). *Regional, Town and Country Planning Act (Chapter 29: 12)*. Harare, Government Printers.
Government of Zimbabwe (2001). *Mines and Minerals Act (Chapter 21: 05)*. Harare, Government Printers.
Government of Zimbabwe (2005). *Environmental and Management Act (Chapter 20:27)*. Harare, Government Printers.
Isik, G., Opalo, K. O. and Toledano, P. (2015). *Breaking out of Enclaves: Leveraging Opportunities from Regional Integration in Africa to Promote Resource-Driven Diversification*. Washington, DC, The World Bank.
Kamete, A. Y. (2012). Of prosperity, ghost towns and havens: Mining and urbanisation in Zimbabwe. *Journal of Contemporary African Studies*, 30(4): pp. 589–609.
Keeble, L. (1969). *Principles and Practice of Town and Country Planning*. London, Estates Gazette.
Kondo, T. (2011). *Beyond the Enclave: Towards a Pro-poor and Inclusive Development Strategy for Zimbabwe*. African Books Collective. Harare, Weaver, Press.
Makore, G. and Zano, V. (2012). *Mining within Zimbabwe's Great Dyke: Extent, Impacts & Opportunities*. Harare, Zimbabwe Environmental Lawyers Association.
Metallon Corporation (2014). *Mazowe Gold Mine*. Accessed online: http://metcorp.co.uk/operations/gold-fields-of-mazowe-ltd.aspx. Accessed 13 January 2021.
Miles, E. M., Berens, G. L., Eppli, M. J. and Weiss, M. A. (2007). *Real Estate Development: Principles and Process*. Washington, DC, Urban Land Institute.
Njoh, A. (2007). *Planning Power: Town Planning and Social Control in Colonial Africa*. London and New York, CRC Press.
Rubbers, B. (2019). Mining towns, enclaves and spaces: A genealogy of worker camps in the Congolese Copperbelt. *Geoforum*, 98: pp. 88–96. doi: 10.1016/j.geoforum.2018.10.005. Accessed 13 January 2021.
Sango, I., Taru, P., Mudzingwa, M. N. and Kuvarega, A. T. (2006). Social and biophysical impacts of Mhangura Copper Mine closure. *Journal of Sustainable Development*, 8(3): pp. 186–204.
The European Union (1997). *The EU Compendium of European Spatial Planning Systems*. Belgium, European Commission.
Vinokurov, E. (2007). *A Theory of Enclaves*. Lanham, MD, Lexington Books.

Wekwete, K. H. (1989a). *Planning Laws for Urban and Regional Planning in Zimbabwe: A Review.* RUP Occasional Paper No. 20. University of Zimbabwe, Department of Rural and Urban Planning, Harare.

Wekwete, K. H. (1989b). Physical planning in Zimbabwe: A Review of the legislative, administrative and operational framework. *Third World Planning Review*, 11(1): pp. 49–69.

Zamasiya, B. and Dhlakama, T. (2016). *An Analysis of the Legal, Institutional and Policy Constraints Affecting the Participation of Men and Women in Local Content Development Outcomes in the Mining Sector in Zimbabwe.* The Zimbabwe Environmental law Association, Harare.

Section V

Methods and Tools for Sustainable and Smart Spatial Planning

15 The Contribution of Spatial Planning Tools towards Disaster Risk Reduction in Informal Settlements in South Africa

Juliet Akola, James Chakwizira, Emaculate Ingwani and Peter Bikam

CONTENTS

15.1 Introduction 213
15.2 Background on Spatial Planning Tools and Disaster Risk Reduction 214
15.3 Disaster Risk Reduction in Informal Settlements as a System 215
15.4 Overview of the Study Area and Methods 216
15.5 Results and Discussion 220
    15.5.1 Spatial Planning Tools in Controlling Developments
           to Reduce Disaster Risks 220
    15.5.2 Disaster Risks in Informal Settlements 221
    15.5.3 Water-Related Risks 221
    15.5.4 Fire-Related Risks 222
    15.5.5 Health-Related Risks 223
    15.5.6 Benefits of Spatial Planning in Informal Settlements in South
           Africa 224
15.6 Conclusion and Recommendations 224
References 225

15.1 INTRODUCTION

As the world is increasingly exposed to disaster risks, spatial planning tools emerge as key to the reduction of disasters particularly in informal settlements (Australian Institute for Disaster Resilience, 2002; UN-Habitat, 2007; Hansford, 2011; Akola et al., 2019). Spatial planning tools aim to improve the welfare of persons and communities by creating sustainable and more convenient, equitable, healthful, efficient and attractive places (Kochtitzky et al., 2006). Thus, spatial planning tools have often been used by public sectors worldwide to influence the distribution of people and activities in spaces of various scales including prevention of disaster risks (UN-Habitat, 2015). Disaster risks refer to the potential disaster losses, in lives, health status, livelihoods, assets and services, which could occur in a community or society over some specified future period (Dewald, 2011).

Studies (Neuvel and van der Knaap, 2010; Vyas-Doorgapersad and Lukamba, 2012; United Nations Office for Disaster Risk Reduction (UNISDR), 2015; United Nations, 2015; Hamza, 2015; GFDRR, 2016; Ran and Nedovic-Budic, 2016; SERI, 2018) have projected that disaster risks will increase in type, intensity and frequency. Therefore, the contribution of spatial planning tools towards disaster risk reduction particularly in informal settlements in developing nations ought to be recognised and given careful consideration in developmental activities (Hansford, 2011).

In the absence of inclusive spatial planning in South Africa, many informal settlements have continued to exist, grow and are exposed to disaster risks (SERI, 2018). This is attributed to unsustainable human practices (Management Institute of Southern Africa (DMISA), 2004), which could have been prevented through adherence to spatial planning tools. While spatial planning tools are regulatory measures used to create orderly developments and reduce disaster risks in urban areas, there is limited literature on their contribution towards disaster risk reduction in informal settlements in South Africa.

This chapter, therefore, reviews the contribution of spatial planning tools and makes recommendations towards disaster risk reduction in informal settlements to promote resilient, liveable, healthy and sustainable communities in South Africa. Following this introduction, the chapter proceeds to present background information on spatial planning tools and disaster risk reduction, systems theory, an overview of the study area and the methods deployed in the study. It further presents spatial planning tools applied for controlling developments to reduce disaster risks related to water, health and fire and the benefits of spatial planning in informal settlements in South Africa. Finally, the chapter provides a conclusion and recommendations.

15.2 BACKGROUND ON SPATIAL PLANNING TOOLS AND DISASTER RISK REDUCTION

Disaster risk reduction through the application of spatial planning tools entails a systematic approach of identifying, assessing and reducing the risks of disasters (UNISDR, 2015). Spatial planning tools that could be used for disaster risk reduction in informal settlements include land use planning regulations, land use schemes (LUSs), zoning, subdivision and building codes and standards (South African Cities Network (SACN), 2017; UN-Habitat, 2010).

As disaster risks are increasing around the world, mostly affecting informal settlements, many regions are experiencing greater damage and heavy losses to property and lives than in the past. Globally, according to GFDRR (2016), there is variability in the annual losses and deaths from disasters but the annual total damage increased tenfold between 1976–1985 and 2005–2014, from US$14 billion to more than US$140 billion. The average population affected each year has risen from around 60 million people (1976–1985) to over 170 million (2005–2014) (GFDRR, 2016). The signs of disaster risks in informal settlements are more evident in the global south than in the global north (Laros, 2014).

It is important to note that most disasters in developed countries happen in general locations and not specifically in informal settlements (DMISA, 2004). On the contrary, in developing nations in Asia, Latin America and Africa, most disaster risks related to health, water and fire are spatially found in informal settlements (ibid).

As such, in Africa, South Africa has taken initiatives to address such disasters. For example, in 2015, the Spatial Planning and Land Use Management Act (SPLUMA) No 16 of 2013 was introduced as a guiding framework for all spatial planning and land use management activities in local municipalities (South African Cities Network (SACN), 2017). The aim of the act is to create a uniform system of spatial planning and land use management throughout the Republic of South Africa (Mashiri et al., 2017). This act mandates local municipalities to prepare spatial development frameworks (SDFs) and land use management schemes as a means of spatial transformation in urban and rural spaces (ibid). Thus, this study focused on ascertaining spatial planning tools applied in controlling developments to reduce disaster risks and analysing water-, health- and fire-related risks and examining the benefits of spatial planning in informal settlements in South Africa.

15.3 DISASTER RISK REDUCTION IN INFORMAL SETTLEMENTS AS A SYSTEM

Assessing the contribution of spatial planning tools in disaster risk reduction in informal settlements requires the consideration of interactions of informal settlements with the surrounding environment from a political, social, economic and environmental context (Coelho and Ruth, 2006). Therefore, the systems theory is adopted to explain how disaster risk reduction in informal settlements is complex in nature and there is a need for a systems approach for the development of appropriate strategies to reduce disaster risks (Pineo et al., 2020).

The systems theory that was developed by Von Bertalanffy in 1968 is likened to the art of wholeness (Cooper et al., 1971), whereby spatial planning tools, disaster risk reduction and informal settlements are regarded as a complex system that is not linear but dynamic in nature (ibid). A system can be defined as a complex of interacting elements and entails three aspects; elements, interconnections and functions that are reasonably organised in a way that achieves intended outcomes (Bertalanffy, 1968). It investigates both the principles common to all complex entities and the models, which can be used to describe them (Meadows, 2009).

In this study, the contribution of spatial planning tools in disaster risk reduction in informal settlements can be looked at as a system as shown in Figure 15.1.

From Figure 15.1, social systems in urban areas particularly the informal settlements receive input from the environment, engage in processes and generate outputs. Reducing disaster risk through spatial planning tools requires long-term systemic thinking (Asian Development Bank, 2016).

From the systems theory perspective, disaster risk reduction in informal settlements through spatial planning tools necessitates a holistic consideration of all components

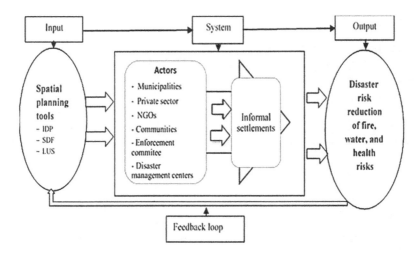

FIGURE 15.1 System interactions in informal settlements.

Source: Figure designed by authors

of the disaster risk reduction system. Based on the current practice of emergency response to disaster risks in informal settlements, it appears that the efforts of the Disaster Management Centres alone may not be adequate. Therefore, all components in the system need to be taken into consideration to reduce disaster risks in informal settlements. This also requires embracing complexity and dynamics and accepting uncertainty in informal settlements. It is a challenging task to deal with dynamic systems and this means planning with change rather than against it. This involves focusing on inclusive planning rather than avoidance as advocated by SPLUMA.

15.4 OVERVIEW OF THE STUDY AREA AND METHODS

South Africa is at the southern tip of the African continent stretching latitudinally from 22°S to 35°S and longitudinally from 17°E to 33°E and covers 1,219,602 km² (Statistics South Africa, 2011). It has a long coastline stretching for more than 3,000 km from the desert border with Namibia on the Atlantic coast southwards around the tip of Africa and then north to the border of subtropical Mozambique on the Indian Ocean as shown in Figure 15.2.

As shown in Figure 15.2, the country shares boundaries with Namibia, Botswana, Zimbabwe, Mozambique and Swaziland and the Mountain Kingdom of Lesotho is landlocked by the South African territory in the south-east.

As of June 2019, the population of South Africa was estimated to be 58.8 million people of which the female gender accounted for 30.1 million representing 51.2% of the total population and the male gender accounted for 28.7 million representing 48.8% of the total population (Statistics South Africa, 2019). Due to a significant lack of affordable housing, many poor and low-income households in South Africa opt

Contribution of Spatial Planning Tools   217

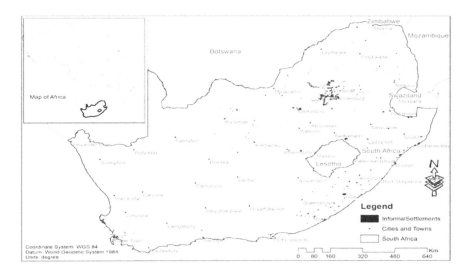

FIGURE 15.2   Map of South Africa.

Source: Figure designed by authors

to live in South Africa's growing informal settlements as indicated in Figure 15.2. According to 2011 estimates, between 2.9 and 3.6 million people lived in informal settlements in South Africa (Statistics South Africa, 2011). However, this number is likely to be significantly higher due to the volatility of residences and insecure tenure arrangements in informal settlements (SERI, 2018). Due to unreadily available data on the status of informal settlements in South African local municipalities, this study provides a highlight of informal settlements in the provinces of South Africa by 2013 (refer to Table 15.1) as compiled by the Housing Development Agency.

The majority of the informal households (52%) as depicted in Table 15.1 have been established in unplanned areas and as such they are more exposed to disaster risks since their settlements are spatially concentrated in the same location in the unplanned arrangement which makes them susceptible to fire, disease outbreaks and water risks.

A systems theory was adopted to explain the concepts of the study through a descriptive research design, which comprised of data that was obtained from secondary sources and analysed thematically. The selection of this theory is built on the argument that informal settlements are part of the spatial planning process and their setting affects the social, economic and environmental aspects of the city. Disaster risks in informal settlements usually occur within social systems due to unsustainable human practices such as constructing buildings in flood plains and wetlands, building on illegally acquired land, encroachment on reserved open spaces and erecting buildings without building plans and permits.

To obtain secondary data used in the study, a literature search involving various stages was conducted as follows: First, the keywords such as 'spatial planning

TABLE 15.1
Proportion of Areas Occupied by Informal Dwellings in the Provinces of South Africa

Provinces	Informal settlement	Urban settlement	Tribal settlement	Farm	Other
Eastern Cape	63%	22%	10%	3%	2%
Free State	47%	46%	3%	2%	2%
Gauteng	58%	37%	0%	3%	2%
KwaZulu-Natal	65%	17%	10%	3%	4%
Limpopo	28%	16%	39%	15%	2%
Mpumalanga	40%	28%	19%	7%	6%
Northwest	19%	32%	38%	7%	4%
Northern Cape	29%	50%	3%	9%	9%
Western Cape	65%	32%	0%	1%	2%
South Africa	**52%**	**32%**	**10%**	**4%**	**3%**

Source: Housing Development Agency (HDA, 2013)

tools', 'disaster risks', 'disaster risk reduction' and 'informal settlements' were used to search for books, papers, reports, news articles and documents that are published in 65 foremost spatial planning, disaster and informal settlements journals, reports, news articles and organisational websites from 2000 to 2020. We considered the search with 22 journals, 29 reports from organisations, 11 news articles and 3 books believing that they provide sufficient, wide-ranging representation of the current information on spatial planning, informal settlements and disaster risk reduction. These sources were selected based on their policy influence and academic significance in spatial planning, disaster risk reduction and informal settlements research. In total, 72 papers, news articles, books and reports were identified at this stage.

Secondly, a reference check of each paper and report obtained from the first stage was conducted. This process provided literature that was thematically analysed based on the objectives of the study. Thirdly, a comprehensive abstract and executive summary review of reports was carried out to disregard inappropriate papers, new articles, books and reports. Inappropriate papers, news articles and reports included analytical studies and studies that assess the contribution of spatial planning in general rather than the contribution of spatial planning tools for disaster risk reduction in informal settlements in South Africa. Finally, 65 papers, reports and news articles were selected and reviewed in this study with dates ranging from 2000 to 2020. Table 15.2 provides a breakdown of studies grouped by source. However, it became inevitable to review books that date back to the 1960s and 1970s due to the need to provide a theoretical underpinning to the systems theory.

TABLE 15.2
Breakdown of the Studies Reviewed

Type of secondary data source	Number
Journals	
Jàmbá: Journal of Disaster Risk Studies	4
Planning Practice and Research	2
Land Use Planning for Public Health	1
Spring Research	1
Tear Fund Roots Resources	1
Planning Theory and Practice	1
Urban Planning and Public Health	1
Sociology and Anthropology	1
International Planning Studies	1
Urban Forum	1
International Journal of Water Resources Development	1
Computers, Environment and Urban Systems	1
Journal of Health and Place	1
Cities Journal	1
Journal of Public Administration	1
South African Medical Journal	1
Policy Sciences	1
Palgrave Communications	1
Reports	
International reports	
United Nations, UN-Habitat and UNDRR	11
ARUP, Asian Development Bank, Australian Institute for Disaster Resilience, USAID, Planning Tank, GFDRR, IFRC, Cities Alliance, Council of Planning Librarians, Monticello Illinois	7
National reports	
Department of Rural Development and Land Reform, Housing Development Agency, Disaster Management Institute of Southern Africa, South African Cities Network, Statistics South Africa	11
News articles	11
Books	3
Total	65

15.5 RESULTS AND DISCUSSION

15.5.1 Spatial Planning Tools in Controlling Developments to Reduce Disaster Risks

The South African spatial planning system is defined by the SPLUMA of 2013 as consisting of the SDFs to be prepared and adopted by the national, provincial and municipal spheres of government. Secondly, it should consist of developmental principles, norms and standards that must guide spatial planning, land use management and land development in the country. Thirdly the management and facilitation of land use through the mechanism of LUS and fourthly procedures and processes for the preparation, submission and consideration of land development applications and related processes should be a part of spatial planning (Department of Rural Development and Land Reform, 2017).

Spatial planning tools such as the integrated development plans (IDPs), SDF and LUS may play an important role in disaster risk reduction in urban areas specifically in informal settlements (Australian Institute for Disaster Resilience, 2002; United Nations, 2002; UN-Habitat, 2007, 2015; United Nations Office for Disaster Risk Reduction (UNISDR), 2015; Ran and Nedovic-Budic, 2016; Planning Tank Admin, 2017; Mashiri et al., 2017; South African Cities Network (SACN), 2017).

Integrated development planning is a tool that not only involves an approach of ascertaining solutions to the challenges affecting communities but also identifies disaster risks that the communities could be exposed to in order to achieve sustainable long-term development (Department of Rural Development and Land Reform, 2017). The process of risk identification allows for the development of disaster risk reduction measures. This approach requires the participation of all stakeholders under the coordination of various municipalities in the preparation of an IDP. The IDP provides an overall framework for development and this presents an opportunity for the disaster risk reduction processes to be incorporated in aspects such as land use, the kind of infrastructure and desirable services and how the environment should be protected in a specific area (Department of Rural Development and Land Reform, 2017). Therefore, an IDP's overall aim is to improve the quality of life of all the people living in an area. It considers the existing conditions, challenges and resources available for development. However, the plan should not only consider spatial development but also the economic and social development of the area, an attribute of the systems theory that requires inclusive planning in urban areas with the input from all actors (Becker, 2009).

An SDF seeks to guide the overall spatial distribution of current and desirable land uses such as commercial, institutional, recreational, residential and special uses within a municipality to give effect to the visions, goals and objectives of the municipal IDP. This tool is used in various municipalities in South Africa to promote sustainable functional and integrated human settlements, maximise resource efficiency and enhance regional identity and unique character of places like informal settlements among other places. Spatially, the SDF aids in mapping disaster risk areas that can be incorporated in the land use scheme of any local municipality (Department of Rural Development and Land Reform, 2011).

The LUS as another spatial planning tool is used to regulate land use in a municipality (Department of Rural Development and Land Reform, 2011). It includes policies related to how land is used on a plot-by-plot basis through the zoning scheme or town planning scheme in the South African context. By regulating land developments through land use zones, zone matrix, bulk regulation (floor area ratio (FAR)), coverage, height, building lines, parking and loading requirements, residential densities, disaster risks related to water, fire and health can be prevented (Department of Rural Development and Land Reform, 2017).

Remarkably, spatial planning has got a long history of addressing public health-, water- and fire-related risks (Fainstein, 2020). To appreciate the contribution of spatial planning tools towards disaster risk reduction, the Department of Rural Development and Land Reform advocates for the preparation of IDPs, SDFs and LUSs to achieve sustainable development across all the municipalities in South Africa that are extended towards disaster risk reduction including in informal settlements (Department of Rural Development and Land Reform, 2017).

15.5.2 DISASTER RISKS IN INFORMAL SETTLEMENTS

In the South African context, informal settlements are unplanned settlements on land that has not been surveyed or proclaimed as residential, consisting mainly of shacks (Housing Development Agency, 2013).

Research shows that disaster risks are likely to increase in South Africa (Anciano and Piper, 2019). Such an increase is demonstrated by the recent fire and disease outbreaks, weather events and natural disasters, as a result of changes in global climate and how human beings interact with the environment (GroundUp News, 2018; Mthuthuzeli, 2020). This poses a great challenge on how local municipalities in South Africa understand and respond proactively to the informal conditions in which millions of poor people live and are exposed to disaster risks related to water, fire and health (Anciano and Piper, 2019).

Findings show that due to the growing number of informal settlements in South Africa with poor living conditions such as poverty, inadequate access to basic services, shortage of housing, poor sanitation and access to water, three categories of disaster risks have emerged from different papers, reports, news and documents reviewed as explained in the subsequent sections (SERI, 2018).

15.5.3 WATER-RELATED RISKS

Communities in informal settlements can be exposed to water-related risks that arise from too much water such as floods, heavy rainfall, torrential rains and cyclones which have often had devastating social and economic effects on society (United Nations, 2015). Table 15.3 presents the documented incidences of disaster risks experienced in some informal settlements in South Africa.

Notably in 2019 as shown in Table 15.3, heavy rain across the country affected informal settlements more than planned settlements. This implies that informal settlements are vulnerable to disaster risks due to their location and the type of building materials used for housing (BBC News, 2019). This also shows that most informal settlements in South Africa are in environmentally sensitive areas such as wetlands

TABLE 15.3
Water-Related Risks that Have Occurred in Some Informal Settlements in Recent Years

Year of disaster occurrence	Kind of disaster risk	Place of occurrence	Location	Impact on lives and property
April 2019	Torrential rains and flash floods	KwaZulu-Natal Province. (Quarry road informal settlement in Durban)	Flood plain (Environmentally sensitive area)	Life loss and property destruction (Singh, 2019a).
April 2019	Torrential rains and flash floods	Eastern Cape Province	Suburb (Not a fragile Area)	Life loss and property destruction (Singh, 2019a)
October 2019	Floods due to heavy rainfall	Gauteng Province (Mamelodi in Pretoria)	Low-lying terrain (Farmland)	People displaced and 700 homes washed away (BBC News, 2019)
June 2018	Floods due to heavy rains	Western Cape Province (In Masiphumelele wetlands, Vrygrond, In Dunoon's New Rest informal settlement, Siyahlala informal settlement,	Wetland (Environmentally sensitive area)	Property destruction (Groundup Staff, 2018)

and floodplains and hence are exposed to water-related risks like floods (Ansumant, 2020).

15.5.4 FIRE-RELATED RISKS

Structures in informal settlements are often built without due consideration to building rules and regulations such as inadequate access, lack of setback limits, no fire extinguishers and illegal connections to electricity (Minnie, 2012). Thus, fire outbreaks have become prevalent in South African informal settlements (ibid). In 2019 alone, as shown in Table 15.4, the recorded incidences of fire outbreaks destroyed property and people lost their lives. In the event of fire outbreaks, because informal settlements have houses built next to each other, narrow access routes, nameless streets or description of locations, firefighters usually find it difficult to save lives and property (Groundup News, 2018).

Devastating fires are frequently happening in informal settlements as indicated in Table 15.4. For example, in March 2017, a large fire destroyed parts of Imizamo Yethu in Hout Bay. In addition, the fire-ravaged Section E, Masiphumelele in July whereby over 250 homes were burnt and left about 1,200 people homeless.

TABLE 15.4
Fire-Related Risks in Some Informal Settlements in Recent Years

Year of disaster occurrence	Kind of disaster risk	Place of occurrence	Location	Impact on lives and property
July 2019	Fire outbreak	Kwa Zulu -Natal Province. (Seacow Lake, north of Durban)	Coastline (Environmentally sensitive area)	100 shacks destroyed (Singh, 2019b)
June 2019	Fire outbreak	Gauteng Province (Zandspruit in the north of Johannesburg)	Suburb (Not a fragile Area)	15 shacks destroyed by fire (Gous, 2019)
July 2019	Fire outbreak	Western Cape Province (Qandu informal settlement in Khayelitsha)	Wetland (Environmentally sensitive area)	One woman died (Pijoos, 2019)
March 2017	Fire outbreak	(Imizamo Yethu in Hout Bay Cape town) The 18-hectare settlement houses approximately 33,600 people	Valley area. (Environmentally sensitive area)	Homes destroyed (GroundUp News, 2018)
July 2017	Fire outbreak	(Masiphumelele section E)	Township (Not a fragile Area)	250 homes were burnt and 1,200 people were left homeless (GroundUp News, 2018)

15.5.5 HEALTH-RELATED RISKS

Poorly planned and managed informal settlements can turn out to be breeding grounds for epidemics like cholera, typhoid, diarrhoea, malaria, influenza, avian flu, SARS and Coronavirus Disease 2019. The poor living conditions in informal settlements are enough to expose residents to public health risks. For example, informal settlements in Cape Town are currently a concern to Government and with the poor health-care systems across the country, there are fears that informal settlement dwellers will bear the brunt of the COVID-19 outbreak (Mthuthuzeli, 2020). Furthermore, people living in informal settlements have poor access to running water to wash their hands and the overcrowded living conditions make the transmission of the virus inevitable (ibid). Families often share communal water taps and ablution facilities, which makes the risk of COVID-19 transmission higher (ibid). Poor waste management systems, inadequate running water and sanitation and inadequate sewerage infrastructure expose residents in informal settlements to health risks.

15.5.6 BENEFITS OF SPATIAL PLANNING IN INFORMAL SETTLEMENTS IN SOUTH AFRICA

The population in informal settlements in South Africa is steadily increasing and due to shortage of housing, people are compelled to settle in haphazardly erected structures that exposes them to fire, health and water risks (Minnie, 2012).

Floods in flood-prone areas according to Ran and Nedovic-Budic (2016) can be mitigated through a spatial planning regulatory role of complying with zoning standards that prevent the location of unauthorised activities on lots such as open spaces, residential, agricultural, commercial and industrial land use. Adhering to zoning regulations, which include location of activities, land use types, scales of development and designs of physical structures may greatly reduce the incidences of flooding and the consequential damage (Neuvel and van der Knaap, 2010).

On the other hand, fire outbreaks can be minimised through spatial planning if land use is left to interact with urban design, building form and construction materials (Wakely and Riley, 2011). The application of building standards in informal settlements aids in the use of appropriate construction methods, use of incombustible construction materials and use of modern heating and lighting sources as opposed to the traditional ways (ARUP, 2018). Adequate infrastructure can be planned for and serviced plots can be provided through subdivision and development applications. In the event of fire outbreaks, spatial planning ensures that settlements need to have access routes, streets names and street lighting for easy accessibility by service providers such as firefighters, police, health officers, spatial planners, environmentalists, social workers and other stakeholders.

Spatial planning issues remain at the basis of some of the most obstinate public health problems resulting in high rates of disease due to high influx of persons and their families from the countryside to the cities (Fallon and Neistadt, 2006). Spatial planning involves assessing and planning for community needs in health-care infrastructure which, in turn, promotes physical activity, better mental health and prevention of infectious diseases through community infrastructures, such as planning for safe running water and sewage systems. It also promotes the protection of persons from hazardous industrial exposures and injury risks through land-use and zoning ordinances (Kochtitzky et al., 2006). Well-planned and functioning urban systems can help prevent the outbreak of diseases, such as COVID-19, malaria and typhoid among others, in highly crowded settlements characterised by poor sanitation.

15.6 CONCLUSION AND RECOMMENDATIONS

This chapter highlights the contribution of spatial planning tools towards disaster risk reduction in informal settlements in South Africa. Adopting spatial planning tools in disaster risk reduction would help to prevent, prepare and mitigate disaster risks and reduce the current practice of relying on emergency responses from the disaster management centres. Unfortunately, the non-consideration of spatial planning tools in informal settlements has left such communities prone to disasters. Applying the IDPs, SDFs and LUSs in reducing disaster risks in informal settlements may require inputs from various disciplines and different stakeholders and a good understanding of the land's environmental, social, economic and political features. Informal settlements that have continued to exist and grow in number are currently experiencing water, health and fire risks in South African local municipalities.

SPLUMA emphasises that land use management systems must include all areas of a municipality and specifically include provisions that are flexible and appropriate for the management of informal settlements (Department of Rural Development and Land Reform, 2017). In addition, the Act requires LUSs to include provisions to deal with informal settlements, a requirement that agrees with the systems theory. Based on the systems theory, to reduce the exposure of informal settlements to disaster risks, these settlements need to become part of an established township so that they are incorporated into the planning scheme of the municipality. For example, recent developments in municipalities such as the City of Johannesburg and the City of Cape Town provide some guidance for modern LUSs. Their processes start with the 'recognition of informal settlements' through a process of regularisation or formalisation. Formalisation entails legal processes whereby townships are established with formal services and residents obtain formal security of tenure. Usually, it may also imply the development of top structures such as Reconstruction Development Programme (RDP) houses.

The aforementioned approach has been successful in reducing disaster risks such as fire, health and water in the City of Johannesburg and Cape Town. This is an incremental approach that allows spatial planning tools to be adopted in an inclusive way that promotes a systematic approach of identifying, assessing and reducing disasters risks in informal settlements to create safe and liveable urban environments.

REFERENCES

Akola, J., Binala, J. and Ochwo, J., 2019. Guiding developments in flood-prone areas: Challenges and opportunities in Dire Dawa city, Ethiopia. *Jamba: Journal of Disaster Risk Studies*, 11(3), pp. 1–8.

Anciano, F. and Piper, L., 2019. South Africa: How cities can approach redesigning informal settlements after disasters [online]. Available at: www.preventionweb.net/news/view/65159%0D [Accessed June 23, 2020].

Ansumant, 2020. What do you mean by sustainable development? *Planning Tank*, [online], p. 20. Available at: https://planningtank.com/sustainable-development/sustainable-development [Accessed August 28, 2020].

ARUP, 2018. A framework for fire safety in informal settlements. Arup International Development, [online] (September), pp. 1–32. Available at: www.arup.com/-/media/arup/files/publications/f/fire_safety_in_informal_settlements_2018.pdf [Accessed March 9, 2020].

Asian Development Bank, 2016. *Reducing Disaster Risk by Managing Urban Land Use: Guidance Notes for Planners.* Mandaluyong: Asian Development Bank.

Australian Institute for Disaster Resilience, 2002. Planning safer communities: Land use planning for natural hazards. *Australian Emergency Manuals Series. Pt. II, Approaches to Emergency Management. v. 2, Mitigation Planning*, [online] 2(manual 7), pp. 1–80. Available at: www.google.com/url?sa=t&rct=j&q=&esrc=s&source=web&cd=1&ved=2ahUKEwi7g9v1kaHmAhWzolwKHQkJD_gQFjAAegQIAhAC&url=https%3A%2F%2Fknowledge.aidr.org.au%2Fmedia%2F1958%2Fmanual-7-planning-safer-communities.pdf&usg=AOvVaw2lsC3MIvcRGzr7WN5zMFR1 [Accessed July 18, 2019].

BBC News, 2019. South Africa hit by floods and power. *BBC*, [online], p. 1. Available at: www.bbc.com/news/world-africa-50726730 [Accessed April 3, 2020].

Becker, P., 2009. Grasping the hydra: The need for a holistic and systematic approach to disaster risk reduction. *Jàmbá: Journal of Disaster Risk Studies*, [online] 2(1), pp. 1–24. Available at: www.preventionweb.net/english/hyogo/gar/2015/en/home/download.html [Accessed June 24, 2021].

Bertalanffy, L. von., 1968. *General System Theory*. 1st ed. New York: George Braziller, Inc.

Coelho, D. and Ruth, M., 2006. Seeking a unified urban systems theory. *WIT Transactions on Ecology and the Environment*, 93, pp. 179–188.

Cooper, A.W.W. et al., 1971. Systems approaches to urban planning: Mixed, conditional, adaptive and other alternatives. *Policy Sciences*, 2(6), pp. 397–405.

Department of Rural Development and Land Reform, 2011. *Guidelines for the Development of Spatial Development Frameworks*. Pretoria: Department of Rural Development and Land Reform

Department of Rural Development and Land Reform, 2017. *Land Use Scheme Guidelines*. Pretoria: Department of Rural Development and Land Reform

Dewald, van N., 2011. *Introduction To Disaster Risk Reduction*. North West University: USAID.

Fainstein, S.S., 2020. Urban planning: The era of industrialisation. *Planning Tank*. [online]. Available at: www.britannica.com/topic/urban-planning#ref10804 [Accessed May 26, 2020].

Fallon, L. and Neistadt, J., 2006. Land use planning for public health: The role of local boards of health in community design and development. *Land Use Planning for Public Health*, [online] 23, pp. 1–33. Available at: www.nalboh.org [Accessed May 26, 2020].

GFDRR, 2016. *The Making of a Riskier Future: How Our Decisions are Shaping Future Disaster Risk*. Washington, DC: Global Facility for Disaster Reduction and Recovery.

Gous, N., 2019. Fire in Zandspruit under control after 15 shacks catch fire. *Sunday Times*, [online] June, p. 1. Available at: www.timeslive.co.za/news/south-africa/2019-06-19-fire-in-zandspruit-under-control-after-15-shacks-catch-fire/%0D [Accessed May 5, 2019].

Groundup News, 2018. Informal settlements flooded by heavy rains. *Sunday Times*, [online], p. 1. Available at: www.timeslive.co.za/news/south-africa/2018-06-15-informal-settlements-flooded-by-heavy-rains/ [Accessed September 4, 2020].

Hamza, M., 2015. *World Disasters Report 2015: Focus on Local Actors, the Key to Humanitarian Effectiveness*. [online] Geneva. Available at: www.ifrc.org [Accessed August 19, 2021].

Hansford, B., 2011. Understanding risk reduction—the theory: Reducing risk of disaster in our community. *Tear Fund Roots Resources*, pp. 15–26.

Housing Development Agency (HDA) 2013. *South Africa : Informal Settlements Status*. Johannesburg: South Africa.

Kochtitzky, C.S., Frumkin, H., Rodriguez, R., Dannenberg, A.L., Rayman, J., Rose, K., Gillig, R. and Kanter, T., 2006. Urban Planning and Public Health at CDC. *MMWR Supplements*, 55(2), pp. 34–38.

Laros, M., 2014. *The State of African Cities 2014: Re-imagining Sustainable Urban Transitions*. Third. Nairobi: United Nations Human Settlements Programme (UN-Habitat).

Management Institutie of Southern Africa (DMISA) 2004. Disaster risk reduction. In *Disaster Risk Reduction Conference held on 13 and 14 October*. Gallagher Estate, Midrand: Management Institutie of Southern Africa (DMISA).

Mashiri, M. et al., 2017. Towards a framework for measuring spatial planning outcomes in South Africa. *Sociology and Anthropology*, 5(2), pp. 146–168.

Meadows, D.H., 2009. *Thinking in Systems*. 1st ed. D. Wright, ed. London: Earth Scan, Publsihing for a Sustainable Future.

Minnie, J., 2012. Informal settlements fire and flooding risk reduction strategy. In *Dmisa Western Cape Provincial Workshop in July 2005*. Cape Town: City of Cape Town South Africa, pp. 1–42.

Mthuthuzeli, N., 2020. Informal settlement dwellers are 'at highest risk' for Covid-19. *IOL News*, [online] March, p. 1. Available at: www.iol.co.za/capeargus/news/informal-settlement-dwellers-are-at-highest-risk-for-covid-19-45031335%0D [Accessed February 17, 2021].

Neuvel, J.M.M. and van der Knaap, W., 2010. A spatial planning perspective for measures concerning flood risk management. *International Journal of Water Resources Development*, 26(2), pp. 283–296.

Pijoos, I., 2019. Woman dies in shack fire in Cape Town. *Sunday Times*, [online], p. 1. Available at: www.timeslive.co.za/news/south-africa/2019-07-14-woman-dies-in-shack-fire-in-cape-town/ [Accessed April 13, 2020].

Pineo, H., Zimmermann, N. and Davies, M., 2020. Integrating health into the complex urban planning policy and decision-making context: A systems thinking analysis. *Palgrave Communications*, [online] 6(1), pp. 1–14. Available at: http://dx.doi.org/10.1057/s41599-020-0398-3 [Accessed November 20, 2020].

Planning Tank Admin 2017. How urban planning can contribute to resilience and disaster risk reduction. *Planning Tank*. [online]. Available at: https://planningtank.com/resilient-city/urban-planning-can-contribute-resilience-disaster-risk-reduction [Accessed January 6, 2020].

Ran, J. and Nedovic-Budic, Z., 2016. Integrating spatial planning and flood risk management: A new conceptual framework for the spatially integrated policy infrastructure. *Computers, Environment and Urban Systems*, [online] 57, pp. 68–79. Available at: http://dx.doi.org/10.1016/j.compenvurbsys.2016.01.008 [Accessed July 19, 2020].

SERI, 2018. *Informal Settlements and Human Rights in South Africa: Submission to the United Nations Special Rapporteur on Adequate Housing as a Component of the Right to an Adequate Standard of Living*. [Online]. Available at: www.ohchr.org/Documents/Issues/Housing/InformalSettlements/SERI.pdf [Accessed February 16, 2021].

Singh, K., 2019a. KZN floods: Death toll up to 85. *New 24*, [online] April, p. 1. Available at: www.news24.com/SouthAfrica/News/kzn-flooding-death-toll-up-to-85-20190425 [Accessed January 29, 2019].

Singh, O., 2019b. 100 shacks damaged as fire rages through Durban informal settlement 25 July 2019–16:12. *Sunday Times*, [online] July, p. 1. Available at: www.timeslive.co.za/news/south-africa/2019-07-25-100-shacks-damaged-as-fire-rages-through-durban-informal-settlement/ [Accessed September 1, 2020].

South African Cities Network (SACN), 2017. *Spluma as a Tool for Spatial Transformation*. Available at: www.sacities.net [Accessed October 9, 2019].

Statistics South Africa, 2019. *Community Survey Data*. Johannesburg: South Africa: Statistics South Africa.

Statistics South Africa, 2011. *Statistical Release Census 2011*. Census 2011 Statistical release—P0301.4 / Statistics South Africa. Pretoria.

UN-Habitat, 2007. *Global Report on Human Settlements 2007: Enhancing Urban Safety and Security*. Nairobi: UN-Habitat.

UN-Habitat, 2010. The state of African cities 2010: Governance, inequality and urban land markets. Nairobi: UN-Habitat.

UN-Habitat, 2015. *International Guideliness on Urban and Territorial? Planning*. [online] Nairobi: UN-Habitat. Available at: www.unhabitat.org [Accssed June 12, 2020].

United Nations, 2002. *Disaster Reduction and Sustainable Development: Understanding the Links between Vulnerability and Risk Related to Development and Environment*. UN/ISDR, revised version 17 May 2002. [online] Johannesburg. Available at: www.unisdr.org [Accessed October 18, 2021].

United Nations, 2015. Water and disaster risk: A contribution by the United Nations to the consultation leading to the Third UN World Conference on Disaster Risk Reduction. In: *United Nations World Conference on Disaster Risk Reduction (WCDRR) 17–18 November 2014*. [online] Sendai Japan: United Nations.pp. 1–6. Available at: www.preventionweb.net/files/38763_water.pdf [Accessed September 8, 2020].

United Nations Office for Disaster Risk Reduction (UNISDR), 2015. *Making Development Sustainable: The Future of Disaster Risk Management.* [online] Geneva. Available at: www.preventionweb.net/english/hyogo/gar/2015/en/home/download.html [Accessed February 6, 2019].

Vyas-Doorgapersad, S. and Lukamba, T.M., 2012. Disaster risk reduction policy for sustainable development in the Southern African development community . . . *Journal of Public Administration*, 47(4), pp. 774–784.

Wakely, P. and Riley, E., 2011. *The Case for Incremental Housing. City Allience Policy Research and Working Paper Series No.1.* Washington, DC: Cities Alliance.

16 Geographic Information Systems for Smart Spatial Planning and Management
Managing Urban Sprawl in Harare Metropolitan, Zimbabwe

Tendai Sylvester Mhlanga and Fiza Naseer

CONTENTS

16.1 Introduction 229
    16.1.1 Urban Sprawl and Its Implication on the Smart City Concept 231
    16.1.2 The Smart Growth and Smart City Frameworks 232
16.2 Linking GIS, Spatial Planning, Smart Growth and Urban Sprawl
    Management 232
16.3 Methodology 233
    16.3.1 The Study Area.. 233
    16.3.2 Data and Pre-Processing.. 233
    16.3.3 Automated Built-Up Area Extraction Using Index Derived
        Built-Up Index 235
    16.3.4 Computation of Spectral Indices 235
16.4 Results and Discussion 237
    16.4.1 Built-Up Area Extraction...239
    16.4.2 Analysis of Urban Sprawl Using Shannon's Entropy 240
16.5 Conclusion 241
16.6 Recommendations 242
References... 242

16.1 INTRODUCTION

Rapid urbanisation is a major challenge to urban planners and policymakers in the development and management of sustainable human settlement. Half of the world's population lives in cities and trends are projected to rise constantly in the future to

70% in 2050 (World Bank, 2010). Eighty percent of the global gross domestic product is produced in cities and enables hundreds of millions to come out of extreme poverty (World Bank, 2019). Hence, this triggers rural–urban migrations in search of better opportunities and living standards. Rural–urban migrations subsequently increase the urban population and exert more pressure on land and on the provision of critical infrastructure and services. The unprecedented and uncontrolled urbanisation causes urban sprawl and informal developments. Most developing countries, particularly Africa as a whole, is threatened by urban sprawl, which is causing large consumption of land and exerting pressure on natural resources. Urbanisation has a direct effect on the environment through huge changes of non-built-up to built-up areas, increase in impervious surfaces and diminishing of wetlands, water bodies and farmland. Urban sprawl in most African countries has seemingly become unmanageable. This justifies the necessity for urban planners and policymakers to adapt to smart growth by implementing the smart city concept.

The successful implantation of the smart city concept in African countries is primarily embedded in the spatial planning and management of human settlements. Urban planners need to continuously monitor and analyse the trends, growth rate, pattern, direction and extent of urban sprawl as well as providing pragmatic measures for spatial planning (Fatih & Bolen, 2009). Intrinsically urban sprawl is one of the major impediments compromising the development of smart cities and sustainable human settlements in most African cities. Unmanaged urban sprawl widens spatial inequalities, increases the costs of service provision and imposes heavy environmental and economic burdens (Mahendra & Seto, January 2019). Henceforth, the contemporary challenges facing cities are most likely to increase dramatically in proportion to urban population growth should the impact of urban sprawl remain unsolved (Fertner et al., 2016).

Urbanisation in Zimbabwe is neither exceptional nor exclusive but has experienced the same phenomenal trends in rapid population growth and housing demand surge, which led to urban sprawl. It is unproportioned to the provision of services, facilities and infrastructure and is a direct effect on urban sprawling (Chirisa et al., 2016). Urban sprawl in Harare Metropolitan dates back to 1980 in the aftermath of suspension of migration restrictions and racial segregation when Zimbabwe got its independence. The population of Harare Metropolitan increased from 7,408,624 in 1980 to 14,862,924 in 2020 (Worldometer, 2020). Post-independence, Zimbabwe has gone through an agrarian to urban-based transition of industries and services. This becomes a pull factor of rural–urban migrations in the metropolitan and caused outward urban growth. The expansion of Harare Metropolitan area has been accelerated by rapid urbanisation which is notable in the land use and land cover changes (Marondedze & Schütt, 2019).

'Like urban sprawl itself, writing about sprawl is scattered in a vast multidisciplinary literature' (Banai & DePriest, 2014). Significant studies have been done on urban sprawl in Southern Africa but most of them are focused on land-use and land cover changes, causes and consequences and policy implications, with only less aimed at management or control of urban sprawl. Considerable effort has been devoted to study the patterns, trends and the drivers of urban sprawl in the Harare Metropolitan Province. One such study focused on the relation between the drivers and the axis of urban growth in the Harare Metropolitan Province (Marondedze & Schütt, 2019). It concluded that urban expansion has been driven by accessibility to transportation (major and secondary roads), water sources (streams and open water sources) and the landscape (Marondedze & Schütt, 2019). A study on the spatial growth patterns in the Harare Metropolitan Province by Kamusoko et al. (2013) indicated that urban growth was mainly in extension, leapfrog and infill developments. Olawole et al. (2011) assessed the encroachment of urban into farmland. This study aims to explore the application of GIS for smart spatial planning and the management of urban sprawl in the Harare Metropolitan Province.

16.1.1 Urban Sprawl and Its Implication on the Smart City Concept

Although several measures and scopes define urban sprawl, there is no universally accepted definition of the process (Krishnaveni & Anilkumar, 2020). It has been defined using landscape patterns, quantitative measures, attitude explanations and qualitative terms (Wilson et al., 2003). A commonly acknowledged definition of urban sprawl or suburban sprawl is the rapid growth of the extent of towns and cities, often characterised by single-use zoning, rigidly separated shops, workplaces and homes, the dominance in private automobile for transportation, a huge block of road network with poor access and low-density residential housing among others (Rafferty, 2019). Urban sprawl is the unplanned outward expansion of the city to the periphery to accommodate growth due to increased land consumption and urban activities (Jaeger, et al., 2010). It leads to the fragmentation of land-use and produces a different spatial form that can be classified as single-use development strip or linear development (normally along major transportation routes such as road/rail), leapfrog development and continuous low-density sprawl (Batty & Besussi, 2003).

Urban sprawl is caused by many factors including but not only urban expansion coupled with unprecedented population growth, lower land values in peripheries, rural–urban migration, high rental prices and lack of affordable housing in the city centre, the development of ITC, the lack of proper planning policies and the failure of enforcement of planning regulation and poor spatial planning (Bhatta, 2010; Basudeb, 2012). These factors vary according to the structure of a society or the country's level of development (Karakayaci, 2016). In Zimbabwe, population dynamics, changes of land tenure policies and political, social and economic factors are often cited as key drivers behind the urban sprawl (Chirisa et al., 2014; Marondedze & Schütt, 2019). All the six pillars of the smart city concept namely, smart living, smart economy, smart mobility, smart governance, smart people and smart environment are entirely subject to the functionality and resilience of the city's soft and hard infrastructure (UN, 2016). It is the hard and soft infrastructure that attributes a city to be an 'intelligent city'. The smart city concept provides new smart strategies to manage the complexity of urban challenges ranging from urban sprawl, environmental pollution and urbanisation to transportation. Therefore, most of these challenges are linked to urban sprawl. The implementation of the smart city concept is fundamental. Smart building, waste management, water, energy and mobility cannot be achieved in sprawl development where these infrastructure and services are poorly provided.

16.1.2 THE SMART GROWTH AND SMART CITY FRAMEWORKS

Urban sprawl is probably the key challenge facing urban planning at the beginning of the twenty-first century (Batty & Besussi, 2003). The smart growth concept emerged in the late 1990s as a proactive urban planning policy tool to mitigate urban sprawl and advocate for new urban planning policies (Raparthi, 2015; Mohammed et al., 2016; Trillo, 2013). Although smart growth and smart city concepts are considered to be proactive to achieve sustainable cities, their implementation in African cities is faced with several obstacles. The major challenges are rooted in the vagueness and fragmentations in institutions both private and public that are responsible for the management and administration of human settlements. Their interaction has diverse agendas and lacks shared goals, visions and principles to achieve smart growth and sustainability. Smart cities often primarily focus on information and communication technology and smart growth narrowly concentrated on planning (Herrschel & Dierwchter, 2018). Urban planners in the developing world are setting their sights on the applicability of technology to redress and digitalise fragmented human settlements that came along with urbanisation to dovetail with the smart city concepts. Geospatial data acquisition and its juxtaposition through GIS images helps the citizen to know the development trend of cities and avoids sprawling. GIS can be used to monitor changes in current and future land use maps to ensure the conformity of the layout plans, master and local development plans with developments that are taking place on the ground.

16.2 LINKING GIS, SPATIAL PLANNING, SMART GROWTH AND URBAN SPRAWL MANAGEMENT

Monitoring and mapping of the spatiotemporal changes is inevitable to understand the rapid changes that are happening in our cities (Basudeb, 2012). Unprecedented population growth coupled with uncontrolled urbanisation results in urban sprawl, which is a threat to building smart cities. Urban growth needs tracking and monitoring through innovative tools such as remote sensing and GIS to constantly manage the urban transition, land-use and land cover changes (Basudeb, 2012). Geospatial technologies as a policy perspective provide the necessary framework for data collection and transforming the acquired observations to facilitate software-based solutions on how best to manage cities smartly. For the successful implementation of smart cities, data collection is the primary step. Existing land uses and infrastructure needs to be mapped using GIS to produce GIS maps for all urban land use. The landuse plans and demographic data, master and local development plans computed in GIS assist policymakers and urban planners in layout planning and site suitability analysis for effective allocation of land-use.

GIS analysis of urban land use and land cover changes in an urban area over different timeframes assists urban planners in monitoring, assessment and detection of squatters. With the application of GIS through regular satellite images, new slum developments and encroachments can be detected early and development control measures can be applied. Digitalised base maps linked with new urban data sets collected from various geospatial technology facilitate the enforcements of redevelopment measures. Real-time GIS is an emerging paradigm for smart cities that enables the acquisition, storage, analysis and visualisation of geospatial data in real-time (Li et al., 2020). Real-time GIS satellite images assist in the identification of land use encroachments, underutilised land and wrongly allocated land use. This facilitates the redevelopment projects such as transit-oriented developments, urban renewal, land-use change and regeneration projects. Real-time GIS is essential in managing urban sprawl. This assists in minimising horizontal expansions through vertical expansions, promoting walkable and compact urban forms. Geospatial data in GIS assists in the identification and prediction of the direction of urban potential growth. This enhances smart transport planning and management. Base maps of existing urban transport networks are provided in the satellite images and this improves accessibility and mobility planning by linking various urban nodes. Unmanaged urban expansion results in disjointed human settlements and depletion and pollution of natural resources. Therefore, GIS can be applied to monitor urban sprawl and ensure sustainable smart growth.

16.3 METHODOLOGY

The study used GIS and the process includes data acquisition, image sub-setting and pre-processing. Index Derived Built-up Index (IDBI), Normalised Difference Built-up Index (NDBI), Modified Normalised Difference Water Index (MNDWI), and Soil Adjusted Vegetation Index (SAVI) were used to map the built-up features from Landsat imageries. Shannon's entropy method of urban sprawl quantification was used to analyse the trend and extent of urban sprawl.

16.3.1 THE STUDY AREA

Harare is the capital city of Zimbabwe. The Harare metropolitan province has an approximate area of 942 km^2 and lies between 17°49'40" S and 31°3'12" E. The study area covers four districts, namely, Harare urban, Harare rural, Epworth and Chitungwiza. It has an average altitude of about 1,500 m above the sea level. The Harare metropolitan province experiences sub-tropical climate conditions with four seasons: hot and dry season between mid-August and mid-November, warm and wet season from mid-November to mid-March, cool dry season between mid-March to mid-August and a post rain season between mid-March to mid-May (Zambuko, 2010). The metropolitan province had an estimated population of 1,435,78 in 2002, 1,485,231 in 2012 and had a projected population of 1,592,369 in 2017 (ZimStat, 2017).

16.3.2 DATA AND PRE-PROCESSING

The study used openly available satellite data that was downloaded from the USGS Earth explorer web portal. The downloaded Landsat imageries were in the GeoTIFF format. The Landsat temporal images of the years 1999, 2009 and 2019 were used as primary datasets of this study. The Landsat Thematic Mapper (TM) of 1999, Landsat Enhanced Thematic Mapper (ETM +) of 2009 and Landsat 8 of 2019 were used

(Table 16.1). The Landsat images of the same period between June and August were selected to reduce the errors from temporal variations. Firstly, the data was pre-processed and finally trimmed around the boundaries of the study area (Figure 16.1). Google Earth was used for mapping, analysis, accuracy assessment and presentation while processing the images and index derivation was done using the ERDAS Imagine software.

TABLE 16.1
Details of Satellite Imageries Used

Date	Sensor	Path/row	Bands width resolution (in metres)
1999 June	Landsat 5-TM	170/072	Blue (30), Green (30), Red (30), NIR (30), SWIR1 (30), TIR (120(30)) and SWIR2 (30)
2009 August	Landsat 7 ETM +	170/072	Blue (30), Green (30), Red (30), NIR (30), SWIR1 (30), TIR (60(30)), SWIR2 (30), SWIR2 (30) and PAN (15)
2019 August	Landsat 8	170/072	Coastal aerosol (30), Blue (30), Green (30), Red (30), NIR (30), SWIR1 (30), SWIR2 (30), PAN (15), Cirrus (30), TIRS1 (100) and TIRS-II (100)

Source: Author

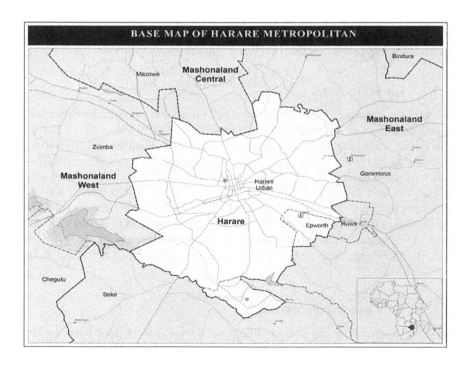

FIGURE 16.1 Harare Metropolitan Province base map.
Source: OCHA (2009)

16.3.3 Automated Built-Up Area Extraction Using Index Derived Built-Up Index

Figure 16.2 summarises the steps and methodology used for mapping built-up areas from Landsat imageries that were downloaded from USGS Earth Explorer. These include Landsat 5-TM of 1999, Landsat 7 ETM + of 2009 and Landsat 8 of 2019. They were all clipped guided by the boundaries of the study area using the shape file of the Harare Metropolitan Province. The data was then processed after the images were accurately clipped in the extent of the shape file.

16.3.4 Computation of Spectral Indices

- Normalised Difference Built-up Index (NDBI)

$$NDBI = \frac{MIR - NIR}{MIR + NIR}$$

In the mid-infrared region (MIR), the built-up features show an increased spectral reflectance as compared to the near-infrared region (NIR) of the electromagnetic

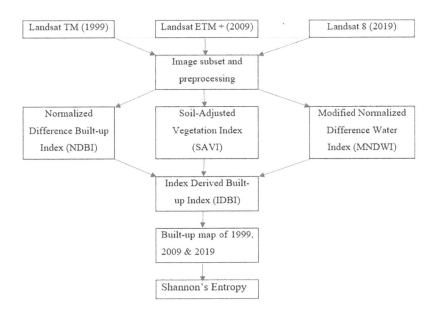

FIGURE 16.2 Chart flow of the methodology.

Source: Adapted from Krishnaveni and Anilkumar (2020)

band. In Landsat TM (1999) and Landsat ETM + (2009), band 4 symbolises NIR and band 5 symbolises MIR, whereas, in Landsat 8 (2019), band 5 and 6 symbolise NIR and MIR, respectively.

- Soil Adjusted Vegetation Index (SAVI)

$$SAVI = \frac{(NIR - RED)(1 + L)}{(NIR + RED + L)}$$

L represents the soil adjustment factor (normal value = 0.5). In Landsat TM (1999) and Landsat ETM + (2009), band 3 represents Red and band 4 represents NIR, whereas in Landsat 8 (2019), bands 4 and 5 represent Red and NIR, respectively. SAVI was chosen over NDVI because of its accuracy in vegetation cover detection. SAVI is more accurate in detecting vegetation cover even in built-up environments where vegetation coverage is less. NDVI is more suitable for mapping vegetation cover in areas with plant cover ≥ 30, whereas SAVI is applicable in mapping vegetation cover in areas with less plant cover to as low as 15%. Therefore, SAVI is more suitable in built-up environments.

- Modified Normalised Difference Water Index (MNDWI)

$$MNDWI = \frac{(GREEN - WIR)}{(GREEN + WIR)}$$

MNDWI is a modified model of the NDWI developed in 1996 by McFeeters (Mcfeeters, 1996). It is obtained by subtracting MIR from the NIR band. In Landsat TM (1999) and ETM + (2009) imageries, bands 2 and 5 represent Green and MIR respectively, whereas bands 3 and 6 represent Green and MIR in Landsat 8 (2019) imagery. MIR has a better performance in sensing water as compared to NIR and therefore MNDWI produces more accuracy than NDWI in extracting water features.

- Index Derived Built-up Index (IDBI)

$$IDBI = \frac{\left(NDBI - \frac{(SAVI + MNDWI)}{2}\right)}{\left(NDBI + \frac{(SAVI + MNDWI)}{2}\right)}$$

IDBI index is developed from SAVI, MNDWI and NDBI indices. There is a substantial reduction in redundancy between thematic and original band imageries and the spectral bands of vegetation, waterbody and built-up area are well isolated. Besides, with IDBI, the image can be used without any geometric modifications for further computations.

16.4 RESULTS AND DISCUSSION

Urban growth and urban sprawl are often debated because both phenomena include urban expansion. However, the key distinguishing feature is the nature or the characteristics of growth which includes type, rate and pattern. Therefore, the growth of Harare metropolitan can be clearly distinguished from the general urban growth. Rapid population growth, the Fast-Track Land Reform Programme (FTLRP) of the year 2000, economic meltdown, political instability and the high unemployment rate in the last two decades has adversely impacted urban spatial growth. As a result of these driving forces, since 2000 the urban sprawl phenomena in Harare metropolitan has become ungovernable. The FTLRP immensely contributed to urban sprawl in Harare metropolitan. Before the FTLRP in 2000, urban expansion in the peripheries was restricted by private land surrounding the metropolitan. The FTLRP changed land ownership and land-uses. White commercial farms surrounding the metropolitan were seized by political elites who later unprocedural subdivided the farms into housing land. The political environment during this period also influenced white farmers who were not affected by the FTLRP to develop their land for urban use fearing expropriation without compensation. As a result, this triggered urban sprawl in the metropolitan.

In addition, the economic meltdown after 2000 can be attributed for urban sprawl. Socioeconomic hardships within the inner city such as high unemployment rates and a surge in rental and land prices drove the urban poor to informal settlements. Informal land sales and markets were the catalysts of the increase in informal housing. In the inner city, housing land was expensive and sold in formal markets whereas in the outer city, housing land was acquired cheaply from private developers and individuals who seized white farms. This growth trend continued and resulted in high urban informality. In 2005, the government implemented *Operation Murambatsvina* (Operation Restore Order) to restore sanity in housing development. However, the directive rendered hundreds of thousands of slumdwellers homeless. In response to the deprivation of housing, the government implemented *Operation Garikai/Hlalani Kuhle* (Operation Live Well), a national housing programme for the displaced slumdwellers. This saw the expansion of the metropolitan towards the northwest and southern direction between 1999 and 2009. Therefore, the increase of high-density housing in the northern and southern sides was due to a government initiative of providing housing for the low-income groups. The programme led to the development of housing schemes, which includes Hatcliffe in the northern side, Whitecliffe along Harare–Bulawayo Road and Hopely in the southern side (UNHABITAT, 2009).

However, the government failed to provide enough housing units to the displaced persons due to economic hardships in 2008 and this reduced the supply of low-cost housing (Chirisa et al., 2014). As a result, an increase in informal settlements was observed between 2009 and 2019 and the city expanded outwards particularly in the southeast and southern directions. In some instances, the FTLRP and the *Operation Garikai* contributed to urban expansion. For example, in the eastern side, the Caledonia housing development is both a result of Operation Garikai and the FTLRP. The findings concur with the findings by Marondedze and Schütt (2019) in their study on the relation between the drivers and the axis of urban growth in Harare metropolitan. As noted earlier, the hyperinflation associated with the devaluation of the local currency after 2000 made it difficult for building societies and land developers to purchase land. Hence, formal housing development decreased which paved the way for high informality. However, a steady economic growth experienced during the Government of National Unity between 2009 and 2013 enabled building societies and land developers to increase their investment in housing developments, thereby increasing the built-up area within the metropolitan. The adoption of the dollarisation monetary policy ushered in a multi-currency system that significantly increased the built-up area from 2009 to 2019.

Changes in non-built-up to built-up urban land was mainly in infill, edge or fringe and leapfrog developments. In medium and high-density zones, fringe and leapfrog sprawls have been observed in the peripheries whereas infill development was experienced in the inner-city. Infill growth is the development of small tracts of vacant land or open spaces mostly encircled by urban built-up land in neighbourhoods (Yue et al., 2013; Heim, 2001). Fringe or edge growth represents the development of the existing urban built-up land patch (Yue et al., 2013). Leapfrog growth refers to development changes from vacant land to developed land separate to or isolated from existing urban built-up land mostly in large tracts (Yue et al., 2013; Wilson et al., 2003). Change of land-use and subdivision of farms into residential properties has caused leapfrog sprawl in the Harare metropolitan province (Kamusoko et al., 2013).

Significant spatial growth was observed between 1999 and 2009 as evidenced by an increase in the urban built-up area and a subsequent decrease in urban no-built-up area. The urban built-up land increased by 24.9% (234.5 km^2), whereas the urban non-built-up land decreased by 75.03% (704.9 km^2) of the study area. Further urban expansion was observed between 2009 and 2019 where urban built-up land increased by 36.56% (343.5 km^2) from 32.30% (303 km^2) while urban non-built-up land decreased by 63.3% (596 km^2) from 65.63% (635 km^2). Figure 16.3 provides a detailed representation of the land-use land cover classification for Harare Zimbabwe over 1999, 2009 and 2019 respectively.

Although urban sprawl was higher in the high and medium density zones, the low-density residential zones also experienced growth changes in the northern side but at a lower level. This can be attributed to the topographical factors and socioeconomic factors. The northern side is hilly with higher land values, whilst the southern side has a gentle slope with lower land values (Olawole et al., 2011). High urban sprawl was observed in the southern and south-eastern directions. Harare is experiencing alarming urban growth and is now merging with Chitungwiza in the south, Epworth in the south-east, Norton in the west and Ruwa in the east. The developmental trend and pattern experienced in the metropolitan since 2000 compromises the applicability of the smart city concept. The distortion in the implementation of masterplans created challenges in the provision of infrastructure and services. Most of the newly built areas are characterised by transportation problems, inadequate services, poor housing and quality of urban life. A cross-dimensional analysis of these indicators against the pillars of smart city concept (*smart people, living, mobility, governance, economy and environment*) shows that neither one of these dimensions can be successfully implemented under the current development trends. If we consider the

Smart Spatial Planning and Management 239

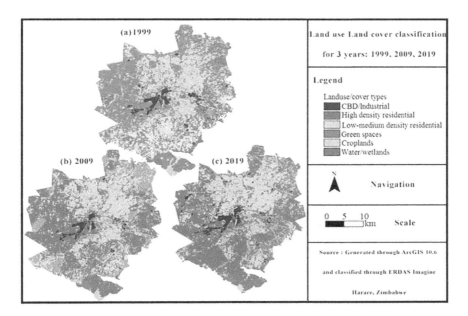

FIGURE 16.3 Land-use land cover classification for the periods 1999, 2009 and 2019.

Source: Author

definition of smart cities to be a 'intelligent city', it is clear that robust infrastructure provision plays a key role. The GIS and RS techniques are now used extensively across various public and private sectors. The application of these techniques includes land-use planning, disaster management, transport and locational analysis, economic planning, demographic analysis and utility and infrastructure planning among other uses. Thus, the application of GIS cuts across all the dimensions of the smart city concept; therefore, GIS and RS have a key role to play in the management of urban sprawl and in the implementation of smart city concept.

16.4.1 BUILT-UP AREA EXTRACTION

The increase in the built-up areas is regarded as an indicator of growth in urban sprawl. IDBI was used for automatic mapping of built-up areas from the Landsat imageries. In the IDBI image (Figure 16.4), the built-up area was coded in light grey to white colour, while water and vegetation features were significantly obscured and outlined in dark grey to black shade. IDBI values for built-up features were positive while those of vegetation and water features were negative. Built-up areas were categorised as with pixel values exceeding the threshold and given a value of 1 whereas the non-built-up features were classified as pixels with a value less than or equal to the threshold. After processing, this results in a binary image with values 1 and 0 representing the built-up and non-built-up classes, respectively.

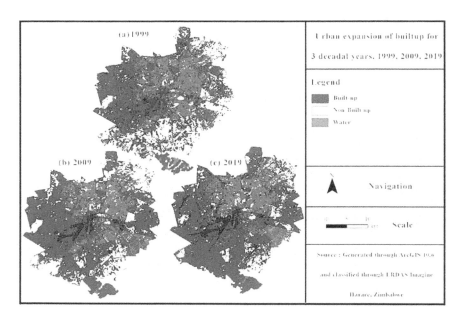

FIGURE 16.4 Built-up area expansion in the years 1999, 2009 and 2019.

Source: Author

16.4.2 Analysis of Urban Sprawl Using Shannon's Entropy

Shannon's entropy method was developed by Shannon in 1948 for calculations of disorder and randomness based on Rudolph Clausius's concept of entropy. It is one of the commonly used methods for measuring patterns of urban sprawl displaced or concentrated. The application of Shannon's entropy method is based on GIS and RS. The results of Shannon's entropy method show an alarming increase in the built-up area and a decrease in the non-built-up area from 1999 to 2019. Green spaces and farmland that are the lungs of the city are drastically decreasing. Shannon's entropy (H_n) is given by:

$$H_n = -\sum_{I=1}^{n} P_i \log_e (P_i)$$

where P_i represents the proportion of the ith variable zone (ward) and n is the total number of zones. P_i refers to the impervious areas in the ith ward, n denotes the total number of wards (12 in this study) and $\log_e(n)$ represents the upper limit of the entropy (2.48 in this study).

In this study, the area of each of the built-up map was divided into 12 grids. Hence, $n = 12$ for Harare metropolitan. Table 16.2 shows the calculations done for Shannon's entropy.

In this research, the built-up area is considered as the indicator of urban sprawl. Shannon's entropy, ΔH_n, is used to calculate the entropy change as shown in Table 16.3.

TABLE 16.2
Computation of Shannon's Entropy over the Years 1999, 2009 and 2019

	Year	1999	2009	2019	1999	2009	2019
		Built-up area (km²)			-pi loge (P_i)		
GRID NUMBER	1	10	10.15	10.95	0.050384446	0.05082227	0.053793747
	2	143.54	143.98	141.98	0.29410575	0.294013596	0.291811595
	3	20	20.19	28.55	0.085212285	0.085570837	0.10974662
	4	50	50.65	60.37	0.161618923	0.162587775	0.181650911
	5	130.579	130.76	135.17	0.281416471	0.28109701	0.28522398
	6	45.87	48.69	49.93	0.152706864	0.158444388	0.160809126
	7	86.7	80.55	83.82	0.226694521	0.216788755	0.221535814
	8	95.5	92.9	91.96	0.239343756	0.235211951	0.233545422
	9	37.33	41.75	51.9	0.132906222	0.143038049	0.164914287
	10	119.73	120.87	82.6	0.269689396	0.270463916	0.219661879
	11	51.87	53.86	55.13	0.165526249	0.169191889	0.171466017
	12	100.01	100.13	104.39	0.245468135	0.245127885	0.250354778
Shannon's entropy, ΔH_n					2.305073018	2.31235832	2.344514175

Source: Author

TABLE 16.3
Shannon's Entropy over the Years 1999, 2009 and 2019

S. No.	Year	Shannon's entropy	ΔH_n
1	1991	2.305	-
2	2009	2.312	0.007
3	2019	2.345	0.033

$\log_e(n) = 2.48490664979$

Source: Author

There is a steady increase in Shannon's entropy's values from 1999–2009 and 2009–2019. The increase in Shannon's entropy values demonstrates an increment in the scattering of developed zones, which indicates the development of urban sprawl in Harare metropolitan. There was a continuous increase in urban sprawl between the two decades.

16.5 CONCLUSION

Urban sprawl is a major hindrance in planning for smart and resilient cities globally, particularly in developing countries. It has become a matter of concern in contemporary urban planning and if left unmonitored and managed, it compromises the sustainability of human settlements. The dynamics of changes in the built-up areas in a particular stipulated timeframe provide significant information about the extent, nature and development of urban sprawl. This will help in building smart, resilient and sustainable cities. A GIS overlay analysis of planned development and existing land-use can be used to monitor and manage urban sprawl. Thus, the application of GIS cuts across all the dimensions of the smart city concept. Therefore, it is clear that GIS and RS have key roles to play in the management of urban sprawl and in the implementation of smart city concept. The application of GIS and RS in urban sprawl management is cost effective compared to the traditional approach of using cartographic methods. Moreover, it enables the manipulation of different spatial datasets in a single map frame. GIS enables the acquisition and analysis of geospatial data which entails evidence-based decision making and planning. Therefore, future land development can be stimulated which provides room for effective implementation of smart cities.

Urban growth patterns and trends can be periodically evaluated to ensure compliance in the implementation of land-use and economic development plans.

The key driving forces of urban sprawl in Harare Metropolitan can be categorised into social, economic, political and land policy factors. Economic policies such as the adoption of the dollarisation in 2009 contributed to new housing developments. The FTLRP of 2000 had a significant impact on the economic recession and the development of informal settlements. The FTLRP caused changes in land ownership of farms that were surrounding the metropolitan. This subsequently caused land-use changes from agricultural to residential land use thereby facilitating urban sprawl.

16.6 RECOMMENDATIONS

From the study findings, the following recommendations are suggested.

- Development of inclusive and flexible urban polices (transport, housing and environmental policies to protect urban open spaces and the natural environment).
- Adoption of mixed land-uses (compact city concept, enabling vertical zoning rather than horizontal expansion).
- Develop digital land administrations platforms.
- Open access of land information and information dissemination (participatory GIS).
- Implement urban renewal programmes and maximise land-use capacity through vertical development (space creation strategy).
- Establishment of an independent institution for infrastructural development that is allocated a certain percentage annually from the national budget.

REFERENCES

Banai, R. & DePriest, T., 2014. Urban Sprawl: Definitions, Data, Methods of Measurement, and Environmental Consequences. *Journal of Sustainability Education*, 7(2151–7452), pp. 455–567.

Basudeb, B., 2012. *Urban Growth Analysis and Remote Sensing: A Case Study of Kolkata, India 1980–2010*.Germany: Springer Heidelberg.

Batty, M. & Besussi, E., 2003. Traffic, Urban Growth and Suburban Sprawl. *Centre for Advanced Spatial Analysis CASA, UCL*, 70(ISSN: 1467–1298), pp. 1–12.

Bhatta, B., 2010. *Analysis of Urban Growth and Sprawl from Remote Sensing Data*. London: Springer.

Chirisa, I., Gaza, M. & Bandauko, E., 2014. Housing Cooperatives and the Politics of Local Organization and Representation in Peri-Urban Harare, Zimbabwe. *African Studies Quarterly*, 15(1), pp. 37–53.

Chirisa, I., Maphosa, A. & Zanamwe, 2016. Past, Present and Future Population Growth and Urban Management in Zimbabwe: Putting Institutions into Perspective. Em: *Population Growth and Rapid Urbanisation in the Developing World 1 edition*. Pennsylvania: IGI Global, pp. 64–81.

Fatih, T. & Bolen, F., 2009. Urban Sprawl Measurement of Istanbul. *European Planning Studies Tandfonline*, 17(10), pp. 1559–1570.

Fertner, C., Jørgensen, G., Nielsen, T. A. S. & Nilsson, K. S. B., 2016. Urban Sprawl and Growth Management—Drivers, Impacts and Responses in Selected European and US Cities. *Future Cities and Environment, Springer*, 2(9), pp. 779–897.

Heim, C. E., 2001. Leapfrogging, Urban Sprawl, and Growth Management. *American Journal of Economics and Sociology*, 60(1), pp. 245–283.

Herrschel, T. & Dierwchter, Y., 2018. *Smart Transitions in City Regionalism: Territory, Politics and the Quest for Competitiveness and Sustainability*. London: Routledge.

Jaeger, J. A. G., Bertiller, R., Schwick, C. & Kienast, F., 2010. Suitability Criteria for Measures of Urban Sprawl. *Elsevier*, 10, pp. 397–406.

Kamusoko, C., Gamba, J. & Murakami, H., 2013. Monitoring Urban Spatial Growth in Harare Metropolitan Province, Zimbabwe. *Advances in Remote Sensing, 2013*, 2, 2, pp. 322–331.

Karakayaci, Z., 2016. The Concept Of Urban Sprawl and its Causes. *The Journal of International Social Research*, 9(45), pp. 815–818.

Krishnaveni, K. S. & Anilkumar, P. P., 2020. Managing Urban Sprawl Using Remote Sensing and GIS. *Int. Arch. Photogramm. Remote Sens. Spatial Inf. Sci*, XLII-3/W11, pp. 69–66.

Li, W., Batty, M. & Goodchild, M. F., 2020. Real-time GIS for Smart Cities. *International Journal of Geographical Information*, 34(2), pp. 311–324.

Mahendra, A. & Seto, K. C., January 2019. *Upward and Outward Growth: Managing Urban Expansion for More Equitable Cities in the Global South*. Washington, DC: World Resources Institute.

Marondedze, A. K. & Schütt, B., 2019. Dynamics of Land Use and Land Cover Changes in Harare, Zimbabwe: A Case Study on the Linkagebetween Drivers and the Axis of Urban Expansion. *Land*, 8(10), pp. 50–89.

Mcfeeters, S. K., 1996. The use of the Normalized Difference Water Index (NDWI) in the delineation of open water features. *International Journal of Remote Sensing*, 17(7), pp. 1425–1432.

Mohammed, I., Alshuwaikhat, H. M. & Adenle, Y. A., 2016. An Approach to Assess the Effectiveness of Smart Growth in Achieving Sustainable Development. *Sustainability*, 8, p. 397.

OCHA, 2009. *Map of Harare Metropolitan*. Harare, Zimbabwe. [Online] Available at: www.ochaonline.un.org/zimbabwe.

Olawole, M. O., Msimanga, L., Adegboyega, S. A. & Adesina, F. A., 2011. Monitoring and Assessing Urban Encroachment into Agricultural Land—A Remote Sensing and GIS Based Study of Harare, Zimbawe. *Ife Journal of Science*, 13(1), pp. 149–160.

Rafferty, J. P., 2019. *Encyclopædia Britannica*. [Online] Available at: www.britannica.com/topic/urban-sprawl [Acesso em 05 05 2020].

Raparthi, K., 2015. Assessing Smart-Growth Strategies in Indian Cities: Grounded Theory Approach to Planning Practice. *Journal of Urban Planning ASCE*, 141(4), pp. 81–100.

Trillo, C., 2013. Urban Sprawl Management, Smart Growth: Challenges from the Implementation Phase. *International Journal of Society Systems and Science*, 5(3), pp. 261–282.

UN, 2016. *Smart Cities and Infrastructre*. Geneva: UN.

UNHABITAT, 2009. *Housing Finance Mechanisms in Zimbabwe*. Nairobi: UNHABITAT.

Wilson, E. H. et al., 2003. Development of a Geospatial Model to Quantify, Describe and Map Urban Growth. *Remote Sensing of Environment*, 86(3), pp. 275–285.

World Bank, 2010. *Cities and Climate Change: Aa Urgent Agenda Urban Development*, Washington, DC: The World Bank.

World Bank, 2019. *World Bank Youth Summit 2019: Smarter Cities for a Resilient Future*. Washington, DC: World Bank.

Worldometer, 2020. *Worldometer*. [Online] Available at: www.worldometers.info/world-population/zimbabwe-population/

Yue, W., Liu, Y. & Fan, P., 2013. Measuring Urban Sprawl and its Drivers in Large Chinese Cities: The Case of Hangzhou. *Land Use Policy*, 31, pp. 358–370.

Zambuko, C., 2010. *Climate Issues and Facts: Zimbabwe*, Harare: Zimbabwe Meteorological Service Departmanet.

ZimStat, 2017. *Facts and Figures*. Harare: GoZ.

17 Appraisal of E-Waste Management Approaches in Zimbabwean Cities

Takudzwa M. Matyatya, Willoughby Zimunya and Tariro Nyevera

CONTENTS

17.1 Introduction 245
17.2 Theoretical Framework and Literature Review 246
    17.2.1 Theoretical Framework 246
    17.2.2 Literature Review 246
17.3 Methodology 249
17.4 Results 249
17.5 Discussion 253
17.6 Conclusions and Recommendations 254
References 255

17.1 INTRODUCTION

The advancement in technology has contributed to the growth of electrical and electronic gadget manufacturing companies, thereby leading to a general increase in their end products (Kalana, 2010). The positive results of these technological advancements are being increasingly enjoyed even in the twenty-first century, making life easy in terms of communication, provision of entertainment and basic household and office devices (Kalk, 2012). However, the challenges of technological advancements in line with e-waste were never anticipated, thereby leading to rising e-waste management problems in the twenty-first century. E-waste refers to electronic and electrical gadgets that are no longer fit for their intended use and those that have expired (Gweme et al., 2016). E-waste management issues in Zimbabwe are not really pronounced because of the reasons that remain veiled in obscurity, some of which can be attributed to the lack of information concerning that field. Existing literature reveals that electrical and electronic gadgets that would have ended their life span and those technologies that could not have worked properly in Europe find their way into Africa as cheap imports (Kasapo, 2013, Theodros, 2010, and Adediran and Abdulkarim, 2012). The illegal dumping of e-waste and the operations of e-waste recyclers is having a negative impact to the environment as it is affecting the amenity of urban areas, thereby undermining the land use planning system.

With globalisation forces at play, electronic gadgets from developed countries continue to be imported into Zimbabwe in the quest for opening doors for economic, social and political interactions (Pellow, 2007). However, this process increases e-waste generated in Zimbabwe and poses e-waste management challenges as well as opportunities. Some e-waste is disposed of in local neighbourhoods at illegal dumpsites, especially on public open spaces that are situated in the high-density residential areas (Gweme et al., 2016). Some of the e-waste is now a source of livelihood for informal e-waste recyclers who also operate mostly in high-density residential areas. Most of the e-waste recyclers occupy land that is not designated for such uses, which could be developed or undeveloped public open spaces and, in some instances, in public utility servitudes (Sthiannopkao and Wong, 2013). This study examines the constraints affecting the adoption of proper e-waste management practices in Zimbabwe using the case study of the Harare city.

17.2 THEORETICAL FRAMEWORK AND LITERATURE REVIEW

17.2.1 Theoretical Framework

The study adopted complexity theory in a bid to understand the complex systems behind e-waste generation and management in urban areas. The theory originated in the mid-1980s and the theory's leading proponent is Stuart Kauffman (Macauley et al., 2003). The basic premise of the complexity theory is that there is a hidden order to the behaviour (and evolution) of complex systems, whether that system is a national economy, an ecosystem, an organisation, or a production line. The theory argues that there is a hidden order to the behaviour of complex systems, even if that system is a national economy (Macauley et al., 2003). Urban areas are complex systems comprising different people with different perceptions, unpredictable dynamics and different age groups and gender (Mattingly, 1995). In e-waste management, there are complexities in handling and disposal of e-wastes such that a multi-stakeholder engagement and a multi-disciplinary approach are necessary to understand the various aspects at play (Widmer et al., 2005). Instead of spending time forecasting a complex environment, it is better to channel the resources and efforts towards a multi-stakeholder engagement aimed at fully analysing e-waste management practices and how to improve them for the benefit of the environment and at the same time not compromising proper land use planning (Macauley et al., 2003). The complexity theory is used to trace and tackle the seemingly insurmountable challenges that urban areas in the twenty-first century are facing. If stakeholder perceptions are shaped by the complexity theory, the strategies that are employed in solving the social, economic and political problems including e-waste management that are bedevilling urban areas can be altered (Widmer et al., 2005).

17.2.2 Literature Review

E-waste is an emerging and fast-growing waste stream with complex traits and impacts. Previous researches show that this type of solid waste is a problem both in developed and developing nations as it is not biodegradable (Gupta et al., 2014;

Gweme et al., 2016). Developed countries felt the negative impacts of e-waste earlier than developing countries because they advanced early in terms of technology development, thereby having faster growing e-waste streams (Macauley et al., 2003).

The world at large is coming together to address the problem of e-waste. The Basil Convention is a global policy that was put in place to restrict the movement of e-waste across borders, especially its dumping in developing nations by developed nations (Sthiannopkao and Wong, 2013). However, the convention is not very articulate in terms of its scope such that developed countries continue to dump e-waste not as real waste but as inferior reusable products and recyclable materials. Nevertheless, developed nations now have their own legal and policy frameworks to deal with the problem of e-waste (Widmer et al., 2005). Chartier (2014) noted that some nations through city authorities have collaborated with electrical and electronic manufacturing companies to try to reduce e-waste through formal recycling and proper disposal that is safe to the environment. However, this is not the case for all developed nations as some are still dumping their e-waste in Africa on the pretext that it is recyclable material or are donating second-hand gadgets (Sthiannopkao and Wong, 2013). An assessment of literature shows that the globalisation process has worsened e-waste dumping by developed nations in developing countries as it opened the doors for economic, social and political interactions (Pellow, 2007). This globalisation process facilitates the movement of e-waste from developed nations to developing nations with serious consequences on land use planning system and the environment (Osibanjo and Nnorom, 2007).

It is important to note that, amongst all the continental regions of the world, Africa is privileged in that it industrialised later than all other continents such that the amount of e-waste it generates is still relatively low (Chartier, 2014). However, it is unfortunate that despite this low generation of e-waste, it has become the dumping ground of e-waste from other continents in the name of cheap imports and donations (Pellow, 2007). Africa is the least industrialised continent in the world such that it embraces and accepts electronic and electrical imports, even second-hand equipment, which few years down the line become obsolete and turn into e-waste (Widmer et al., 2005). Ghana and Nigeria are good examples of African countries that are victims to dumping of electrical and electronic gadgets from other developed countries (Robinson, 2009). The dumping of e-waste leads to unhealthy environments, compromises the amenities of land and since the regions are connected, negative externalities manifest in other areas if not attended to in time (Osibanjo and Nnorom, 2007; Robinson, 2009). Thus, this means the interconnectedness of countries in Africa needs proper waste management system so as to avoid negative externalities of e-waste manifesting in other countries (Osibanjo and Nnorom, 2007).

Rapid urbanisation in Zimbabwe has seen a tremendous increase in urban population (increasing at a rate of 5% per annum since 1980) and extensive urban developments in urban areas (Mangwende, 2015). Urban areas are the ones with greater ownership and consumption of electrical and electronic gadgets when compared to rural areas, thereby making the problem of e-waste an urban challenge (Gweme et al., 2016). Harare, as the capital of Zimbabwe with a population of about 2 million, is vulnerable to this challenge of e-waste (ZimStat, 2012). The increased volume of e-waste associated with high population is ending up in illegal dumpsites, which are mainly public open spaces and servitudes that are situated in populous high-density residential areas, thereby causing a threat to the environment, amenity of land and human health, especially groundwater contamination (Osibanjo and Nnorom, 2007). Informal e-waste recyclers occupy open spaces where they undertake their operations and also dump the residue thereon (ibid). E-waste is generated internally in the country by local industries and externally through imports into the country. Fast-growing technology industry and innovations are making existing electronic and electrical gadgets obsolete, thereby adding them to e-waste (Gweme et al., 2016). Thus, the problem of managing e-waste is increasing in Zimbabwe.

Proper waste management is a critical component of good land management. Land use planning is critical in managing the competition between various uses and optimising the use of land (Robinson, 2009). Land use planning is an iterative process based on dialogue amongst all stakeholders aimed at the negotiation and decision for a sustainable form of land use in urban areas and initiating and monitoring its implementation (Zimmerman, 1999). E-waste management practices that were evident in Ghana show that informal activities are being done on undesignated land, thereby affecting land use patterns and negatively impacting the environment (Osibanjo and Nnorom, 2007). E-waste management practices happen within urban land and if not formally planned, they negatively affect the environment, as they will not be integrated with citywide land uses.

E-waste management involves several stakeholders who play various roles as far as it is concerned. The actors include local authorities, the public, the private sector, the government and parastatals (Macauley et al., 2003). Each one of them belongs to either the formal or the informal sector. The central government sets the theme of any development and supports it through creating a conducive environment (Gupta et al., 2014). It aligns policies with e-waste management and its role is crucial because without policies no progress takes place. The Indian government subsidised companies involved in e-waste management and aligned its budget to fund e-waste awareness campaigns and strategies (Sthiannopkao and Wong, 2013). Local authorities have the mandate to manage all types of wastes in their areas of jurisdiction. They also have the power to regulate players and stakeholders in the management of e-waste. The private sector works as manufacturers, traders, collectors, refurbishers and recyclers (Theodros, 2010). In addition, as manufacturers, the private sector contributes to e-waste generation since their products over time eventually become waste (ibid). Besides, the private sector also collects e-waste for re-cycling and use the recycled material for manufacturing of other goods. As such, the private sector can aid the waste management efforts of local authorities, thereby becoming a major stakeholder in the management of e-waste (Kalk, 2012).

Zimbabwe is a signatory of the Durban Declaration of 2008 on e-waste management in Africa (Gweme et al., 2016). The declaration clearly pointed out that every African country needs to map its strategy on how to manage e-waste being guided by the declaration recommendations, which include improving cooperation amongst stakeholders, establishing an institutional framework and creating general awareness and legal framework (Chartier, 2014). In this case, the country already has a starting point as the Durban Declaration of 2008 provides guidelines on how to manage e-waste in Africa (Gweme et al., 2016). At the global level, the Basil Convention restricts the trans-boundary movement of hazardous waste, in particular, e-waste from the developed to the less developed countries as this is causing the informal sector to thrive as it continues to get hold of e-waste in huge volumes (Sthiannopkao and Wong, 2013). While Zimbabwe is a signatory to these international policies on e-waste management, it does not have explicit and sound policies on e-waste management. The continental and international laws pertaining to e-waste management create a base for managing e-waste in all countries, including Zimbabwe (Gweme et al., 2016). In this case, local policies and laws pertaining to e-waste management are supposed to align with regional and international legal and policy frameworks.

17.3 METHODOLOGY

The study adopted the constructivist paradigm and the qualitative methodology linked to a typical case study approach. Qualitative approach was adopted to explore the behaviour, perspectives and experiences, attitudes of people involved in the e-waste management practices and how they are affecting land use planning and environment. Six semi-structured interviews were conducted with respondents from the City of Harare (Town Planning Department and Waste Management Division), environmental management agency, private sector organisations and individuals who are involved in recycling, refurbishing, dismantling and electronic manufacturing. The interview respondents were purposively selected based on their experience in planning and regulating waste management as well as the recycling, refurbishing, dismantling and electronic manufacturing of electronic gadgets. The study also used unobtrusive observation constituted primary data collection methods. The literature review of peer reviewed journals, reports, text books and statutes was undertaken with the aim to draw lessons and insights from global, regional and local practices and trends. The Mbare township in Harare was used as the focus of the study because of the prevalent cases of e-waste management recycling and dumping in that area. Content and thematic analyses were used as major qualitative data analysis tools.

17.4 RESULTS

The study noted that the players that are involved in general e-waste management in the Mbare township are the Harare city council, Environmental Management Agency (EMA) and private formal and informal players. The study established that the institutional framework for general solid waste management in Harare is backed by a proper legal and policy framework that is comprised of legislations such as the EMA Act, Public Health Act, Regional Town and Country Planning Act and Urban Councils Act. In addition, Zimbabwe is a signatory of the 2008 Durban Declaration on e-waste management in Africa and this demonstrates the commitment of Zimbabwe to properly manage e-waste. However, the study also revealed that there is no formalised and functional institutional framework for e-waste management. For instance, one of the key informants from the Harare city council indicated that there is no formal system for management of e-waste in the city, despite the council being the principal player in waste management. Instead, the Harare city council is using the general solid waste management system to handle the bulk of e-waste. Some of the e-waste is being handled by private sectors. These private players are licensed to collect waste for recycling. In addition, in a bid to improve general solid waste management from source, the council is also authorising individuals and other companies to collect solid waste including e-waste from the neighbourhoods free of charge. Mostly informal collectors collect the other e-waste from the landfill sites and neighbourhoods. Formal e-waste collection is not yet in existence. The study also noted that in cases where private players are involved in waste collection, the Harare city council is supposed to supervise them but this is not being effectively done resulting in illegal dumping of e-waste. In addition, EMA as the regulatory authority on environmental issues is also involved in waste management by ensuring the compliance of the council and other waste collectors with environmental standards.

The study observed that the failure by the Harare city council waste management division to collect solid waste throughout the city including Mbare has led to the emergency of informal waste management set-ups in Mbare. The informal set-ups are involved in e-waste management collection and they are unregulated by the EMA and the Harare City Council. The study observed that the informal sector collects most of the e-waste from the landfill sites and surrounding neighbourhoods and takes it to Mbare for re-processing. This practice is worsened by the fact that there are no mechanisms to regulate the informal and illegal re-collection and re-use of waste from landfill sites such as Pomona. The study also established that the town planning division of the council is involved in waste management in the city by selecting landfill sites. However, the division is not facilitating the creation of a proper working space for informal e-waste recyclers resulting in them occupying undesignated lands to undertake their operations. As such, these e-waste recyclers have occupied public open spaces and other undesignated sites in the Mbare township. The hotspots for e-waste recycling and dumping include the sites between the back of Shawasha hostels facing the Cripps Road, the whole open space comprising recreational grounds and road servitudes along the Simon–Mazorodze Road opposite the Mbare Police station, along the Adbernnie road servitude opposite the Mbare Old Bus Terminus and along the Mukuvisi River.

The study also revealed that one of the constraints facing the Harare City Council in handling e-waste is the lack of proper legal and policy frameworks. E-waste is an emerging waste stream and not much ground has been covered in terms of having a legal and policy framework for its management in Zimbabwe. Thus, the council is managing e-waste based on the general legal and policy frameworks for managing general wastes in the city. This general waste management policy is not articulate on e-waste issues such that e-waste is managed as any other solid waste. The policy is deficient in that it is allows e-waste to be disposed of in dumpsites, despite its negative impacts on the environment. While Zimbabwe has a legislation on general waste management like the EMA Act, it does not have an explicit e-waste management policy or legal framework. In view of the emerging and growing e-waste streams, EMA has crafted draft guidelines for the management of e-waste. Once finalised, this policy will assist local authorities including the Harare Municipality in the management of e-waste.

The town planning division is responsible for allocating and controlling land uses to ensure that any development complies with the master plan and local development plan (LDP) of Mbare. Yet, e-waste is being managed informally in a study area in violation of the Mbare North LDP. Table 17.1 presents proper land uses as indicated in the Mbare North LDP against the current informal uses occupied by e-waste handlers.

As depicted in Table 17.1, the land uses that are provided in the Mbare North LDP are being violated by e-waste handlers and these include recreational and public spaces that have been occupied illegally. One typical example is the e-waste that is gathered on public space between the Shawasha Flats and Cripps roads as depicted in Figure 17.1. The informal e-waste management activities on these sites are affecting the adjacent land uses in terms of land value, amenity and perception. The informal e-waste handlers do not adhere to environmental quality regulations and hence end up disposing the residue of e-waste on illegal dumpsites as shown in Figure 17.2.

TABLE 17.1
Contrast between Planned and Existing Land Uses in Mbare Township

Sites	Designated use in the LDP	Current use on the site
Along the Cripps road	Road reserve	SMEs including informal e-waste handlers
Around Shawasha flats	Recreation, open space and Freeway reservation	Chain of e-waste handlers occupying the space
Along the Adbernnie road	Road reserve	Occupied by e-waste handlers
Along Simon Mazorodze	Road reserve	SMEs including informal e-waste handlers

Source: Matyatya (2018)

FIGURE 17.1 E-waste gathered on an area designated as an open space in Mbare township.
Source: Matyatya (2018)

FIGURE 17.2 E-waste dumped at undesignated dumpsites in Mbare township.

Source: Matyatya (2018)

E-waste is different from any other solid waste because it has got toxicants in the form of heavy metals and it is not biodegradable, thereby polluting the environment and land. The informal e-waste handling activities lead to land pollution in areas, where e-waste recycling activities are taking place. Town planners are responsible for development control and are required to ensure that the provisions of the land use plans are strictly adhered to. However, the study revealed that as far as the town planners and the development control are concerned, the guard over illegal e-waste management practices is not effective as exemplified by the rampant operations on these hotspots. E-waste activities have been persisting for over six and ten years in some hotspot areas and nothing has been done, despite the existence of the Regional Town and Country Planning Act which provides for developmental control by the town planning department.

If not disposed well, e-wastes pollute the soil, air and water because electrical and electronic equipment such as mobile phones and laptop batteries contain heavy metals such as lead, barium, mercury and lithium. Officials from EMA indicated that the effects of e-waste on the environment in Mbare are already manifesting but it is unfortunate that no explicit tests were done to verify e-waste impacts on water and soil. Despite the lack of comprehensive soil, air and water tests to support the claim of pollution from e-waste, the available preliminary analysis by EMA shows that there is pollution on land, water and air from e-waste. However, this will require comprehensive tests and evidence-based conclusions.

Current e-waste handling practices in Mbare can be categorised into formal handling practices and informal handling practises. Of the formal practices, the ones that include players registered and operating within stipulated regulations to protect the environment and amenity, there is still a lack of formal system in e-waste management in Zimbabwe from collection, dismantling, refurbishing and recycling. The study noted only one company, Enviroserve, that operates in Harare and specialises in the collection of e-waste and then exports it to Dubai for recycling. With only one formal company in e-waste collection, the opportunity for informal e-waste management practices increases and thus violates the town planning provisions and damages the environment. The absence of an e-waste legal and institutional framework, low levels of awareness, high demand for second-hand electronic appliances such as refrigerators, stoves, laptops and cell-phones, the selling of e-waste to individuals and collectors has given rise to a boom in informal e-waste recycling. Some Mbare residents are earning a living through e-waste handling and this shows the significance of the matter when people operate informally realising the economic potential of the activity. The prevalent informal practices of handling e-waste that were observed in the Mbare area include collection, dismantling, refurbishing and recycling. Most of the people that are into e-waste handling in Mbare do not have the capacity to recycle it and hence resort to the aforementioned methods. Collection is done mostly by individuals who go for scavenging at the Pomona dumpsite and throughout the city. These collectors sell the items to dismantlers, refurbishers and recyclers.

The EMA has provisions that guide the proper disposal of waste without causing pollution to the environment. The informal practices are a violation of this act and pollute the environment and amenity of urban areas. EMA has taken a note of this policy gap and has come up with an e-waste draft policy to deal with such issues. The draft policy when approved and functional is targeting to phase out the informal activities and maintain the zoning depicted in land use plans.

17.5 DISCUSSION

E-waste is an emerging and fast-growing waste stream that needs pro-active strategies before it causes adverse effects to the environment and the land begins to manifest. Experiences in developed countries have shown that there has been improvements in the management of e-waste, owing to the technology developments that necessitated fast-growing e-waste streams (Gweme et al., 2016). The study has shown that even some of the responsible authorities lack knowledge on e-waste management, let alone the general public. This is the first step that impedes proper e-waste management because people do not know what e-waste is and how it can be dealt with in a way that does not compromise the health state of the urban environment. The study revealed that there is no formal institutional structure to deal with e-waste. The absence of a formal e-waste management system leads to informal e-waste management practices. A formal institutional structure is important in that it organises a proper management chain of e-waste and closes the loopholes within the system. A formal system is capable of putting an end to rudimentary e-waste management practices that are not environmentally friendly. However, this framework can only work when the right policies are put in place to support it. When policies are non-existent, the institutional framework can cease to work.

The findings of the study indicate that informality in e-waste management such as in Mbare is a reality and the practices being employed are negatively affecting land use planning goals and the good health state of the environment. Informality is there because there is a policy gap that needs to be closed through adequate and functional e-waste management policy. However, policies can be introduced but it does not guarantee that informal e-waste management will stop as informality is described as a viral trait of African cities (Widmer et al., 2005). This is in line with the complexity theory, in which behaviour and outcomes are influenced by the underlying complex systems. The findings of the study converge with the complexity theory in the sense that there are complexities at policy levels and practice levels and these complexities involve different actors that include the central government, local government, parastatals, other countries trading with Zimbabwe, the private sector and the informal traders.

The welcoming development is that EMA has realised the policy gap and drafted a policy aimed at addressing e-waste management problems. However, the draft is taking too long to be finalised; yet, the situation is worsening. It is evident that the prolonged delay in adopting the policy means continual operation of informal e-waste handlers with negative effects on the environment. The unfortunate part is that Mbare does not exist in isolation and the persistence of these informal activities is going to affect other neighbourhoods. There are unavoidable externalities that are brought about by informal practices such as water and air pollution. The study revealed that e-waste management in the informal sector has got environmental implications as the heavy metals and toxic chemicals found in e-waste affect water, soil and air qualities.

In town planning, land designated as open spaces, freeway and road reservations are recognised as legal land uses and breaching such stipulations affect the land use pattern with detrimental effects on the amenity of localities. As e-waste is being managed and disposed informally, the goals of planning that include maintaining order and functional pleasant urban environments are jeopardised. This is largely a result of the lack of pro-active planning approaches and also the slow adaptation of urbanisation levels within the country. The existing literature has shown that the increased volume of e-waste is associated with high population increases in the country (Osibanjo and Nnorom, 2007). As a result, there are illegal dumpsites that are emerging and threaten urban sustainability as they are a threat to the environment, amenity of land and human health, especially when the groundwater is contaminated (Gweme et al., 2016).

17.6 CONCLUSIONS AND RECOMMENDATIONS

Sustainable and smart spatial planning is about anticipating to future challenges. New challenges in the form of e-waste management have emerged and this calls for innovative approaches in response to these new challenges. E-waste is an emerging and fast-growing waste stream that has disastrous public health and environmental consequences if not handled and disposed of properly. Experiences in Mbare indicate that Zimbabwe is failing to manage e-waste properly. This is evidenced by the rampant illegal dumping of e-waste, which is worsened by the operations of illegal e-waste recyclers. As a result of poor management of e-waste, threats to public health, environmental degradation and deterioration of amenity are looming. It is concluded from the study that the lack of proper e-waste management frameworks supported by planning is posing serious threats to sustainability in Zimbabwe.

The study recommends the formulation of proper legal and policy frameworks to guide e-waste handling through import controls, safe disposal and recycling.

The starting point will be import controls by the central government. At the local authority level, the waste management by-laws will need to be revised to include e-waste and how it is disposed. The private sector in conjunction with the government may need to formulate partnerships that work out the opportunities for recycling of e-waste where possible. This reduces the volume of e-waste that needs to be disposed. In addition, since e-waste recycling in Mbare has already taken an informal stance, it is imperative to accommodate the informal players through regularisation of these operations. Individual collectors, for instance, can still be engaged to supply to formal refurbishers, dismantlers and recyclers. The informal collectors, refurbishers and recyclers can receive proper training in handling e-waste in safe and healthy ways. There is also need for the town planning department to be acquainted with e-waste management aspects relating to proper handling and monitoring practices. This will inform them about the need to create working space for such recycling activities. In this case, the invasion of open spaces and servitudes can be abated when such properly designated working space is provided for e-waste activities. The workspace has to be formally designated and serviced for the various e-waste management operations starting from the neighbourhood level.

REFERENCES

Adediran, Y.A. and Abdulkarim, A., 2012. Challenges of electronic waste management in Nigeria. *International Journal of Advances in Engineering & Technology*, 4(1), p. 640.

Chartier, Y. ed., 2014. *Safe management of wastes from health-care activities*. Geneva, Switzerland: World Health Organization.

Gupta, S., Modi, G., Saini, R. and Agarwala, V., 2014. A review on various electronic waste recycling techniques and hazards due to its improper handling. *International Refereed Journal of Engineering and Science (IRJES)*, 3(5), pp. 05–17.

Gweme, F., Maringe, H., Ngoyi, L. and van Stam, G., 2016. E-waste in Zimbabwe and Zambia. In *1st institute of lifelong learning and development studies international research conference* (pp. 1–16). Chinhoyi, Zimbabwe: Chinhoyi University of Technology.

Kalana, J.A., 2010. Electrical and electronic waste management practice by households in Shah Alam, Selangor, Malaysia. *International Journal of Environmental Sciences*, 1(2), pp. 132–144.

Kalk, B.B.C., 2012. *A critical analysis of e-waste policies*. Eindhoven: Eindhoven University of Technology.

Kasapo, P., 2013. *E-waste management in selected institutions of higher learning in the Klang Valley, Malaysia* (Doctoral dissertation, University of Malaya).

Macauley, M., Palmer, K. and Shih, J.S., 2003. Dealing with electronic waste: Modeling the costs and environmental benefits of computer monitor disposal. *Journal of Environmental Management*, 68(1), pp. 13–22.

Mangwende, P., 2015. Urbanisation and its impact on sustainable development in Harare: A case of Harare south. http://hdl.handle.net/11408/2044

Mattingly, M., 1995. *Urban management in less developed countries*. London: University College, Development Planning Unit.

Matyatya, T., 2018. *An investigation of e-waste management practices and impacts on the environment: A land use planning perspective, the case of Mbare township*. A Dissertation Submitted in Partial Fulfilment of the Requirements of a BSc (Honours) Degree in Rural and Urban Planning, University of Zimbabwe, Harare.

Osibanjo, O. and Nnorom, I.C., 2007. The challenge of electronic waste (e-waste) management in developing countries. *Waste Management & Research*, 25(6), pp. 489–501.

Pellow, D.N., 2007. *Resisting global toxics: Transnational movements for environmental justice.* Cambridge: MIT Press.
Robinson, B.H., 2009. E-waste: An assessment of global production and environmental impacts. *Science of the Total Environment, 408*(2), pp. 183–191.
Sthiannopkao, S. and Wong, M.H., 2013. Handling e-waste in developed and developing countries: Initiatives, practices, and consequences. *Science of the Total Environment, 463,* pp. 1147–1153.
The Government of Zimbabwe, Regional Town and Planning Act (RTCP) (Chapter 29:12) Government Printers, Harare.
Theodros, G., 2010. *Assessment of electronic waste management: Case study of some TVET colleges in Addis Ababa* (Doctoral dissertation, A Thesis Submitted in Partial Fulfilment of the Requirements for the Degree of Masters of Arts in Development Studies).
Widmer, R., Oswald-Krapf, H., Sinha-Khetriwal, D., Schnellmann, M. and Böni, H., 2005. Global perspectives on e-waste. *Environmental Impact Assessment Review, 25*(5), pp. 436–458.
Zimbabwe National Statistics Agency (ZimSat), 2012. *Census 2012: National report.* Harare: ZimStat
Zimmerman, W., 1999. *Land use planning methods, strategies and tools.* Eschborn: Deutsche Gesellschaft fur Technishe Zusammenarbeit.

18 Three-Dimensional Layout Planning in the Context of Zimbabwe's Planning Profession
Scope, Fears and Potentialities

Brilliant Mavhima

CONTENTS

18.1 Introduction	257
18.2 Literature Review	258
18.2.1 Understanding Shapes and Dimensions	258
18.2.2 Use of 3D in Town Planning Design	258
18.2.3 Introduction of 3D Modelling in Planning	260
18.3 Methodology	260
18.4 Findings	260
18.4.1 Layout Design Approach in Zimbabwe	261
18.4.2 3D Design from a Central Government Perspective	262
18.4.3 3D Design from a Local Authority Perspective	262
18.4.4 3D Design from a Private Sector Perspective	263
18.5 Discussion and Synthesis	264
18.6 Conclusion and Recommendations	265
References	266

18.1 INTRODUCTION

The Minister of Local Government Public Works and National Housing woke up one morning and requested three-dimensional (3D) layout plans upon the submission of 2D blueprints. The question that followed was does it matter to design a 3D layout plan in addition to the conventional 2D blueprint? The literature points to the improvement of one's understanding of an object with the increase in the number of dimensions provided (Couprie and Bertrand, 2008). Urban problems have become so complex, thus requiring a system understanding. The systems can only be understood following the availability of as many dimensions as possible on the blueprint of the proposed development (Hu et al., 2000). This points to the need of adding dimensions to the designs of the layout plans. This does not seem to be the case in Zimbabwe. The Ministry of Local Government Public Works and National Housing with its mandate to approve layout plans that are implemented in Zimbabwe has decided to force roll out a paradigm shift in the design of settlement plans.

However, the use of 3D designs in planning seems to be ritualistic, merely fulfilling the requirements of layout submission for approval. This follows the issues that surround Zimbabwean-planning education. Most spatial planners that operate within the realm of Zimbabwe graduated before 2013; yet, the full engagement of software packages for design only came to life in the year 2016 (Chigudu, 2021). Furthermore, the design of settlements using technology is affected by the technology lag that exists in Zimbabwe. The design of three-dimensional layout plans requires computer hardware that is hard to find in Zimbabwe and if found is very expensive. This creates a challenge for the design of settlements in a 3D format. The chapter tries to identify how a 3D layout plan is perceived and done by Zimbabwe's professional spatial planners. The chapter is organised as follows: Introduction, literature review, methodology, findings and analysis, discussion and synthesis and then conclusion and recommendations.

18.2 LITERATURE REVIEW

The use of 3D is becoming prevalent in most countries, particularly the developed nations. The use of 3D in design can only be achieved by understanding the various meanings of the concept of dimensions; that is, how they are used in the context of town planning and the importance of design in town planning.

18.2.1 Understanding Shapes and Dimensions

Dimensions are the different values that are required to locate points on a shape. The first stage of understanding what 3D is about is looking at the dimensions that are in existence. The starting position is a point which in general gives a position and has no dimensions. The next is connecting two points and creating a line (Couprie and Bertrand, 2008). A line has just one dimension known as the length. From the line, there is a plane. Planes have two dimensions that are the length and width. A 3D shape is a solid shape that has a length, width and height. This is how people view things. The fourth dimension is 4D, which is the time component to the 3D shapes (Hu et al., 2000). For general sketches of the dimensions (Figure 18.1).

When drawing or designing, the dimensions play a significant role. The main purpose of a dimension is to provide a clear and complete description of the object being designed. Having a complete set of dimensions will make it easy to implement the design. Therefore, the more dimensions one can provide, the better the implementation.

18.2.2 Use of 3D in Town Planning Design

In town planning, the use of 3D in the design and development of cities has been utilised in most developed nations. The 3D design is used to develop virtual models that

Three-Dimensional Layout Planning

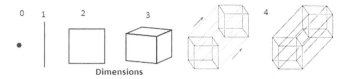

FIGURE 18.1 Dimensions of shapes.

Source: Figure by Author

represent proposed urban settlements. The virtual models are simply the surfaces of the terrain that is being planned for the proposed buildings, vegetation, infrastructure and all the elements of the landscape that are being proposed (Marshall et al., 2019). This helps settlement planners to perceive their proposal in the context of relationships. The designed model assists a spatial planner to identify how their proposed model fits in the existing context, thereby highlighting any possible planning issues.

3D design models can be used in the context of spatial analysis. In this case, spatial planners do an analysis and communicate the impacts of both the existing and new developments in space (Aharon-Gutman et al., 2018). The models help in looking at sun movement, zoning regulations, traffic gravity and other potentials that each of the proposed development has to offer (Polys et al., 2018). While these can help the planner, it also helps in the engagement of citizens even more as they can understand the proposed development easily compared to the way they would in the context of a 2D blueprint (Bajpai, 2019).

In terms of sun movement, the model helps the spatial planner to carry a sun movement analysis. In this context, the model that is developed is placed in context. Upon placement in context, the designer then simulates the movement of the sun relative to the orientation of the proposed development (Chandrasekaran, 2019). This process helps the spatial planner to understand how the settlement will perform following to access of sun in the design of buildings. The process places in questions planning zones, bulk factors and building lines as it identifies buildings that will not have access to sunlight when development occurs.

The other purpose of a three-dimensional drawing in spatial planning is to identify and put in question the existing zoning codes (Altintas and Ilal, 2021) Designing in 3D creates a settlement that in a 2D design is imaginary. The development of the settlement helps to identify if the proposals are functional and feasible in the context of the zones. Land-use zones partition land into sections of broader chunks of permitted land uses in the area (Foster and Brostoff, 2016). Placed within a model, one can identify if the proposed zoning code can help in developing the settlements that one intends to design. For instance, if a spatial planner is looking into the development of a walkable compact city, developing a model helps in identifying areas with large chunks of single-use zones. This will indicate the feasibility and areas with zoning codes that do not permit the development of the intended city/settlement.

Another issue that 3D models assist in is the identification of traffic gravity (Haag, 1999). Traffic gravity identifies the various sources of traffic and the volume per traffic route. The designing of settlements as 3D models helps in identifying possible traffic sources and the intersections that are likely to be the most congested.

This position assists a spatial planner in determining the road sizes of the areas with the largest catchment areas and thereby developing a sustainable and functional settlement.

18.2.3 Introduction of 3D Modelling in Planning

The use of 3D in planning has been occurring in the world. The Brisbane City Council in Australia has a virtual Brisbane that was designed using 3D modelling software packages. This computer-generated model of Brisbane is used for strategic planning, development assessment and control as well as community engagement. The 3D model is used in making changes and determining the degree of impacts of the proposed change in the context of transport and population dynamics among other considerations (Krasnovyd, 2019).

In the year 2018, the city planners in Singapore developed a 3D replica of the city-state. The model was developed by replicating every aspect of the city and is amazingly detailed (Yan et al., 2019). The model was developed to assist in planning, management, monitoring as well as controlling of development in Singapore (Jusuf et al., 2017). With the model, the proposals are first applied to the model before actual development, simulations are done and then the results are used in decision making within the city (Thompson et al., 2017).

The use of 3D has slowly become integrated into planning education across the globe. In Australia, the University of Sydney has integrated virtual modelling into the Urban and Town Planning curricula, thereby developing 3D modelling skills (Burry and White, 2020). Students are introduced to design software packages including ESRI's City Engine which is one of the standard software packages in the profession of town planning (Evans-Cowley, 2018). The process has allowed students to understand how to visualise, plan, communicate and model the developments that are proposed during the planning of the city in question.

18.3 METHODOLOGY

Understanding the significance of 3D planning in Zimbabwe's professional planning field requires one to engage professional planners. As such, the study sought to get the views of only spatial planners that are carrying out spatial planning practice in Zimbabwe. The author engaged professional planners who operate from the government, local authorities, private sector and academic institutions. This was done through WhatsApp platform focus groups. A focus group discussion guide was engaged to collect data with regards to the approach and attitudes towards 3D planning and discussions were done. A document review was also done. This involved reading government and local authority publications. These were purposively chosen. To derive meanings, the author used thematic content analysis. No names or anything linked to the respondents was presented.

18.4 FINDINGS

This section focuses on the findings of the study. It is divided into four subsections that include the layout design approach in Zimbabwe, the opinion of planners in government, planners in local authorities and planners in the private sector. The section indicates the status and perspectives of spatial planners in Zimbabwe concerning 3D designing.

18.4.1 Layout Design Approach in Zimbabwe

The design of layout plans in Zimbabwe is guided by various instruments. The subdivision of land is provided for by the Regional Town and Country Planning Act [Chapter 29: 12]. The design process is guided by Circular Number 70 of 2004 as well as the layout design manual. Circular No 70 of 2004 is a government instrument that was developed to facilitate the development of affordable housing in Zimbabwe. The circular attempted to lower the design and planning standards in Zimbabwe. It superseded Circular Number 20 of 1992. On the other hand, the layout design manual is a manual developed by the Ministry of Local Government Public Works and National Housing to guide spatial planners with the guidelines of how to draft settlement layout plans. The existing legislative instruments do not provide for the 3D design of layout plans. The whole idea behind 3D layout plans was championed by the Minister and is not backed by any instrument.

The existing legislation does not provide for the design of layout plans in 3D. This is so as it does not stipulate what to consider when coming up with a layout plan in 3D form. This implies that in some cases, layout plans would be submitted in 3D form but without any relevance to the actual 3D form. In a discussion session with a senior planner in central government, a question was asked: Will the Department of Spatial Planning and Development (DSPD) pick it if a planner was to download images of drawings off the internet and submit them in compliance with the 3D requirement: Here are some of the responses:

Response 1: I doubt.
Response 2: As long as it is matching the planning idea.
Response 3: This they will pick it.
Response 4: Will you be doing yourself justice? This work will be your legacy.
Response 5: How will they see it, as long as it is in line with the 3D?

The responses from the five planners picked for the discussion all pointed towards difficulty in the identification of the difference between an actual context-relevant 3D design and an internet image. The follow-up question sought to establish why was this impossible to pick the difference? The responses indicated that the 3D requirements were of parts of the settlement and not the entire settlement. One planner said that:

> If I say I am going to put a house I can just put a house and pull the picture off the internet. Will I be wrong? The same applies to a shopping mall.

The statement by the planner indicated that the whole idea of introducing 3D in layout planning was not well-conceived and influences the quality of 3D designs being submitted for the minister's approval. To understand how the different groups viewed the issue of 3D responses from planners in different sectors are outlined in the following sub-sections.

18.4.2 3D Design from a Central Government Perspective

The central government has used 3D designs for layout presentation and analysis since the 1980s. Some of the commercial centre layout plans were accompanied by a 3D drawing. This was done using drawing pens and ammonia paper coupled with colour washing. In discussion with a central government planner, it was highlighted that:

> I joined the DPP in 1991 as a cadet student. 3Ds were there but mainly for commercial layouts. Check the town centre layouts for some of our towns. They have 3D elements.

The focus of the 3D component in the design of layout plans seems to be on commercial nodes. The major drive behind the design of commercial nodes in 3D was to create aesthetically pleasing shopping centres, particularly the central business districts (CBDs). One planner indicated that:

> *If you look at the 3D's that exist, they are mostly for huge commercial centres, or may I say CBD. These were done because when designing a CBD, we want a centre that is beautiful and functional. This could only be understood by designing a 3D view of the layout.*

The statement by the planner working in the central government highlighted that 3D layout designing was mostly to ensure that planners get a view of the beauty associated with the proposals they have made. Furthermore, the design was also made to view the functionality of the city.

The plans that planners in the central government started to have 3D designs drawn using drawing pens. Furthermore, in all of the designs, one name always appeared accompanied by different names as co-designers. This highlighted that the concept of the 3D design software package was not yet part of the central government planning system. Furthermore, the 3D designs seem to have been understood by one person who assisted others to grasp the concept. This indicates a gap in terms of 3D design expertise.

18.4.3 3D Design from a Local Authority Perspective

3D planning in the context of local authorities in Zimbabwe is not as clear as that of the government. Local authority planners design 3D plans for plan approval and not for any other use. One of the local authority planners highlighted this by stating that:

> 3D's are a waste of time. What aspect of a 2D layout plan will make it weak because it does not have a 3D design.

The driving force behind the development of 3D layout plans then becomes more of a mandate than a need. The planners indicated that 3D designs of layouts or any other design is done specifically for layouts only. One planning technician within a local authority stated that:

What the 3D's are doing for us is to add extra work. Do you know how hard to use those 3D software packages are?

The technician simply stated what most planners echoed. To local authority employees, 2D dimensions were enough to do proper planning and come up with a well-organised plan which is easy to implement and control. In a different view, one local authority employee, though speaking in his capacity emphasised the significance of 3D in the planning of Zimbabwean cities. He stated that:

In my view, I think 3D's are important. We just need to develop comprehensive standards for the design of 3D's [. . .]. It helps in the identification of the use of vertical space. Issues of building heights and building lines become easy to understand.

The points raised by the planner were possibly linked to two things: First, the local authority that the planner worked for did not use 3D in planning, though some of the employees understood its significance. Furthermore, the planner works for a politically volatile local authority that does not welcome change as such indicating a whole lot of resistance to ideas.

The issue of resistance to the introduction of 3D design by local authorities can be ascribed to the limited technical capacity—3D design using drawing pens and pencils as art. As such, only those that are talented in art could come up with a meaningful 3D layout plan. This meant that most planners shunned the idea. Furthermore, the old generation of planners in local authorities cannot use the design software packages. In an interview, one of the design academics stationed at a Zimbabwean University that trains spatial planners said:

It is only in 2014 that 3D software packages and design components became important and taught at the university. . . . Before that, the emphasis was on 2D layouts drawn on paper.

Planners that were trained before 2013 do not have an appreciation of most design software packages unless one had a personal interest or was trained on the job by the employer. The issue of the employer trying to update the skills of the employee was raised by one spatial planner who said that:

We were once trained by this person who was paid by the council to do so. Most of us do not even remember how to use the ArchiCAD software that the trainer was emphasising [chuckling]. That's what you get for training old horses.

18.4.4 3D Design from a Private Sector Perspective

Spatial planners operating in the private sector had views similar to their local authority counterparts. One of the planners stated that they are focusing on 3D just because of the DPP requirement for layout approval. The statement that the planner gave highlighted that the sole purpose of 3D designs was to satisfy the requirements of the government.

Another planner who works for a different organisation believed that 3D designs were very useful for marketing development proposals. The visualisation is very convincing for clients and investors to undertake development. The planner stated that:

> 3Ds play a role in convincing buyers as they can visualise development before it comes to existence. It helps people in arriving at investment decisions.

The view that 3D plans are designed for marketing and convincing clients to buy into ideas indicates that 3D designing is used as a visual impression tool. That is the end. Planners in the private sector are interested in profitability, which can be can be achieved through 3D design.

While private-sector planners focused on 3D for marketing and plan approval, one planner indicated that from experience with implementing private developments, they sometimes have to design layouts in 3D for the development proposal to realise its potential. The spatial planners stated that 2D designs are adequate in layout designs. However, the failure to realise the potential is largely a matter of different interpretations of the same design and hence highlighting the need for 3D layout designing. The planner stated that:

> *In my view, it's not about presentation but visualisation. It's good for a planner or urban designer to share his vision not in 2D but in 3D which is close to reality. Sometimes the potential of good layouts is being failed by architects and engineers who are not good with massing and triangulation.*

In this context, the other driving force behind the design of 3D layouts was the implementation of projects that had a high return on investment. The idea was that for a design to realise its full potential, it had to be interpreted by the designer, not anyone. This introduced the need for a 3D design.

18.5 DISCUSSION AND SYNTHESIS

Planning using the third dimension has been highlighted in the literature as having various advantages that include introducing a better analysis capability and a better option for citizen's participation through simulation of other planning-related features that are proposed for the development (Onyimbi et al., 2018). A simulation of settlements requires a broader set of data, which ranges from population data, socio-economic data and traffic data to existing policies that guide the areas. Having the data and engaging them in simulating settlements indicates the functionality and feasibility of settlements. This gives the 3D modelling approach a better and wider scope within the planning profession. The scope of 3D layout designing Zimbabwe seems to be obscure. Each planner from each department has a different opinion as to the purpose and the reason behind the designing of 3D layout plans. The planners that operate within the context of the government argue that the idea of 3D planning was always in existence since the 1980s. The problem, however, was that the idea was not enforced. Only planners with an interest in designing the settlements in 3D would spend time designing in 3D.

In the context of local authorities, the 3D design does not show a ray of focus. Planners who work for local authorities only prepare layouts in 3D for purposes of satisfying central government requirements. The approach is satisficing. This implies that local authorities are yet to realise the benefits that are associated with the use of 3D layout designs and the management of urban settlements.

For private-sector planners, the scope of the use of 3D designing is clear—getting the job done. The 3D designs are done to guard the return on investment Furthermore, the purpose is also to convince investors as well as satisfy the requirements of the DPP.

The issue that comes up is, what is the purpose of 3D designs? The DPP seems to request it as a ritual. Local authorities and private-sector planners follow the process to ensure that the ritual is followed. Scholarship has highlighted the importance of 3D designs in the planning of cities (cite the scholars). Simulation as well as an understanding of issues before implementation of the actual plan are highlighted as some of the greatest strengths of adding another dimension on 2D drawings (Kormann et al., 2016). The advantages of 3D design indicate a limitation in the design fraternity in Zimbabwe.

The issues associated with the layout plans in Zimbabwe open an opportunity for the introduction of 3D design. The first opportunity is the absence of an organised 3D design system. This opportunity opens avenues for the development of context-relevant 3D design standards that can be used as a benchmark for the development of settlements in Zimbabwe. The interest by the minister to introduce the 3D designs during the submission of layout plans for approval provides a champion for steering change. This provides a huge opportunity for a leapfrog within the design fraternity.

Another opportunity that comes within the context of 3D layout designing is planning schools. This widens the scope for schools to review and update their curricula. Furthermore, there is an opportunity for planning schools to introduce town planning relevant design software packages. The study highlighted that the 3D software package that was being used is ArchiCAD. The software is an architectural software meaning using the software in the context of town planning makes the work tedious and cumbersome to partake.

18.6 CONCLUSION AND RECOMMENDATIONS

The chapter focused on understanding the 3D layout design approach in Zimbabwe. It emanated from the proclamation by the minister that all layout plans be submitted together with 3D inserts of the design. The study identified that 3D layout design has existed since the 1980s but only for commercial centres. The technical capacity for 3D designing is very limited as the professional planners are very important in such a process. The layouts are done as a ritual to fulfil the requirements of the minister. Planners in local authorities and the private sector use 3D designs to fulfil government requirements for approval. The private sector has since engaged the 3D designs for layout marketing and investment promotion. This is contrary to the advantages of using the 3D design approach for cities. For the planning profession to fully utilise the concept of 3D layout design, there is a need to develop an instrument that guides the design of 3D layouts. Layout designing in Zimbabwe is guided by the layout design manual and circular no 70 of 2004. These instruments are silent on 3D layout design. This creates a gap in the development of a functional design manual that informs on what 3D design is, how it is done and the aspects that need to be considered by spatial planners who are the designing as well as planning officers that inspect the submitted spatial planners.

Furthermore, planning schools should emphasise training on the use of town planning software and reduce an over-emphasis on architectural software as it makes the 3D layout design process cumbersome. The spatial planners who were designing settlements had a focus on ArchiCAD which most of them learnt on their own. This indicates a gap in planning education to teach spatial planning software packages that make it easier to design settlements that can easily be simulated and be understood.

More so, there is a need for courses on vertical design to give planners an understanding of the use of 3D layout planning processes. These classes will be used to refresh the minds of settlement planners that are already in the spatial planning field. This will help in covering the gap in planning organisations that requires them to train staff and enhance their technical capacities to ensure that the full benefits of 3D layout designs are realised.

REFERENCES

Aharon-Gutman, M., Schaap, M. and Lederman, I., 2018. Social topography: Studying spatial inequality using a 3D regional model. *Journal of Rural Studies*, *62*, pp. 40–52.

Altintaş, Y.D. and Ilal, M.E., 2021. Loose coupling of GIS and BIM data models for automated compliance checking against zoning codes. *Automation in Construction*, *128*, p. 103743.

Bajpai, S., Kidwai, N.R. and Singh, H.V., 2019. 3D wavelet block tree coding for hyperspectral images. *International Journal of Innovative Technology and Exploring Engineering (IJITEE)*, *8*(6C), pp. 64–68.

Burry, J. and White, M., 2020. From 'To 'π'(pi)-shaped people: Better urban practice through dual depth in architecture and planning education. *Architectural Design*, *90*(3), pp. 114–121.

Chandrasekaran, V., 2019. A study on sun path of Coimbatore and its irradiation effects on buildings using ecotect software and design recommendations for minimizing the heat gain. *International Journal of Architecture and Infrastructure Planning*, *5*(2), pp. 36–47.

Chigudu, A., 2021. Influence of colonial planning legislation on spatial development in Zimbabwe and Zambia. *Journal of Urban Planning and Development*, *147*(1), p. 04020057.

Couprie, M. and Bertrand, G., 2008. New characterizations of simple points in 2D, 3D, and 4D discrete spaces. *IEEE Transactions on Pattern Analysis and Machine Intelligence*, *31*(4), pp. 637–648.

Evans-Cowley, J.S., 2018. Planning education with and through technologies. In *Urban Planning Education (editors)?* (pp. 293–306). Springer, New York, City.

Foster, S. and Brostoff, J., 2016. Digital doppelgängers: Converging technologies and techniques in 3D world modelling, video game design and urban design. In *Media Convergence Handbook-Vol. 2* (pp. 175–189). Springer, Berlin, Heidelberg.

Haag, M. and Nagel, H.H., 1999. Combination of edge element and optical flow estimates for 3D-model-based vehicle tracking in traffic image sequences. *International Journal of Computer Vision*, *35*(3), pp. 295–319.

Hu, H., De Angelis, A.A., Mandelshtam, V.A. and Shaka, A.J., 2000. The multidimensional filter diagonalization method: II. Application to 2D projections of 2D, 3D, and 4D NMR experiments. *Journal of Magnetic Resonance, 144*(2), pp. 357–366.

Jusuf, S.K., Mousseau, B., Godfroid, G. and Hui, V.S.J., 2017. Integrated modelling of CityGML and IFC for city/neighbourhood development for urban microclimates analysis. *Energy Procedia, 122*, pp. 145–150.

Kormann, M., Katsarou, S. and Katsonopoulou, D., 2016. The Digital Helike Project in the Early Helladic Period: Further Insights from Archaeological and Geological Data Through Combined Modelling, 3D Reconstruction, and Simulation. In *2nd Conference Computer Applications and Quantitative Methods in Archaeology (CAA-GR)*", Athens, Greece, 20–21 December 2016.

Krasnovyd, V., 2019. The importance of 3D modelling. In *Modern Technologies: Improving the Present and Affecting the Future*. Dnipro, Russia: Dnipropetrovsk National University of Railway Transport.

Marshall, S., Hudson-Smith, A. and Farndon, D., 2019. Digital participation—taking planning into the third dimension. *Town and Country Planning, 88*(1), pp. 11–14.

Onyimbi, J.R., Koeva, M. and Flacke, J. 2018. Public participation using 3D web-based city models: Opportunities for E-participation in Kisumu, Kenya. *ISPRS International Journal of Geo-Information, 7*(12), p. 454.

Polys, N., Newcomb, C., Schenk, T., Skuzinski, T. and Dunay, D., 2018, June. The value of 3D models and immersive technology in planning urban density. In *Proceedings of the 23rd International ACM Conference on 3D Web Technology* (pp. 1–4). Association for Computing Machinery, New York, NY.

Thompson, R.J., Van Oosterom, P. and Soon, K.H., 2017. LandXML encoding of mixed 2D and 3D survey plans with multi-level topology. *ISPRS International Journal of Geo-Information, 6*(6), pp. 171–193.

Yan, J., Jaw, S.W., Soon, K.H. and Schrotter, G., 2019. The Ladm-based 3D underground utility Mapping: A case study in Singapore. *International Archives of the Photogrammetry, Remote Sensing & Spatial Information Sciences*. Academic Editors: Efi Dimopoulou and Wolfgang Kainz. Received: 31 March 2017; Accepted: 5 June 2017; Published: 12 June 2017.

19 From Two-Dimensional to Four-Dimensional Layout Design
A Necessary Leapfrog in Zimbabwean Urban Planning

Brilliant Mavhima

CONTENTS

19.1 Introduction and Background 269
19.2 Conceptual Framework 270
    19.2.1 Leap Frogging 272
19.3 Literature Review 272
    19.3.1 Visualisation 273
    19.3.2 Simulations and Cities 273
    19.3.3 Simulation and Design in Africa 276
19.4 Research Methodology 277
19.5 Results and Discussion 277
    19.5.1 Results 277
    19.5.2 Discussion 279
19.6 Conclusion and Recommendations 279
References 280

19.1 INTRODUCTION AND BACKGROUND

Urban design in Zimbabwe has been largely done as a process of creating space for various uses. This is usually associated with designing two-dimensional (2D) layouts and writing layout reports and then implementing them. Layouts created through this process have been implemented without much knowledge of how the introduction of the human factor can distort the intention of the plan. This has seen the emergence of problems like traffic congestion, white elephants and vending. In some countries, layout design has progressed from 2D to 3D, while others are already doing 4D layout planning. This implies different qualities in urban spaces that are created during the process.

    The planning and design of plans in Africa remain lagging (Dodo, 2018). The plans are often criticised for partially addressing issues at the territorial level (not covering the entire city) and addressing only the needs of the investors in the society

(Nair et al., 2020). The other issue that African plans face is the existence of a paucity between planners, policymakers and the residents. This comes from the fact that planning idea approaches and tools are often Eurocentric and are not properly adapted to suit the contexts in which they are to be implemented. This is often so as the African nations will not have the financial capacity to implement urban planning in most parts of sub-Saharan Africa. They are still lagging behind developments in America, Europe and Asia. The developed parts of sub-Saharan Africa are using 3D urban designs and GIS-based software for collecting data (Chipambwa and Chimanga, 2018). In South Africa and Botswana, urban planning and layout design is slowly moving towards 3D planning. This is largely done through software like ArchiCAD and Google Sketch-up (Sante et al., 2010). 3D designs have been used largely as the representation of how a development proposal will look like, particularly in, marketing ideas (Gwamuri and Pearce, 2017). As a result, the full potential of advances in design have not yet been realised. In less-developed parts of the region, layout planning is still 2D. Consultancy companies have been trying to introduce new design technology but have encountered government inertia. This is linked to the issue that most government employees do not have the capability to use the relevant software (Benguigui et al., 2001).

Urban design and planning in Zimbabwe have been advancing from drawing boards and food colours to the use of computer-aided drawing. However, this has not changed the type of plans under production, as they are largely 2D. 3D layout design is still in its primitive stage with consultancy companies spearheading work on the designs (Chipambwa and Chimanga, 2018). A 4D layout design is still a marketing technique used by land developers but its potential has not being fully exploited (Gwamuri and Pierce, 2017). Government offices are still lagging with some even using drawing boards for layout design (Gwamuri and Pierce, 2017). As such, urban layout design in Zimbabwe is still largely 2D. The chapter is organised as follows: the introduction and background, the conceptual framework, the literature review, methodology, results, discussions and synthesis than the chapter conclusion and recommendations.

19.2 CONCEPTUAL FRAMEWORK

Urban planning and design can be done using various ways and angles. The data that can be acquired from a design largely depends on the number of sides one can perceive from a drawing. As such, there is a need to increase the number of sides to be able to predict the potential threats and opportunities of a design before it is implemented. This chapter uses various concepts like 2D, 3D and 4D designs as well as leapfrogging. As such, there is a need to understand the various meanings of the concepts.

When looking at shape lines and forms, the dimensions come into play. The dimensions can be defined as visible sides of a shape (Hiraiwa, 2005). For instance, a shape's height or a shape's length. When looking at a 2D design, the designs are based on plane shapes with length and width but without height. The shapes can be regular or irregular (Bänsch, 1991). So, when discussing a 2D layout design, roads and blocks

From 2D to 4D Layout Design

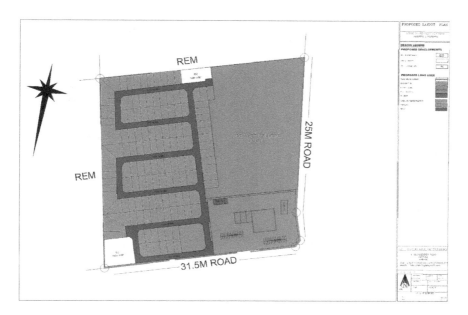

FIGURE 19.1 Two-dimensional layout.

Source: Figure made by author

are shown in terms of length and width. As such understanding the dynamics of the area planned for is limited (Hiraiwa, 2005). Therefore, adding extra dimensions can improve the understanding of the area that is being planned. Figure 19.1 shows an example of a 2D layout.

Adding dimensions to a shape with length and width involves adding height. Adding height to a design that is two-dimensional makes it three-dimensional (Faugeras and Hebert, 1986). Three dimensions are the simplest way of conceptualising the real world. A 3D layout design entails adding blocks and their heights on a plan. Therefore, a 3D layout design shows the skyline of an area (Thurston, 2014). 3D plans help understand the area planned better than a plan drawn in the 2D form. Figure 19.2 shows a 3D layout. However, the weakness is that it does not allow an understanding of the effect of time and activities on an area. This requires the addition of extra dimensions.

From three dimensions, one cannot fully comprehend the effects of a plan, as such, there is a need to add movement and space (Boulos et al., 2017). This is more complex but enhances the conceptualisation of the real world (Hinton, 1880). This idea was introduced by Joseph Louise Lagrange in the 1700s and was popularised in 1880 by Hinton (Hinton, 1880). Adding the fourth dimension improves one's understanding as one can now perceive the hidden dimensions. In urban layout planning, the fourth dimension helps one to see how the layout will look and affect the existing environment in different aspects (Rucker, 2014).

FIGURE 19.2 Three-dimensional layout.
Source: Figure made by author

19.2.1 LEAP FROGGING

When developing from one stage to another, it is expected to follow a stage, for instance, Amartya Sen's development theory, Weber's modernisation theory among other theories. The same appears to be the norm even when developing in terms of technology (Benguigui et al., 2001; Brezis et al., 1993). Therefore, when improving in urban design, one moves from mushrooming developments to 2D plans, then 3D plans and so on (Fong, 2009). The question that arises is if we are all going to the same destination, is it necessary to follow the same route. From this question emanates the idea of leapfrogging. This is a process in which a nation can jump some stages into a current and dominant position (World Bank, 2016). Thus, it is possible to move from 2D urban designing, skip 3D designing and jump into 4D urban designing.

19.3 LITERATURE REVIEW

Cities are among the most complex structures created by human societies and unlike other complex manufactured objects, like computers, scholars do not pretend to understand them fully (Falconer et al., 2018). Furthermore, most parts of the world are currently undergoing rapid and philosophical changes that influence the quality of life for hundreds of millions of people—changes which should be managed to preserve or enhance the quality of life and to ensure economic and environmental sustainability. Effective planning and management require both data on current conditions and an ability to foresee the likely consequences of proposed projects and policies. Using advanced technology to improve the quality of cities that are under production is essential. One can achieve several positive as well as negative yields through the engagement of virtual reality. As such, one needs to fully grasp how one can harness the positive benefits as well as curb the negative externalities of virtual realities.

19.3.1 Visualisation

The world is filled with mathematical models that have been used to try to reduce the complexity of the world. The theoretical models are critical in understanding cities and urban development; the models are too simplified for cities to yield the results that are needed for the planning and designing of functional urban settlements. This calls for the concept of visualisation. The power of visualisation in urban planning is very essential (Christodoulou et al., 2018). The ability to plan, project and understand plans help in managing urban problems before the plan is implemented. This creates an advantage of viewing the future before its existence (Florio et al., 2018). This has been termed virtual reality (VR). The potential of virtual reality in urban planning is huge.

Cities and urban spaces are complex areas. The method of communication that urban planners and urban designers use in presenting ideas for solving urban planning problems is usually through visuals. Visualisation helps in understanding an object by perceiving it from various viewpoints (Dodge et al., 2011). This, in turn helps urban planners, the policymakers and the public understand complex urban information (Christodoulou et al., 2018). Communication and visualisation form the basis of urban planning (Bhunua et al., 2002). This is usually done in 2D and 3D forms. Recently, the 3D form has been introduced which is helping in viewing the plan from more sides than the conventional 2D form (Boulos et al., 2017).

Planners can now model urban environments in a 3D electronic space, using Computer-Aided Design (CAD) or advanced visualisation techniques (Bhunua et al., 2002). Virtual reality models of entire cities are possible: a model of Bath, produced by researchers from CAD models, is one such example of a large-scale urban model (Bourdakis, 1997). Virtual models can be viewed from different viewpoints: both bird's eye views, which quickly give a survey information about the city and eye-level perspectives from the vantage point of pedestrians and motorists can be generated.

Utilising VR, planners can map the site and its surrounding census data, land use and behavioural diagrams, climate and pollution simulations and other data enhancing expert's assessment and most importantly reducing the need for site familiarity (Schindler et al., 2018). Furthermore, planners can use virtual reality models to assess the impact of new housing schemes concerning transportation patterns and access to schools, shopping facilities, parks and other amenities. From the ability to perceive proposed developments in the added third dimension, models can also be given the aspect of life, which is called simulation (Rocha et al., 2021).

19.3.2 Simulations and Cities

Simulations and attempts to understand the physical aspects have been classified into two groups of models that include static models and dynamic models. The static models are the models that have been discussed under visualisation. These include models like the physical and environmental models, social models and urban models. These are 3D models of the issues as they are on the settlement or the proposed development. On the other hand, there are dynamic models. These include optimisation models, control models and simulation models. The specific focus is on simulation models that can be used to assess the spatial evolution of settlements, transport dynamics as well as environmental issues that may arise.

The concept of simulation has been brought into necessity by the complex nature of city problems. The chapter engages the definition by White and Ingalls (2009) which indicates that

A simulation is a particular approach to studying models, which is fundamentally experiential or experimental. In principle, simulation is much like running field tests, except that a physical or computational model replaces the system of interest. Simulation involves creating a model, which imitates the behaviours of interest; experimenting with the model to generate observations of these behaviours; and attempting to understand, summarize, and/or generalize these behaviours. In many applications, simulation also involves testing and comparing alternative designs and validating, explaining, and supporting simulation outcomes and study recommendations.

From the definition of simulations, the critical aspect is the model, which is developed through a 3D design. The model is then placed under an experiment with the supposed possible aspects that will influence the day-to-day running of the settlement under study. The way simulations are done is two-fold. This can be in a man-in-loop simulation or analysis of design artefacts. The man-in-loop simulation is often done to replicate reality and is done for training or entertainment, that is, development of movies or training firefighters (White and Ingalls, 2009). The second aspect of the simulation looks into how the proposed design can work under perceived or predicted challenges (ibid). These are the kind of simulations that are done for settlement analyses.

The most common simulation strategy that has been used in city simulation models has been that of a game called SIMCITY. This game was developed by Will Wright and published by Maxis in 1989 as a strategy game of making settlement decisions in urban planning and law to ensure that citizens are happy. The simulation process is based on four broad urban indicators that include: *desirability*—this aspect looks at the norms and values of the people as well as the economic aspect of the proposed development. *Geographical information views*—this looks at zoning all hazards, traffic movements, land values, age, health and garbage collection and treatment. *Time behaviour graphs*—this looks into issues of crime, power and water supply, pollution, employment, education, life expectancy and traffic volumes. *Monthly budget*—incomes and outcomes. The considerations of the simulation cover aspects of city development that are available in regions with good governance and access to information (Davison et al., 2000).

The city authorities in America have engaged in simulations in urban planning through what is called the SimCity (Devisch, 2008). This gaming idea of urban planning provides one with the rules of human factors, economic factors, survival factors and political factors that are both opportunities and constraints to the planner (ibid). In addition, there are numerous maps to monitor land-use patterns, zoning, demography, pollution and other factors as the simulation progress (Frasca, 2013). The simulator also allows one to manipulate tax rates as well as public infrastructure funding levels as the simulation progresses. This allows the replication of the real world in a game and problems that are likely to occur if the plan is implemented can be identified. As a result, a plan may not be implemented. Thus, in America, certain plans are replicated in the games and data are manipulated following existing situations leading to an analysis of how the plan will perform (Dooley, 2002). However, as good as the SimCity simulator is, some weaknesses and situations do not exist in the real world like the over-reliance on mass transit systems. Simulations do not end in America. Other organisations are looking into transport simulation models as well as city models like UrbanSim that try to replicate real-life scenarios.

In Europe, simulations are becoming the norm and the strategy of testing ideas before full implementation (White et al., 2000). Cities that are being developed in Europe largely depend on data, interactive analysis, simulation and visualisation of projected futures (Sante et al., 2010). This has been done through various tools; one of them is the interactive analysis, simulation and visualisation tools for Urban Agile Policy Implementation (urban API). This is led by Germany but involves countries like Austria, Italy, United Kingdom and Spain among others (Strauch et al., 2005). The application is based on three urban planning contexts that are: First, the 3D Scenario Creator application directly addresses the issue of stakeholder engagement in the planning process through the development and provision of enhanced 3D VR visualisations of neighbourhood development proposals. Second, Mobility Explorer provides mobile phone-based information and communication technology (ICT) solutions that permit the analysis and visual representation of socio-economic activities across cities and concerning the various land-use elements of the city. Third, the Urban Development Simulator prototype provides ICT simulation tools for interactive city-region development simulation addressing urban growth and densification because of planning interventions.

European urban design approaches have been advancing with advancements in technology. Urban design software advanced from simple computer drawing boards to more complex software packages with simulation capabilities (Sante et al., 2010). European governments, collaborating with research institutions, have been researching how urban changes can affect the performance of cities through creating virtual reality (Benguigui et al., 2001). In Britain, for instance, traffic management and building capacities are tested using simulation software (White et al., 2000). This has improved the efficiency of towns that are currently under design.

Simulation capability is proving to be essential in American cities (Axelrod, 1997). This is so as cities have been employing GIS-based software to collect data from the cities and apply it into prototypes leading to the prediction of possible opportunities and threats (Adams, 1998). As such, 4D urban planning has been employed to try to improve the quality of plans under production in America. The Massachusetts Institute of Technology has been developing various strategies for improving the quality of urban spaces under production (Frasca, 2013). Recently, the institution launched a free urban data collection application that can be employed in simulations.

Four-dimensional presentations were forced into a necessity by the natural environment of the Asian continent (White et al., 2000). Earthquakes, volcanoes and tremors have made it necessary to try to replicate events to understand them even more. Therefore, planning in developed parts of Asia is done through various simulations to see how the proposed developments will react in case of various emergencies (Sante et al., 2010). The use of such simulations has seen them advancing towards the development of 4D city planning. In the less developed parts of Asia, for instance in India and Malaysia, city planning still depends on 2D planning. The result of the discrepancies in planning approaches can be viewed in the qualities of the urban spaces that are created.

19.3.3 SIMULATION AND DESIGN IN AFRICA

African municipalities have implemented urban planning for the past 50 years and often with no noticeable results in socio-economic development (Albuquerque and Guedes, 2021). The city plans are often well designed but suffer from the issues of lack of realism (Das, 2020). This comes from the fact that the planning of cities in Africa continue to operate in a limited environment where decision-making is done without enough information. The implementation of ICT in urban planning in Africa is marred with deficiencies because of issues like lack of information by the city and inhabitants, unavailability of inadequate land management and a lack of approaches to implement ICT. This is coupled with issues of political will, which often determine the implementation of proposals. Sub-Saharan Africa is not in a unique situation.

Design and modelling of cities have been attempted in the context of Sub-Saharan Africa, particularly in South Africa. This has been done following the increased access to data and availability of GIS. The Gauteng City-Region following its neverending growth has necessitated the use of models to anticipate developmental issues (Mubiwa and Annegarn, 2013). This has seen the development of transport models done by the Gauteng Department of Road and Transport. This has been done using software packages like EMME2. The attempts, however, have been affected by a mismatch of attempts at local and national levels leading to the differences in results of the models (Davison et al., 2000). The difference in the efforts at the local and metropolitan levels saw the development of the Gauteng Integrated Transport Modelling Centre (GITMC). The centre was focused on integrating the modelling efforts towards the development of a functional transport model. The centre utilised software applications like UrbanSim and MatSim to generate scenarios based on different transport initiatives and infrastructure (Cich et al., 2017).

From transport, the region has also developed the Gauteng Infrastructure Planning Tool (GIPT) (Kleynhans, 2012).). This tool indicates the location of all public facilities like schools and health centres to help spatial planners in making infrastructure-planning decisions (Engelbrecht, 2012). Cape Town has also utilised modelling in planning. The city's growth has been predicted and monitored using the Urban Growth Model as well as the Urban Growth Monitoring System. The GIS-based tools predict where development is likely to occur and how it will happen.

While various urban models have guided the development of cities in South Africa, the systems have been facing some challenges. The major challenge is that of coordination. The efforts by the government of South Africa does not align properly with the efforts of the local authorities. The current efforts of the Government of

South Africa have been to develop a National Planning Commission that focuses on aligning all national development attempts. The lack of coordination has seen the duplication of efforts across various departments. An example is the GIPT (developed by the City of Gauteng) and the National Human Settlement Index (developed by the Housing Development Agency) identified land suitable for different public infrastructure. The availability of the opportunities for urban simulations and the challenges that affect the potential to leapfrog into the design of simulated cities requires a further understanding. The next section focuses on the methodology utilised by the chapter.

19.4 RESEARCH METHODOLOGY

The chapter is informed by case studies and document reviews. Data on urban planning and design was collected from various sources. This included scholarly works, newspaper articles and videos among other data sources. Data was collected from the sources and placed against the variables indicated in the SIMCITY modelling game. Cases of how urban design has advanced in different contexts were also reviewed to borrow best practices. The aspects of how simulation operates were taken from the considerations taken in a game called SIMCITY. This helped in understanding how simulations can be done and why they are hard to implement in Zimbabwe. Data collected from all the sources was then organised and analysed through thematic content analysis.

19.5 RESULTS AND DISCUSSION

19.5.1 Results

Urban planning improvements largely depend on the prevailing environment. For a nation to advance into sophisticated urban planning approaches, the governments and the local authorities should be prepared to improve their operations. Central and local government officers should be trained in the use of software to enhance their capabilities.

19.5.1.1   Layout Designing in Zimbabwe

Designing of layouts in Zimbabwe is done under the guidance of circular 70 of 2004, the layout design manuals and the Regional Town and Country Planning Act [29:12]. This has seen the creation of layout plans following the existing planning ordinances. The plans that are produced are largely 2D plans (Chitekwe-Biti et al., 2012). In the delivery of housing, the Government of Zimbabwe receives approximately 180 layout plans per province per year with all of them being 2D plans and having reports that are based on the number of residential stands created as well as the existing features of the sites that are being planned for (Musandu-Nyamayaro, 1993).

Data collected proved that local authorities and governments only require hard copy layout plans that are 2D in nature. This leaves the developer with the discretion of creating whatever environment he or she perceives from the 2D designs. The design approach and the plans under production have also led to the uncoordinated accumulation of layout plans.

In the context of layout design technology, Zimbabwe is slowly going digital. Government offices still use drawing boards, tracing paper and technical pens, advanced offices and large format printers. This state of affairs shows that even government officers are not capable of harnessing technology in designing layout plans. With officers showing limited capabilities, some urban problems have become so advanced that government and local authorities are short of solutions. The City of Harare is failing to deal with touting and vending. The city of Mutare has run out of developmental ideas and so is the Masvingo City as evidenced by their call for ideas in the local press. This is all evidenced in the existence of efforts by the Government of Zimbabwe to rationalise the planning and development of settlements through the development of a Provincial Development Committee that seeks to plan and rationalise cities that are viewed as being chaotic and unmanaged (Mlambo, 2020).

In the context of Zimbabwean urban problems, there is an issue of an increased percentage of urban habitats. This means that the increased human pressure and human behaviour ought to be considered when designing for a place that has to be vibrant and functional (Potts, 2006). It can also be noted that the current problems that most urban centres are facing are a result of the human factor (pirate taxi drivers and politicians). Congestion in the cities is increasing partially because of increased traffic volume and human behaviour. All these cannot be predicted or managed in a 2D layout.

19.5.1.2 Potential to Leapfrog

Urban planning technology has been increasing in Zimbabwe, particularly in consultancy companies in a bid to stay relevant (Gwamuri and Pierce, 2017). This has seen the introduction of 3D layout designs as a marketing strategy. Recently, various advertisements of cluster homes were featured in newspapers. Accompanying the advertisements were 3D projections of urban planning. This shows the existence of 3D skills within the body of planning professionals in Zimbabwe.

Furthermore, local authorities have been clamouring for enhancing service delivery through the introduction of GIS-based software. The City of Bulawayo has engaged in such a project. The local authority was aided by the Australian Aid, which was initiated by an assessment of the local authority capacity. The Kariba Municipality was also assisted in developing a GIS database developing from existing municipal systems. This saw the local authority employees being trained to use GIS-based equipment to map existing infrastructure, thereby creating a GIS-based database. This presents an opportunity for advancement into 4D layout design as it is useful in predicting potential problems that may hinder flawless service delivery.

Another significant opportunity for the introduction of 4D urban design is the problem that has affected most of Zimbabwe's urban spaces. The problems have left most local authorities helpless and desperate for any method that can help in curbing the challenges (Potts, 2006; Chirisa, 2014, Kawadza, 2017). Problems like vending, traffic congestion, water shortage, air pollution and encroachment into environmentally sensitive areas, all create a headache for city leaders. The problems present a chance for 4D urban planning as it helps in reducing their potential impacts through virtually identifying them.

The already existing information regarding city budgeting, zoning codes and traffic movements is also a critical opportunity for zoning codes. The Urban Councils

Act Chapter 29:15 of 1996 indicates that local authorities are mandated to make annual budgets of their areas of jurisdiction. These budgets are perceived as public information that can be accessed at any time upon request. Land use zones are available in local development plans that can also be accessed from local authorities upon payment of a small administrative fee. The existence of such information makes it easy to simulate settlements budgeting, zoning relations in the context of a proposed development.

19.5.1.3 Challenges to Simulation in Zimbabwe

The simulation of a settlement according to the specifications of a SIMCITY game involves the aspects of desirability, Geographical information view, time behaviour graphs and monthly budget. Besides the broad and complex nature of seemingly simplified variables, the information required to simulate Zimbabwe is in short. The design, modelling and simulation of a settlement require updated geographical information, which is found on maps and diagrams from the surveyor general. These are not available in Zimbabwe as the maps are still found as hard copies and yet to be digitised. Furthermore, information about social interests is not easy as public interest is not easy to define.

19.5.2 Discussion

Urban layout design is still generally 2D and the development of models and simulations is still in its primitive stages as such the layouts that are produced are implemented without much consideration of the potential or the threats that proposed developments may cause. Technological innovations that can reduce this are yet to be established by relevant departments. The planning environment in Zimbabwe is not developing fast enough to catch up with the problems that are emanating from the cities that have been created. This has seen most of the urban planning problems occurring without being anticipated. Planning departments in Zimbabwe are using obsolete planning technology and comparing the methods to the rate at which planning problems are developing, layouts are becoming more of organised disorder. This calls for the development of robust digital planning techniques and methods.

The Government of Zimbabwe and local authorities exist as parallel structures as such, the modelling of harmonised city models remain challenges as the information is hard to acquire. This makes it hard to develop city models that require the availability of comprehensive information. As such, the current city designs and plans are still based on one's imagination of how it will work as well as predictions and development of plans based on skeletal information. This position can explain how certain layout plans were approved regardless of their impacts in different cases (Mbare flea markets, various wetland developments among other plans that have failed to work or have been criticised).

19.6 CONCLUSION AND RECOMMENDATIONS

This chapter intended to identify the opportunities for leapfrogging from 2D to 3D urban planning. It identified that the planning environment in Zimbabwe is divided into two, the private and the public sector. In the public sector, there is little to no technological innovation on how planning is done. Technology is implemented by the private sector in trying to increase its competitiveness. The public sector has the most influence in shaping urban planning and design in Zimbabwe. Therefore, its use of obsolescent design techniques implies the failure of urban plans in Zimbabwe.

A leapfrog from 2D to 4D layout design in Zimbabwe requires investment in urban planning technology. The government and local authorities have to cease depending on drawing offices for their urban planning. There is a need to come up with city laboratories where simulations and designs are done, visualised, analysed, synthesised and implemented. This will improve the quality of pans that are under the production of Zimbabwe.

Furthermore, there is a need to engage universities and research institutions in developing relevant software that is useful in simulating Zimbabwean urban scenarios. This will help in fully understanding the performance of a proposed plan in the context of Zimbabwe. This will also help in the further development of planning ideas through research and analysis.

REFERENCES

Adams, P.C., 1998. Teaching and learning with SimCity 2000. *Journal of Geography*, 97(2), pp. 47–55.

Albuquerque, N. and Guedes, M.C., 2021. Cities without slums and the right to the city: Slums in Sub-Saharan Africa. *Renewable Energy and Environmental Sustainability*, 6, p. 24.

Axelrod, R., 1997. Advancing the art of simulation in the social sciences. In Rosario Conte, Rainer Hegselmann and Pietro Terna (eds.), *Simulating Social Phenomena*. Berlin: Springer, pp. 21–40.

Bänsch, E., 1991. Local mesh refinement in two and three dimensions. *IMPACT of Computing in Science and Engineering*, 3(3), pp. 181–191.

Benguigui, L., Czamanski, D. and Marinov, M., 2001. City growth as a leap-frogging process: An application to the Tel-Aviv metropolis. *Urban Studies*, 38(10), pp. 1819–1839.

Bhunua, S.T., Ruthera, H. and Gainb, J., 2002. 3-Dimensional virtual reality in urban management. *International Archives of the Photogrammetry, Remote Sensing and Spatial Information Sciences*, 34(Part 6), p. W6.

Boulos, M.N.K., Lu, Z., Guerrero, P. et al., 2017. From urban planning and emergency training to Pokémon Go: Applications of virtual reality GIS (VRGIS) and augmented reality GIS (ARGIS) in personal, public and environmental health. *International Journal of Health Geographics*, 16, 7. doi: 10.1186/s12942-017-0081-0.

Bourdakis, V., 1997. The future of VRML on large urban models. *The 4th UK VRSIG Conference* Centre for Advanced Studies in Architecture (CASA), University of Bath, Claverton Down, BA2 7AY, Somerset. UK

Brezis, E.S., Krugman, P.R. and Tsiddon, D., 1993. Leapfrogging in international competition: A theory of cycles in national technological leadership. *The American Economic Review*, pp. 1211–1219.

Chipambwa, W. and Chimanga, T., 2018. Creative design software: Challenges and opportunities to the graphic designer in Zimbabwe. *Journal of Graphic Engineering and Design*, 9(1), pp. 29–45.

Chirisa, I., 2014. Building and urban planning in Zimbabwe with special reference to Harare: Putting needs, costs and sustainability in focus. *Consilience*, (11), pp. 1–26.

Chitekwe-Biti, B., Mudimu, P., Nyama, G.M. and Jera, T., 2012. Developing an informal settlement upgrading protocol in Zimbabwe—the Epworth story. *Environment and Urbanization*, 24(1), p. 131148.

Christodoulou, N., Papallas, A., Kostic, Z. and Nacke, L.E., 2018. Information visualisation, gamification and immersive technologies in participatory planning. *Extended Abstracts of the 2018 CHI Conference on Human Factors in Computing Systems*. April 2018 Paper No.: SIG12, pp. 1–4. doi: 10.1145/3170427.3185363

Cich, G., Knapen, L., Maciejewski, M., Bellemans, T. and Janssens, D., 2017. Modelling demand responsive transport using SARL and MATSim. *Procedia Computer Science*, 109, pp. 1074–1079.

Das, D.K., 2020. Perspectives of smart cities in South Africa through applied systems analysis approach: A case of Bloemfontein. *Construction Economics and Building*, 20(2), pp. 65–88.

Davison, R., Vogel, D., Harris, R. and Jones, N., 2000. Technology leapfrogging in developing countries—an inevitable luxury. *The Electronic Journal of Information Systems in Developing Countries*, 1(1), pp. 1–10.

Devisch, O., 2008. Should planners start playing computer games? Arguments from SimCity and Second Life. *Planning Theory & Practice*, 9(2), pp. 209–226.

Dodge, M., McDerby, M. and Turner, M. eds., 2011. *Geographic Visualization: Concepts, Tools and Applications*. West Sussex, England: John Wiley & Sons.

Dodo, M.K., 2018. Why is Africa lagging behind in economic development? A critical review. *Journal of Asia Pacific Studies*, 5(1), 93–124.

Dooley, K., 2002. Simulation research methods. *Companion to Organizations*, pp. 829–848.

Engelbrecht, G., 2012. Infrastructure planning tool. *GISSA Ukubuzana 2012 Conference*, Ekurhuleni, 2–4 October 2012, pp. 68–80, GISSA.

Falconer, R.E., Isaacs, J.P., Gilmour, D. and Blackwood, D.J., 2018, Indicator modelling and interactive visualisation for urban sustainability assessment. In IRMA (ed.), *E-planning and Collaboration: Concepts, Methodologies, Tools, and Applications*. vol. 1, IGI Global, Hershey, PA, pp. 486–508. doi: 10.4018/978-1-5225-5646-6.ch023

Faugeras, O.D. and Hebert, M., 1986. The representation, recognition, and locating of 3-D objects. *The International Journal of Robotics Research*, 5(3), pp. 27–52.

Florio, P., Probst, M.C.M., Schüler, A., Roecker, C. and Scartezzini, J.L., 2018. Assessing visibility in multi-scale urban planning: A contribution to a method enhancing social acceptability of solar energy in cities. *Solar Energy*, 173, pp. 97–109.

Fong, M.W., 2009. Technology leapfrogging for developing countries. In: *Encyclopaedia of Information Science and Technology*, Second Edition, Hershey, PA: IGI Global, pp. 3707–3713.

Frasca, G., 2013. Simulation versus narrative: Introduction to ludology. In: *The Video Game Theory Reader*. New York: Routledge, pp. 243–258.

Gwamuri, J. and Pearce, J. M., 2017. "Open Source 3-D Printers: An Appropriate Technology for Building Low Cost Optics Labs for the Developing Communities," in *ETOP 2017 Proceedings*, X. Liu and X. Zhang, eds., (Optical Society of America, 2017), paper 104522S.

https://www.osapublishing.org/abstract.cfm?URI=ETOP-2017-104522S.

Hinton, C.H., 1880. What is the fourth dimension? *The University Magazine, 1878–1880*, 1(1), pp. 1534–1623.

Hiraiwa, K., 2005. Dimensions of symmetry in syntax: Agreement and clausal architecture (Doctoral dissertation, Massachusetts Institute of Technology).

Kawadza, S., 2017. Cry my beloved town planning profession. *The Herald*. www.herald.co.zw/cry-my-beloved-town-planning-profession/ Accessed 22/08/2018.

Kleynhans, P.J., 2012. The provision of short-term technical assistance services to the Gauteng provincial treasury: Closure report—version 2. Mogwariepa Consulting CC, Johannesburg, South Africa.

Mlambo, N., 2020. An overview of the local government system of Zimbabwe. *Journal of African Problems and Solutions*, 2(1), pp. 11–45.

Mubiwa, B. and Annegarn, H., 2013. *Historical Spatial Change in the Gauteng City-Region* (GCRO Occasional Paper 4), GCRO, Johannesburg, South Africa.

Musandu-Nyamayaro, O., 1993. Housing design standards for urban low-income people in Zimbabwe. *Third World Planning Review*, 15, pp. 329–354.

Nair, K.K., Pillai, M.M., Lefophane, S. and Nair, H.D., 2020, August. Adaptation of smart cities in the South African Internet of Things context. In *2020 International Conference on Artificial Intelligence, Big Data, Computing and Data Communication Systems (icABCD)*, pp. 1–6. IEEE, Durban, South Africa.

Potts, D., 2006. 'All my hopes and dreams are shattered': Urbanization and migrancy in an imploding African economy—the case of Zimbabwe. *Geoforum*, 37(4), pp. 536–551.

Rocha, F.W., Francesquini, E. and Cordeiro, D., 2021, May. An approach inspired by simulation points to accelerate smart cities simulations. In *Anais da XII Escola Regional de Alto Desempenho de São Paulo*, pp. 49–52. SBC, Sao-Paolo, Brazil

Rucker, R., 2014. *The Fourth Dimension: Toward a Geometry of Higher Reality*. Mineola, New York: Courier Corporation.

Sante, I., García, A.M., Miranda, D. and Crecente, R., 2010. Cellular automata models for the simulation of real-world urban processes: A review and analysis. *Landscape and Urban Planning*, 96(2), pp. 108–122.

Schindler, M., Dionisio, R. and Kingham, S., 2018. A multi-level perspective on a spatial data ecosystem: Needs and challenges among urban planning stakeholders in New Zealand. *International Journal*, 13, pp. 223–252.

Strauch, D., Moeckel, R., Wegener, M., Gräfe, J., Mühlhans, H., Rindsfüser, G. and Beckmann, K.J., 2005. Linking transport and land use planning: The microscopic dynamic simulation model ILUMASS. *Geodynamics*, pp. 295–311.

Thurston, W.P., 2014. *Three-Dimensional Geometry and Topology*, Volume 1. Princeton: Princeton University Press.

White, K.P. and Ingalls, R.G., 2009. Introduction to simulation. In *Proceedings of the 2009 Winter Simulation Conference (WSC)*, Austin, Texas, 13–16 December 2009, pp. 12–23, doi: 10.1109/WSC.2009.5429315

White, R., Engelen, G., Uljee, I., Lavalle, C. and Ehrlich, D., 2000, June. Developing an urban land use simulator for European cities. In *Proceedings of the Fifth EC GIS Workshop: GIS of Tomorrow*. June 2018. Ispra, Italy.

World Bank Group, 2016. *World Development Report 2016: Digital Dividends*. World Bank Publications, Washington DC.

20 An Evaluation of the Effectiveness of Material Waste Management Techniques in the Construction Industry of Zimbabwe
A Case of Harare and Bulawayo

Crytone Kusaziya and Yvonne Munanga

CONTENTS

20.1 Introduction 283
20.2 Theoretical Framework... 285
20.3 Literature Review 285
    20.3.1 Forms and Sources of Construction Material Waste........................ 285
    20.3.2 Construction Material Waste Management Techniques................... 287
    20.3.3 Effects of the Legislative and Institutional Frameworks................. 289
20.4 Methodology... 290
20.5 Results and Discussion 290
20.6 Conclusion, Policy Options and Recommendations.................................... 294
References.. 295

20.1 INTRODUCTION

Material waste management has become a challenge all over the world due to population increase, unplanned urban development and improvements in standards of living (Mabaso and Hewson, 2015). Zimbabwe has not been spared the challenge as it has been faced with rapid urbanisation and infrastructure development, which has resulted in enormous construction material waste being generated from the construction activities. Wastes are materials that are discarded after use at the end of their intended life span where residuals recycled or reused are excluded from it (Kharamova et al., 2016). Hence, construction material wastage can be defined as the difference between the values of materials delivered and accepted onsite and those properly used as specified and accurately measured in the work after deducting the cost saving of substituted materials transferred elsewhere in which unnecessary cost and time may be incurred by the material wastage (Shen et al., 2004).

The causes of construction material waste can be measured and evaluated using a large number of construction phase–related factors such as design and documentation, materials procurement and management, site management practices and site supervision including environmental conditions (Adewuyi and Otali, 2013). Construction waste may be generated due to various reasons that are demolition and dismantling of structures, damage or excess construction material, non-use of material due to alteration in specification, changes in design during construction, designers' inexperience in methods or sequence of construction and improper planning of the required quantity amongst others. Rapid population growth in most developing nations has been the source of pressure on urban infrastructure and services; as a result it has led to an increase in construction activities and hence the generation of huge amounts of construction waste. Managing this waste is one of the challenging issues in the fast-developing world due to inadequate manpower, financial resources, implements and machinery that results in environmental pollution (Tsiko and Togarepi, 2012; Adewuyi and Otali, 2013).

Waste management works towards reduction, application and use of all resources. Effective waste management encourages the reduction of energy consumption, conservation and acquisition of reused and recycled products. Statistics have also revealed that less than 20% of urban solid waste is collected and disposed of properly. As such there is a need to effectively manage material waste at all stages of a construction project to minimise costs on the construction project and the environment at large. Waste management for construction activities has also been promoted with the aim of protecting the environment and the recognition that wastes from construction and demolition works to significantly contribute increased project costs and pollution of the environment. Generation of material waste means loss of profits to contractors due to extra overhead costs and delays in work execution which lead to lower productivity (Skoyles and Skoyles, 1987; Bossink and Brouwers, 1996; Shen et al., 2004).

Zimbabwe produces an average 2.5 million tonnes of solid wastes (household and industrial combined) per annum. In general, a very high level of waste is assumed to exist in construction. This kind of waste typically accounts for between 15% and 30% of urban waste. Studies have shown that Zimbabwe still faces material waste management challenges, despite the existence of laws dealing with material waste management such as the Environmental Management Act of 2003 (Chimhowu et al., 2002; Tsiko and Togarepi, 2012). Construction material waste continues to be dumped along major highways, in soccer grounds and open spaces near residential areas. In light of those statistics, there is a need for comprehensive and sustainable material waste management techniques in the Zimbabwean construction industry. The construction industry continues to look for ways to improve the construction material waste management techniques. This research will evaluate the effectiveness of the construction material waste management techniques in reducing the amount of construction material waste being generated. The chapter will look at the theoretical framework the study is hinged on, literature review, methodology, results and discussions and lastly the conclusion, policy options and recommendations.

20.2 THEORETICAL FRAMEWORK

This study is grounded on the sustainability theory, which forms the theoretical framework guiding this research. In its literal rudiments, sustainability means a capacity to maintain some entity, outcome, or process over time. Sustainability is also the ability of a society, ecosystem, or any such ongoing network to remain functional into the indefinite future without being exposed to malfunctioning due to exhaustion or excessive use of key resources which that network or system relies on. In relation to material waste management, sustainability entails the ability of the construction industry to implement effective material waste management techniques that protect the environment as well as ensure effective resource utilisation. This can be achieved by promoting implementation of effective waste management techniques in the built environment. Sustainability also refers to a beneficial change in the urban form that would accommodate the socio-economic and environmental needs of the society, thus leading to corresponding lasting outcomes and impacts in people's lives (Mabaso and Hewson, 2015).

20.3 LITERATURE REVIEW

20.3.1 Forms and Sources of Construction Material Waste

Material waste has been recognised as a major problem in the construction industry that has implications both for the efficiency of the industry and for the environmental impact of construction projects. The waste results from the various construction processes and construction stages comes in different types (Formoso and Eduardo, 1999; Garas, 2001). Construction waste can be categorised into two principal types, direct waste or total loss of materials and indirect waste (Skoyles and Skoyles, 1987). Indirect waste is distinguished from direct waste in that the materials are not usually lost physically and only the payment for part or the whole of the value is lost. Table 20.1 summarises the forms of direct and indirect wastes.

There are five main sources of material waste in construction that are design, procurement, material handling, operation and residual. Project design and specifications can contribute significantly to the amount of material waste generated during the construction of a project particularly when uneconomical design solutions are selected or when unsuitable materials are specified. Improper designs result in excessive cut-offs and this is one of the major causes of material waste. To reduce or avoid the generation of construction material wastes from design, there is a need to pay attention to dimensional co-ordination and changes made to design while construction is in progress. Designers should be fully experienced in the method and sequence of construction. Designers should also be familiar with alternative products, complexity and errors and also pay attention to standard sizes available in the market so as to avoid excessive cut-offs which have been attributed as a major cause of construction material waste (Alarcon, 1994, Bossink and Brouwers, 1996; Ekanayake and Ofori, 2000).

TABLE 20.1
Forms of Indirect and Direct Waste

Type	Forms
Indirect Waste	– **Substitution**, where materials are used for purposes other than those specified. – **Production waste**, where materials are used in excess of those indicated or not clearly defined in contract documents, e.g. additional concrete in trenches, which are dug wider than was designed, because no appropriately sized digger bucket is available. – **Operational waste**, where materials are used for temporary site work for which no quantity or other allowances have been made in the contract documentation, e.g. tower-crane bases, site paths, temporary protection. – **Negligent waste**, where materials are used in addition to the amount required by the contract owing to the construction contractor's own negligence.
Direct Waste	– **Deliveries waste** comprises all losses in transit to the site, unloading and placing into the initial storage. – **Site storage and internal site transit waste** comprise losses due to bad stacking and initial storage, including movement and unloading around the site, to stack at the work place or placing into position. – **Conversion waste** comprises losses due to cutting uneconomical shapes, e.g. timber and sheeted goods. – **Fixing waste** comprises materials dropped, spoiled, or discarded during the fixing operation. – **Cutting waste** includes losses caused by cutting materials to size and to irregular shapes. – **Application waste** includes materials such as mortar for brickwork, paint spilled or dropped during application similarly and materials left in containers or cans that are not resealed. Mixed materials like mortar and plaster left to harden at the end of the working day. – **Management waste** includes losses arising from an incorrect decision or from indecision and not related to anything other than poor organisation or lack of supervision. – **Waste caused by other trades**. This includes losses arising from material or damage by succeeding trades. – **Criminal waste** covers pilfering, theft from sites and vandalism. – **Waste due to the incorrect type or quality of materials**. This includes waste stemming from materials wrongly specified, waste due to errors, particularly in the bills of quantities and specification. – **Learning waste** that is usually caused by apprentices, unskilled 'tradesmen' and tradesmen on new operations.

Source: Urio and Brent (2006)

Procurement of materials is also a critical point at which contractors and subcontractors can influence material waste for a project as this activity determines the materials that are to be supplied to site. Construction material waste at this stage has been attributed to the purchase of products that do not comply with specifications. Ordering of materials that do not fulfil project requirements is reported to be the 86% cause of construction material waste (Gihan et al., 2010; Bhosale, 2015). In this light, procurement of materials plays a critical role in construction material waste management and hence there is a need to accurately procure the required materials at the required quantities.

Material handling is another factor that causes material waste. Effective material handling is using the right method, amount, material, place, time, sequence, position, condition and cost. This involves handling, storing and controlling of the construction materials. It has been noted that an average material waste volume of 27.6% spreading across sources such as lack of quality control, effective material handling, off-cuts, and labour error is generated on construction sites (Lingard et al., 2000, Navon and Berkovich, 2005). In addition, the way construction activities are carried out during the construction process, that is, operations also impact the quantity of material waste produced. Material waste production in construction sites is often due to inadequate storage and protection, poor or multiple handling, poor site control, over ordering of material, bad stock control, lack of training and damage to material during delivery. Other causes of construction material wastage include errors by tradespersons and accidents due to negligence (Bossink and Brouwers, 1996).

Residual-related material waste arises from issues such as uneconomical shape, off cuts, over mixing of material and waste from the application process (Urio and Brent, 2006). This type of material waste is common where there is lack of knowledge about construction process during design activities hence designers producing uneconomical shapes hence leading to a lot of off cuts hence material wastage. All players in the construction process have a critical role in the prevention of material wastage and hence there is a need for consultative meetings during the design of projects so that uneconomical shapes will be avoided at all costs and that the designers are kept in light of the construction process.

In addition, material waste can also be attributed to cultural factors because waste generation is not only a technical issue but also a behaviouristic one. Construction is labour intensive in nature and henceforth behavioural impediments are likely to influence the level of material waste generation. To effectively manage the generation of material wastes in the construction industry, the construction team has to change attitude and behaviour towards material waste issues. Cultural attributes that contribute to material waste include lack of training of the participants in the construction industry and lack of incentives for the participants and hence rendering them not motivated to carry out their duties effectively. If the behavioural tendencies and attitudes of the construction industry are aligned correctly then effective material waste management can be achieved (Lingard et al., 2000).

20.3.2 Construction Material Waste Management Techniques

Materials management is an important function for improving productivity in construction projects. Hence, concern over construction material waste is becoming a prevalent part of any construction project. Several approaches to construction waste management have been proposed and these will be discussed in this chapter. The process of managing construction waste goes far beyond the disposal of the wastes itself. Research has shown that waste management techniques cannot be uniform across regions and sectors because individual waste management methods cannot deal with all potential construction waste materials in a sustainable manner. Varying conditions of regions and sectors also mean that the procedures of material waste management should also vary for them to be effective. However, material waste management must remain flexible in light of changing economic, environmental and social conditions (Chimhowu et al., 2002; Kharamova et al., 2016).

The three R principles have become the central principle in sustainable material waste management. The principle denotes that waste should be reduced, re-used or recycled depending on alternative uses the waste can be utilised on. It has been noted that the most effective environmental solution may often be to reduce the generation of waste, where further reduction is not practicable, products and materials can sometimes be re-used, either for the same or a different purpose and also the value should be recovered from waste, through recycling, composting or energy recovery (Macozoma, 2002; Polat and Arditi, 2005). Only if none of these solutions is appropriate should waste be disposed of, using the best practicable environmental option.

20.3.2.1 Reduce

Waste minimisation includes source reduction. Source reduction is defined as any activity that reduces or eliminates the generation of waste at the source, usually within a process. The concept of waste reduction or waste minimisation involves redesigning products or changing societal patterns of consumption, use and waste generation to prevent the creation of waste and minimise the toxicity of waste that is produced (Macozoma, 2002).

20.3.2.2 Reuse

Reuse means the recovery of useful materials from the waste stream for immediate use as secondary materials. Reusable materials account for a significant proportion of the total material waste generated during site activities; therefore, it is the contractor's responsibility to adopt the re-use technique to recover material (Ekanayake and Ofori, 2000; Macozoma, 2002). Reuse offers greater environmental advantage than any other material waste management technique. Research indicates that the construction industry faces barriers in the implementation of the reuse technique and amongst others the barriers are lack of standardisation of components, ensuring and warranting the performance of reused components, lack of detailed knowledge of the product's properties and in-use history, quality assurance of reused products, robustness of products in the deconstruction process, that is, many lighter products do not intactly survive the deconstruction process and practicalities of economic deconstruction including deconstructing composite components.

20.3.2.3 Recycle

Recycling refers to the reprocessing of salvaged useful waste materials that cannot be put into direct re-use to produce secondary materials and products. Recycling involves separating waste into recyclable and non-recyclable waste materials. Recycling is a preferred option to landfill disposal; however, it ranks below reuse on the waste management hierarchy due to its energy requirements for reprocessing before reapplication. There are various advantages of recycling waste materials in the construction industry and amongst other advantages is the protection of environment as it reduces the amount of energy consumed in manufacturing new materials by processing raw materials. Recycling material also has the advantage that it reduces the amount of waste ending up in landfills, thereby also increasing the profitability of the construction industry (Polat and Arditi, 2005; Kharamova et al., 2016).

20.3.3 Effects of the Legislative and Institutional Frameworks

Developing countries suffer from the effects of poor material waste management and efforts in place to curb the growing concern of construction material wastes. There is a need for a legislative framework in developing countries dealing with material waste management and this has been a chronic problem. Zimbabwe is no exception as there is no legislation that guides material waste management specifically in the construction industry. At the moment, the industry is only guided by the Environmental Management Agency (EMA) through the EMA Act that covers all sectors (Tsiko and Togarepi, 2012; Mabaso and Hewson, 2015). There is a need for a separate arm or agent that focuses specifically on construction waste. EMA seems to be overwhelmed as it focuses on all sectors, which then affects the enforcement of the law. Most of the waste produced is being disposed in landfills and this is harmful to the environment. Despite the environmental effects that include but is not limited to air pollution, land degradation, land pollution, landfills have however provided a convenient and cost-effective solution to wasteful practices for the construction industry across the globe. However, there has been a surge of illegal landfills; hence, landfills cannot be seen as a sustainable way of waste materials disposal. Contractors just choose the nearest convenient dumping place (which may be a landfill) for them, despite legality issues for the material wastes hence causing the environment to deteriorate.

20.3.3.1 Benefits of Construction Material Waste Management

Effective material waste management has benefits to the construction industry. First, the benefits can be financial benefits to the contractor, subcontractors and the client. The financial benefits manifest in the form of economic advantages that result in lower project costs, increased business patronage and lower risk of litigation regarding wastes. Second, are environmental benefits. These include the minimisation of the risk of immediate and future environmental pollution and harm to human health (Coventry and Guthrie, 1998). Reduced waste means less quantity of landfill space used and reduced environmental impacts associated with extracting, transporting and manufacturing or processing of the raw materials of construction products.

20.3.3.2 Challenges in Effective Construction Material Waste Management

Despite the benefits of proper waste management, implementation of effective material waste management remains a challenge in both developed and developing world (Garas, 2001). Some of the barriers to effective material waste management include lack of awareness in the industry, lack of interest from clients, lack of proper training and education, lack of skilled labour, lack of market competition, lack of government intervention and lack of appreciation of waste reduction approaches by architects and designers.

20.4 METHODOLOGY

The study was carried out in Harare and Bulawayo which is the capital city and the second largest city of Zimbabwe, respectively. There has been an upsurge of construction and demolition activities being carried out in these economic hubs of Zimbabwe. Moreso, according to the 2019 CIFOZ national categorisation register, 77% of the registered general building contractors and general civil engineering construction companies have their administrative offices in Harare, 21% have offices in Bulawayo and 2% in other cities and towns. So, this justifies the choice of the study area. At the time the research was carried out, there were 136 registered building and civil contractors in Harare and Bulawayo, according to CIFOZ register. The sample size was calculated using a mathematical formula and a sample size of 30 was considered for analysis purposes. In descriptive research, anything from 10% to 20% of the population in question is representative enough to warrant the generalisation of results (Creswell, 2007).

The research used a triangulation of literature review, document review and key informant interviews. For data collection, a desk study was used for secondary data and primary data was gathered mainly through interviews, field observations and questionnaires. Sources consulted included contractors, professionals in the built environment, government officials, environmentalists and policy-makers. These were considered as they are stakeholders in the construction industry and appreciate construction waste management issues. A combination of purposive and snowball sampling was used to select the respondents. Data collected was analysed mainly through content analysis as it was mostly qualitative data.

20.5 RESULTS AND DISCUSSION

From the data collected, there was a clear indication of the lack of understanding of what construction material waste constitutes which stemmed from the various perspectives from the respondents. It is therefore evident that indeed there is a need to raise awareness as to what material waste is about. The Zimbabwe construction industry mainly uses materials such as concrete, bricks, mortar, steel, timber, glass, plastics, bitumen, stones and sand. Having an understanding of the types of materials commonly used in the construction industry and the amount of material waste generated due to their usage helps in implementing effective waste management techniques.

With regards to sources of material waste in the construction industry in Harare and Bulawayo, reworks, variations and negligence were ranked the highest by contractors. The other sources in the descending order of ranking include the procurement of poor-quality materials, design changes, onsite storage of materials, lack of training, poor coordination between trades, design complexities, lack of awareness, poor product knowledge, poor coordination and sequencing of works, material delivery methods, lack of incentives, site ordering process and poor delivery schedules. Table 20.2 summarises the field results of the sources of construction material waste.

Reworks, variations and negligence are ranked highest as the source of construction material waste with a mean of 3.86. Site managers posited that most reworks

TABLE 20.2
Sources of Construction Waste

Source of construction material waste	Mean	Rank
Rework variations and negligence	3.86	1
Poor quality of materials	3.62	2
Design changes	3.59	3
Onsite storage	3.41	4
Lack of training	3.32	5
Poor coordination between trades	3.07	6
Design complexities	3.07	7
Lack of awareness	3.03	8
Poor product knowledge	3.00	9
Coordination and sequencing of works	2.93	10
Material delivery methods	2.79	11
Lack of support from senior management	2.76	12
Programme constraints	2.74	13
Lack of incentives	2.55	14
Coordinated site ordering process	2.52	15
Delivery schedules	2.45	16

Source: Kusaziya field survey

were a result of lack of drawings and standards at the time of construction whereas variations were mainly attributed to changes issued and directed by the client. Procurement of poor-quality materials also contributes greatly to material waste and has a mean of 3.62. Site personnel noted that in most instances, poor materials are rejected by the project team and there is no take back facility for some of the materials and hence they end up being material waste. They also noted that related to this, lack of coordinated site ordering process, poor delivery schedules and delivery methods contribute to construction waste.

Project managers and site managers indicated that design changes were major contributors to the generation of material waste on construction sites since procurement of required materials would have been done and change of designs usually renders some materials unusable therefore ending up accounting for material waste. This had a mean of 3.59. Designers interviewed concurred with this perspective but noted that in some instances design changes are beyond their control as the client is the one to decide on which design he or she wants. They also claimed that clients are mostly not sure of what they want until actual construction takes place. Design changes contribute to alterations and in some instances demolitions to allow the modification to be made hence contributing to material waste. In these instances, the interviewees agreed that there is a need to have an active involvement of the client at the design stages to avoid procuring materials that usually end up being waste.

Relating to design changes is design complexity which had a mean of 3.07 and was also pointed as one of the causes of material waste. Complex designs usually result in some instances working in confined areas or acute areas that are difficult to work on and hence result in breakages that form a part of material waste (Mabaso and Hewson, 2015).

Observations on most construction sites indicated that behavioural tendencies and lack of support were a major observable attribute causing material waste as the labourers were observed dropping the bricks while handling attributing to lack of personnel protective equipment such as gloves and this is lack of management support. In addition, onsite workers were observed to occasionally sit whilst the mixed mortar and concrete were left to dry unused and hence contributing to material waste. This can also be attributed to the lack of awareness as to how much cost will that add to the overall project cost. This is in line with Skoyles and Skoyles (1987) observations that the problem of material wastage was more dependent upon the attitudes and behavioural tendencies of individuals involved in the construction process than upon the technical processes it employed and hence the need to impart construction labours with the right attitude in order to minimise material waste occurring in the construction industry.

Effective material handling is using the right method, amount, material, place, time, sequence, position, condition and cost. This involves handling, storing and controlling of the construction materials (Tsiko and Togarepi, 2012). Awareness and training on effective material handling will go a long way in minimising construction waste as waste cannot be avoided totally. Since construction waste is unavoidable completely, waste management techniques should be implemented and there is need to evaluate the effectiveness of these techniques in the Zimbabwe construction industry. There may be a need to tailor make techniques to suit the Zimbabwean environment. From the sample population, a remarkable 65% of the respondents had material waste management techniques within their organisations, whilst the remaining 35% had no material waste management techniques in use.

The results revealed that the most used waste management technique in Harare and Bulawayo is the reduce technique that had a frequency of 14. Waste minimisation includes source reduction. Source reduction eliminates the generation of material waste at the source, usually within a process (Adewuyi and Otali, 2013). The reduce technique is the most widely used technique as the contractors strive to eliminate losses through material waste in their construction processes. The respondents noted that the criteria that they used for the selection of the reduce technique was mainly due to the fact that they did not have any other material waste management techniques hence the option was the reduction of material waste through the processes. Effectiveness of the method and feasibility due to the economic conditions was one of the factors that the respondents cited as to why they chose the reduction principle of material waste management.

Cost and benefit were also cited by the respondents as to why they chose the reduction technique as a material waste management technique on their projects. The type of the process being conducted also influenced the choice of material waste management technique to use, for example, material from demolitions could be used on some projects elsewhere. Reduction of material waste can also be achieved through accurate estimating of materials and ordering (Bhosale, 2015). Material waste reduction needs to be catered for during the design stages of the projects and also during the construction stage where modern construction methods can be employed in order to reduce the generation of waste.

The reuse technique came second on ranking on the waste management techniques employed in the construction industry with a frequency of 10. Reusable materials account for a significant proportion of the total material waste generated during site activities, thus explaining why most of the respondents would use the reuse technique to minimise waste generated due to site activities. It is the contractor's responsibility to adopt the reuse technique to recover material (Ekanayake and Ofori, 2000; Macozoma, 2002). Cost–benefit analysis was cited by the respondents as to why they would choose the reuse technique. Effective reuse preserves the present structure of material and does not require additional time or energy for utility. If the reuse technique is implemented effectively, it may yield many benefits for both the contractor and the client. Reusable materials account for a significant proportion of the total material waste generated during site activities and hence it is the contractor's responsibility as well as his best interest to adopt the re-use technique to recover materials (Ekanayake and Ofori, 2000; Macozoma, 2002

Recycling ranked last with a frequency of 6. Recycling is a preferred option; however, it ranks below reuse on the waste hierarchy due to its energy requirements for reprocessing before reapplication (Kharamova et al., 2016). In this view, although the recycling technique is a preferred option, the cost associated with it discourages contractors from using the method. Another issue in Zimbabwe is the lack of the requisite technology for recycling waste. Therefore, there really is a need for awareness and investment in this area so that waste can be recycled. It was also worrying to discover that some organisations (10 out of the 30 contractors) actually do not make use of any waste management techniques.

The recycling technique presents great challenges in Zimbabwe even though some nations have adopted this technique very well. Although there are various advantages associated with recycling waste materials in the construction industry which include the protection of environment, reduction of the amount of energy consumed in the manufacturing of new materials by processing raw materials, reduction of the amount of waste ending up in landfills, increasing the profitability of the construction industry, among other advantages, it is the least used technique. There is need for government support towards the adoption of the recycling technique. This support can come in the form of supporting policies and financial assistance. Awareness programmes can be held by the construction industry boards and the relevant government departments as another form of support.

It is well appreciated that material waste management techniques cannot be uniform across regions and sectors because individual waste management methods cannot deal with all potential construction materials waste in a sustainable manner. Varying conditions of regions and sectors also means that the procedures of material waste management should also vary for them to be effective. However, material waste management must remain flexible in light of changing economic, environmental and social conditions (Navon and Berkovich, 2005).

The research also indicated some of the challenges facing the construction industry in implementing effective construction material waste management. Amongst the challenges put forward were weakness in legislation, lack of incentives, lack of proper training, lack of awareness and lack of waste reduction approach by designers amongst others. These issues need to be addressed to ensure effective waste management. Bhosale (2015) asserts that lack of awareness about material waste management techniques and approach is the major barrier in the construction industry among local contractors, construction labour and architects. There is therefore need for training amongst the construction labour and amongst all the players of the built environment on effective material waste management techniques and approach.

There are remarkable impacts of poor material waste management practices in the construction industry. These include increased transport costs of waste to appropriate dump sites, cost of material waste to the client, material excess or shortage on site, environmental impacts, construction time delay, low productivity due to material waste and loss of profit to contractors. On the other hand, effective material waste management offers both financial and environmental benefits. The financial benefits include lowering overall project costs, reduced cost of disposal, increased profit margins and reduced project cost. The environmental benefits include a cleaner environment, conserving natural resources, reduced quantity of material waste generated minimisation of the risk of immediate and future environmental pollution and harm to human health (Coventry and Guthrie, 1998; Gihan et al., 2010).

20.6 CONCLUSION, POLICY OPTIONS AND RECOMMENDATIONS

From the research, it was evident that material waste in the construction industry occurs at various stages of a project lifecycle and that minimisation of material waste on construction projects lies within the supply chain stakeholders, that is; the clients, designers, contractors and suppliers. More so, contractors play a pivotal and critical role in the management of material waste on construction projects. It was also concluded that the three most important sources contributing to construction material waste generation in the Zimbabwean construction industry are rework, variations and negligence. Use of poor-quality materials, design changes, lack of incentives, uncoordinated site ordering process and delivery schedules were also considered common sources of construction material waste.

On techniques implemented in managing waste, waste reduction and reuse are the most viable options being utilised by the construction industry in Harare and Bulawayo. There is therefore need for coordination between professionals from the project inception to completion so as to raise awareness as to which construction methods to use and also to optimise design in accordance to the skills available. Effective waste management yields environmental and financial benefits. Recycling is not a viable method in the Zimbabwean construction industry at the moment.

From the findings and the conclusions, it is recommended that there is a need to promote material waste recovery for reuse and recycling in accordance with waste hierarchy before final waste disposal. There is also need to create awareness amongst project team members and they should be educated on the effects and benefits of construction material waste management. Training sessions for the construction phase team should be done, as most material waste is perceived to occur during the construction phase of the project. In addition, greater importance should be for planning and controlling of materials to ensure that the right quality and quantity of materials are available when needed thus minimising material waste due to material handling and storage.

Moreover, to promote sustainable and smart spatial planning, a legislative framework that deals specifically with material wastes management in the construction industry is recommended. Infrastructure development is delivered through construction projects and this justifies having a separate agency dealing with waste management issues in the construction sector. With the current national objective of achieving a middle-income status by 2030, there will definitely be a lot of construction works and hence waste management issues should be addressed effectively.

REFERENCES

Adewuyi, T. O. and Otali, M. 2013, *Evaluation of Causes of Construction Waste: Case of River State Nigeria*, University of Uyo Press, Uyo.

Alarcon, L. F. 1994, *Tools for the Identification and Reduction of Waste in Construction Projects*, Catholic University of Chile, Santiago.

Bhosale, A. 2015, *Plastic Waste Management: A Survey at an Institute*, LAP LAMBERT Academic Publishing, Trenton, NJ.

Bossink, B. A. G. and Brouwers, H. J. H. 1996, *Construction Waste: Quantification and Source Evaluation*, ASCE Publishers, Amsterdam.

Chimhowu, A., Tevera, D., Chimbetete, N., Gandure, S., Conyers, D. and Matovu, G. 2002, *Urban Solid Waste Management in Zimbabwe*, Municipal Development Partnership Eastern and Southern Africa, Harare.

Coventry, S. and Guthrie, P. 1998, *Waste Minimisation and recycling in construction*, John Wiley and Sons, London.

Creswell, W. J. 2007, *Qualitative Inquiry & Research Design: Choosing among Five Approaches*, 2nd ed., Sage Publications, Thousand Oaks, CA.

Ekanayake, L. and Ofori, G. 2000, *Construction Material Waste Source Evaluation*, AOSIS Publishing, Cape Town.

Formoso, C. T. and Eduardo, L. I. 1999, *Method for Waste Control in Building Industry*, University of California, Los Angeles, CA.

Garas, G. L. 2001, *Minimizing Construction Material Wastes*, Cairo University, Cairo.

Gihan, L. G., Ahmed, R. A. and Adel, E. G. 2010, *Material Waste in Egyptian Construction Industry*, Cairo University, Cairo.

Kharamova, M. D., Mada, S. Y. and Grchev, V. A. 2016, *Landfills: Problems, Solutions and Decision Making to Waste Disposal in Harare, Zimbabwe*, Peoples Friendship University of Russia, Moscow.

Lingard, H., Graham, P. and Smithers, G. 2000, *Employee Perceptions of the Solid Waste Management System Operating in a large Australian Contracting Organisation: Implications for Company Policy Implementation*, Royal Melbourne Institute of Technology, Melbourne.

Mabaso, C. H. and Hewson, D. S. 2015, *Employees' Perception of Food Waste Management in Hotels*, University of Johannesburg, Johannesburg.

Macozoma, D. S. 2002, *Construction Site Waste Management and Minimization*, CIB Publications, Ontario.

Navon, R. and Berkovich, O. 2005, *Development and On-Site Evaluation of an Automated Materials Management and Control Model*, Blackwell Publishers, London.

Polat, G. and Arditi, D. 2005, *The JIT Materials Management System in Developing Countries*, Blackwell Publishers, London.

Shen, L. Y., Tam, V. W., Tam, C. M. and Drew, D. 2004, *Mapping Approach for Examining Waste Management on Construction Sites*, Mitchell Publishing Company Limited, London.

Skoyles, E. R. and Skoyles, J. R. 1987, *Waste Prevention on Site*, Mitchell Publishing Company Limited, London.

Tsiko, R. and Togarepi, S. 2012, *A Situational Analysis of Waste Management in Harare, Zimbabwe*, Marsland Press, New York.

Urio, A. F. and Brent, A. 2006, *Solid Waste Management Strategy in Botswana. The Reduction of Construction Waste*, Routledge Taylor and Francis Group, London.

Section VI

Future of Sustainable and Smart Spatial Planning in Africa

21 Furthering Sustainable and Smart Spatial Planning in Africa

Charles Chavunduka, Walter Timo de Vries and Pamela Durán-Díaz

CONTENTS

21.1 Introduction 299
21.2 Sustainable and Smart Governance..300
21.3 Sustainable and Smart Infrastructure..301
21.4 Sustainable and Smart Technological Change 302
21.5 Conclusion 303
References...303

21.1 INTRODUCTION

As the goal of this book is to provide various critical assessments of how, where and why spatial and settlement planning is currently changing, synthesising the key messages of the chapters of this book in a uniform way is not evident nor provides justice to the variations that currently exist.

Global changes affecting spatial planning worldwide include rural–urban migration and global human mobility, climate change and unpredictability of weather patterns, spatial disparities and injustices, Internet of Things, interconnectedness of technologies, open data and open source technologies, multiple active and passive data sensors and data surveillance technologies and the use of open and big global data. The pace of these changes remains rapid whilst the uncertainty coming along with the impacts of these changes has increased.

Experiments with smart city concepts and policies in different parts of the world are gradually turning into actual smart city implementation strategies and elements of daily life. With these experiments have also come a number of expectations from policy makers, planners and citizens. These include a more efficient use of land, more effective monitoring and compliance mechanisms, better access to government and public information and a more direct relation between planning measures and sustainable development.

In Africa, a big divide remains however visible in the adoption of strategies for urban areas (or cities) versus those for rural areas and rural communities. Compared to urban areas, rural areas lag behind in information communication technology

DOI: 10.1201/9781003221791-27

infrastructure, data access and human resources for smart concept adoption and innovation.

On the theoretical and conceptual side, there are both converging and diverging definitions and practices of sustainable and smart spatial planning. In southern Africa, sustainability is especially evident through the active pursuance of sustainable development goals (SDGs) in the strategies of multiple countries. They manifest themselves as the empirical evidence of the uptake of renewable energy projects

Smartness in most African strategies strongly builds on international examples. Typically, the national strategies refer to the six aspects of smartness, that is, smart economy, smart people, smart governance, smart mobility, smart environment and smart living. There are indeed emerging technical innovations in Africa. Adoption of Internet of Things based smart spatial planning is still in its infancy stage in Africa, but there are some promising examples and experiences in Rwanda and Nigeria, which could act as a gateway for the rest of Africa. In southern Africa, in particular, one can observe the first signs of the adaptation, revision and extension of the theory of sustainable and smart spatial planning in and for Africa and a gradual adoption, revision and extension of the spatial planning tools and methodologies. The use of GIS 2D and 3D technologies, for example, in managing and administrating properties, the monitoring and assessment of public safety and 3D layout planning, is also gradually developing. In terms of broadness, however, the smart developments in Africa tend to focus on smart mobility, smart laboratories and smart urban labs, manifested in the adoption of new sensor and surveillance technologies, supported by e-government services, e-financing and banking and the use of mobile apps. Moreover, so far, the adoption of bottom-up spatial co-design and co-construction initiatives, smart environment and smart living are still lagging behind. Participatory approaches, involving both communities and smart technologies, are still limited despite their appeal.

The next sections discuss the emerging themes and ideas, which the preceding chapters have developed. Sustainable and smart spatial planning needs to be embedded in sustainable and smart governance, supported by a sustainable and smart spatial institutional, social and technical infrastructure and being pro-actively open to adopting sustainable and smart spatial planning technologies.

21.2 SUSTAINABLE AND SMART GOVERNANCE

The various chapters in the book have made it clear that sustainable and smart governance in the context of Africa needs to be better aligned with the spatial planning and socio-political realities of African cities and rural towns. There are a number of constraints that hinder any progress so far. Some of the chapters specifically note that political instability and corruption in many countries in southern Africa often disrupt achieving sustainable and smart goals, whilst economic instability and unrest causes smart city initiatives to fail. At the same time, the examples of smart village initiatives demonstrate that community residents and their visionary chief were able to make the best out of the new relocated residential sites and thus were able to transform their daily practices of earning a living.

Concepts of what a city or a rural town constitutes need to be redefined. Both spatial and social boundaries between urban and rural are fluid and dynamic and contain a political economic character (Chavunduka and Chaonwa-Gaza, 2021). Informality and illegal land occupation is simply a rule rather than an exception, given that some 60% of the population live in such areas, and given that policies to combat informality sternly relate to the broader context of socio-political forces.

Moreover, the movement of people in search for livelihoods creates continuously changing patterns of densities of people, structures and social and economic activities. This implies that the rules guiding both the spatial planning processes and practitioners, copied from the colonial epistemologies of rational spatial planning are no longer valid in such a context. Bounded rationality, accepting spatial uncertainty and dynamic and fast changes are the fundamental principles for a new type of spatial planning geared towards sustainable spatial governance. This implies employing more facilitation in reaching pragmatic and fit-for-purpose solutions that are valid and appropriate for shorter times, enhancing the ability to change, recycle and upgrade progressively, more regular rethinking and re-evaluating the appropriateness of given solutions and faster upgrading of professional staff members in the uptake of smart technologies that can handle uncertainties. Spatial planning, in other words, is then much more an accompanying socio-spatial innovation process than a process relying on the design, implementation and monitoring of master plans. Instead, when spatial relations change continuously monitoring should probably be the first task of effective spatial planning. This requires effective informational infrastructures and expertise.

One of the crucial aspects of spatial planning also concerns the creation of accessible and affordable housing. Yet, findings from Zimbabwe show that this remains problematic. The persistence of informal settlements is a manifestation of this difficulty and despite the existence of opportunities for using urban land readjustment processes to enable land development and rationalise urban land use, it remains complicated in practice to mobilise public participation and execute such processes accordingly (Chavunduka, 2018). This would imply that either the rules need to be changed or the rules will need to be applied in a different manner and from a transnational approach.

21.3 SUSTAINABLE AND SMART INFRASTRUCTURE

Handling these new spatial planning requirements also necessitates an adaptation and extension of the socio-institutional, informational and technical infrastructures. To a large degree, the increase of open data and open-source algorithm repositories have resulted in a situation where the global informational and technological infrastructures are becoming rapidly available and accessible, even in the African contexts. There is increasing evidence that remote sensing data are becoming more accessible for public use, also for African users, but the employment of these data is still far below its potential. Therefore, the socio-institutional infrastructure, upgrading knowledge, knowledge networks, e-learning and development and recognition of technical skills needs to be further developed and maintained.

The extension and enhancement of technical infrastructures are crucial but also necessitate significant funding. One way to fund such infrastructures could be public value capture (Chavunduka, 2020). However, implementing such practice still requires a set of fundamental conditions. In the African context, city governments require transparency and accountability structures that can ensure that the public benefit from the public land value capture. Using land value capture as a capital mobilisation instrument implies having a legal framework, for example, that can enable and secure multiple forms of value capture in addition to endowment fees and property taxes. In addition, a better alignment between the central and local governments is required, whereby financial capacities and capturing mechanisms interrelate in order to provide for urban infrastructure investment. Currently, local governments are insufficiently able to rely on traditional revenue, such as from property taxation, electricity generation and vehicle licensing and are thus incapable of funding major infrastructure investments.

21.4 SUSTAINABLE AND SMART TECHNOLOGICAL CHANGE

Technologies do not develop by themselves but rely on persuasive adoption and perceived benefits. According to de Vries et al. (2020), technologies are persuasive if they can change the socio-economic relations and perceptions and expectations. If this occurs, they can be sufficiently disruptive such that they can displace and replace existing socio-organisational structures and work flows, inter-personal and inter-institutional relations and societal situations. For spatial planning, this implies that geospatial technologies need to generate fit-for-purpose approaches, connecting geographic information with open, big and linked data and employing new visualisation, representation and simulation technologies that can enhance interaction and participatory processes in a much more convenient and effective manner. This would include better conversions of 2D static images of cities and landscapes into more dynamic 3D real-time representations of specific phenomena and clusters of changes of cities and rural landscapes. Ideally, one would like to make use of advanced image processing technologies that combine virtual, augmented, immersive and mixed reality with dedicated interactive devices, such that one can truly visualise and perceive simulated environments. If combined with the new possibilities of artificial intelligence (AI), machine learning and neural networks, the possibilities would even be wider (Wagner and de Vries, 2019; de Vries et al., 2014).

Spatial data remain crucial. Collecting data should not be just for the sake of having abundant volumes of data at hand but it should enable practitioners as well as citizens to make the data adapted to particular contexts and make these accessible in multiple forms and shapes and in all sorts of application fields. Taking into account a time dimension and the third dimension would allow better information descriptions and management of subsoil—below the ground—resources and infrastructures—and managing the space above soil—to secure rights in spaces which are currently still unused, but which may be rapidly used in the future is important. The accomplishment of such developments—even in the current African context—can certainly support to build scenarios for responsible use of space and to find more usability for new and old spaces. Examples of application areas include investigating the reasons and causes for the existence of vacant land and depleted land, monitoring of land use and land ceiling, assessing which land is used ineffectively or inappropriately, determining which land requires which energy sources and determining the relations between land tenure and outbreaks of diseases of people, plants and animals.

21.5 CONCLUSION

Prospects and implications for sustainable and smart spatial planning research and practice in Africa are multiple. First, Africa needs new paradigms of space, occupation, allocation and use of space and design of new spaces. Holistically, a new paradigm of the African cities is needed, one that is not bound by colonial legacies. The African city is dynamic and increasingly a co-creation of many factors rather than a creation by technocrats. It has thus taken on complex and specific forms. This paradigm should rely on the idea of dynamic and fluid settlements without geographic or social borders and classes. There should be a governance continuum of social and territorial space. Secondly, transnational cooperation and exchange could support the development of smart initiatives and strengthening land governance. The example of twinning agreements of border towns of Zimbabwe and South Africa given in one of the chapters justifies that planning for an effective and just use of space does not stop at borders.

COVID-19, in particular, has created new challenges but also revealed a number of shortcomings and opportunities that may have an impact for future sustainable and smart developments. It has revealed that:

- Technical and social vulnerabilities can be global and inter-connected.
- There is a global need for reliable, transparent and accurate data collection and data access in order to take swift and appropriate action. Dashboard type of apps, containing both background information and action-oriented information are needed. The use of similar transparent, open and inter-linked data at the global level has become crucial to design new vaccines and to combat the outbreaks in all parts of the world.
- An emerging heated debate of pro-active protection of citizens and strong government rules and interventions versus individual freedoms and choice, protection of privacy and alternative realities and policies pends resolution.

REFERENCES

Chavunduka, C. 2018. Land readjustment: The missing link in urban Zimbabwe. *African Journal on Land Policy and Geospatial Sciences*, 1, 1–17.

Chavunduka, C. 2020. The use of land value capture instruments for financing urban infrastructure in Zimbabwe. *Responsible and Smart Land Management Interventions*. CRC Press, Boca Raton.

Chavunduka, C. & Chaonwa-Gaza, M. 2021. The political economy of urban informal settlements in Zimbabwe. *Urban Geography in Postcolonial Zimbabwe: Paradigms and Perspectives for Sustainable Urban Planning and Governance*, 287.

de Vries, W. T., Bennett, R. M. & Zevenbergen, J. A. 2014. Neo-cadastres: Innovative solution for land users without state based land rights, or just reflections of institutional isomorphism? *Survey Review*, 47, 220–229.

de Vries, W. T., Bugri, J. & Mandhu, F. 2020. Advancing responsible and smart land management. *Responsible and Smart Land Management Interventions*. CRC Press, Boca Raton.

Wagner, M. & de Vries, W. T. 2019. Comparative Review of methods supporting decision-making in urban development and land management. *Land*, 8, 123.

Index

0–9
3D, 258, 302
    layout, 265, 269
    modelling, 259
    planning, 260, 269
4D, 269

A
ADLAND project, 3
affordable housing, 5, 125, 145, 216, 231, 261, 301
Agenda 2063, 45

B
brown fields, 26, 147

C
city master plan, 57, 107, 110, 113, 250
city strategic plan, 112, 113
complexity theory, 246, 254
corruption perception, 93, 94, 97, 98, 104
cost-benefit analysis (CBA), 116, 172

D
data-driven city, 14, 109
data mining, 144, 149
decentralisation, 36
disaster, 213–215, 221
    Disaster Management Act of 1998 of South Africa, 47
    disaster risk (reduction), 214–215, 218, 220, 224

E
e-governance, 12, 17
enclave (theory), 200, 204, 208
entrepreneurship, 17, 28, 46, 50, 132, 199
Environmental Impact Assessment (EIA), 172
Environmental Management Act (EMA Act), 200, 284
e-waste, 245–249, 254

G
governance
    good, 21, 43, 55, 70, 73, 90, 102, 113, 117, 274
    indicators, 90, 93–95, 104
    participatory, 6, 42, 50
    smart, 16, 28, 116, 181, 231, 300
    transnational (land), 90, 102, 104
    weak (poor), 20, 90, 93, 101, 111, 117, 180
green fields, 16, 26, 148

H
housing, 30, 34, 47, 116–117, 127, 143, 146, 148, 202, 224, 231, 237
Housing Act, 34, 200

I
infocracy, 143
informal settlements, 30, 113, 121–123, 125, 133–136, 214–216, 221–225
infrastructure, 18, 42–43, 57, 79, 109, 111, 123, 147, 169–170, 174, 201, 239, 276, 301
    ICT, 16, 18, 28, 33, 35
    physical, 18, 159, 199, 201
institutions, 15, 50, 90, 93–94, 98, 111, 115, 232
    land administration, 55
    public, 35, 36, 98
Internet of Things (IoT), 14, 27, 50, 300

L
land administration, 44, 47, 55, 242
land and property rights, 4, 44, 47–49, 95, 104
land development, 56, 90, 220, 242, 301
    sustainable, 90
land governance, 44, 90, 93, 104, 303
land grabbing, 48
land management, 3, 90, 105, 248, 276
    responsible, 3, 94, 106, 302
    smart, 3
    sustainable, 89–90, 94
land use planning, 5, 26, 55–56, 61–62, 65, 199, 214, 246, 248
liveability indicators, 124–125, 134
living labs, 111, 115
local development plan (LDP), 59, 61, 65, 232, 251, 279

M
mining, 25, 101, 111, 129, 133, 198–203, 207–208
    Mines and Minerals (MM) Act, 198, 204, 208
modernism, 110, 113

N
NELGA (the Network of Excellence on Land Governance in Africa), 4
New Urbanism, 56–58, 62, 65
Normalised Difference Built-up Index (NDBI), 233, 235

P
Paris Climate Agreement, 105, 154
participatory planning, 61, 65, 98

peri-urban settlements, 43
public space (concept), 184–187

R
redevelopment, 56–57, 62, 64, 233
Regional Town and Country Planning (RTCP) Act, 108, 198, 249, 252, 261, 277
relocation, 46–47, 49, 56, 62, 64
resilience, 155, 165, 231
right to the city, 184–186, 189
rural landscape, 43, 302
rural-urban linkages, 49, 170, 230, 299

S
scarcity of land, 11, 161
Shannon's entropy, 233, 240
simulation, 264–265, 274–275, 302
sister city, 70, see also twinning city
small and medium entrepresies (SMEs), 6, 169
smart city
    concept, 4, 12–15, 19, 26, 34, 108, 117, 123, 156, 185, 201, 230–231, 238
    development, 108–109, 111
    human smart city model, 184–185
    smart city framework (SMELTS), 16, 22, 28, 33, 232
    smart city theory, 13, 15
smart energy, 20, 157
smart growth, 6, 13, 26, 34, 56–57, 62, 155, 232
smart infrastructure, 169, 181, 301
smart planning (theory), 26, 184
smart village, 42–43, 45–46, 300
solar energy, 115, 154, 158
South Africa, 20–21, 29, 42–43, 71, 184, 201, 214–217, 276
Southern Africa, 4, 70, 82, 198, 230
Southern African Development Community (SADC) Treaty of 1992, 70
spatial data infrastructure, 32
spatial development (framework), 64, 70, 73, 79, 188, 215
spatial justice, 6, 108, 117, 184–185, 193
spatial planning (framework), 200, 202
    layout planning, 232, 261, 266, 269, 300
    settlement planning, 4, 117, 124
    spatial planning and land use management Act (SPLUMA), 44, 215
    spatial planning tools, 214, 220, 300
    sustainable spatial planning, 5, 59
    trans-border spatial planning, 77, 80
sustainability
    sustainable city concept, 27, 65, 201
    sustainable energy, 46, 157

    sustainable governance, 300
    sustainability theory, 285
sustainable development goals (SDGs)
    SDG1 (reducing poverty), 5, 94
    SDG7 (universal access to energy), 154, 159
    SDG11 (sustainable cities and communities), 5, 45, 46, 94, 198, 201
    SDG15 (life on land), 5, 94
systems theory, 15, 214–215, 225

T
theory of space production, 186
Town and Country Planning Act, 34, 108, 249, 252, 261, 277
traditional authorities, 44, 47, 51
transparency, 17, 95, 102, 111, 148, 302
twinning
    agreements, 70, 73–75, 84
    city, see sister city
    indicators, 75

U
unplanned settlement, see informal settlement
Urban and Regional Planning Act, 34, 35
urban design, 58, 65, 124, 224, 270, 275
urban growth (theory), 17, 26, 57, 109, 198, 231, 237, 276
urbanisation, 6, 34, 109, 122, 143, 230, 232
urban laboratory (UrbLab's), 141, 142
urban liveability, see liveability indicators
urban planning, 13–14, 28, 34, 59, 113, 232, 270, 273, 277
urban poor, 108, 114, 237
urban redevelopment, 56–57
urban spaces, 51, 185, 273, 278
urban sprawl, 57, 230–233, 240
urban systems theory, 15
user-centred systems design (UCSD), 171

V
virtual reality, 273

W
waste management, 245–246, 248, 284, 287, 289
    material waste, 284–286, 292

Z
Zambia, 30–33, 202
Zimbabwe, 20, 58, 71, 108, 112, 122, 149, 154, 156, 170, 201–203, 230, 233, 247, 261, 277, 284, 290